# Survival or Extinction?

Bridget Martin

# Survival or Extinction?

## How to Save Elephants and Rhinos

 Springer

Bridget Martin
Lancashire Law School
University of Central Lancashire
Preston, Lancashire, UK

ISBN 978-3-030-13292-7          ISBN 978-3-030-13293-4    (eBook)
https://doi.org/10.1007/978-3-030-13293-4

Library of Congress Control Number: 2019933565

Cover illustration: Bornean Pygmy elephants seeming to pose for their picture to be taken. Copyright Marc Ancrenaz, Elephant Family

This Springer imprint is published by the registered company Springer Nature Switzerland AG
The registered company address is: Gewerbestrasse 11, 6330 Cham, Switzerland

*For Sudan, the last male northern white rhino, who, unlike so many other rhinos, was fortunate enough to pass away peacefully, in the company of loving friends and carers.*
*Sudan 1972–2018. Copyright Daniel Stiles*

*A great many animals and plants on our planet are, like northern white rhinos, threatened with extinction, even though their future is not quite so bleak. So, the author pledges 80% of all her income from the sales of this book shall go to protecting these animals and plants.*

# Preface

Elephants and rhinos are in trouble; they need your help. Poaching and illegal trade in ivory and rhino horn have caused their numbers to plummet, and it is debatable whether sustainable use in these commodities is now viable as some species and populations are teetering on the brink of extinction. In recent years, organized crime has not been slow to take advantage of the opportunities offered by skyrocketing prices for elephant ivory and rhino horn, bringing with it death and destruction to both animals and the people who protect them. Law and its enforcement have never been more important.

In this book, I have tried to explain what has gone wrong and how we are trying to put things right. The story, which is made up of material from a great many sources, each footnoted to show its origin and linked together by a guiding narrative, is complicated, so sections signpost the route. You do not need to read the footnotes to follow the story, but I have made them easier to use by adapting OSCOLA. All the material has been selected by me, and although I have done my best to produce as accurate and rounded a picture as possible, if I have failed, I apologize.

People power and the results achieved by working together form a key element, and now that the United Nations has recognized the seriousness of the situation and votes have been taken in the UN General Assembly, it is encouraging to see so many countries keen to face the challenge and try to bring an end to such a cruel and unnecessary trade.

Whether you live in a producer country (e.g. South Africa or Namibia) or a consumer country (e.g. China or Viet Nam), you can see where the problems have arisen and how they are gradually being overcome, because so many people care about these iconic animals, care that elephants and rhinos are teetering on the edge of extinction and are determined to save them. This book will give you some ideas about how it is being done.

This continues to be a fast-moving story, so a line was drawn not to include most of the reports published after October/November 2018.

Preston, Lancashire, UK                                    Bridget Martin

# Acknowledgements

A great many people have helped to make this project possible. Wherever I turned for advice, to ask questions and to bounce ideas, I always found kindness and enthusiasm. I was deeply moved. Sadly though, there is only room to mention a few.

The following people and organizations generously gave their permission to use their photographs and figures:

Photos: Marc Ancrenaz and Elephant Family, Dr. Daniel Stiles, Guy Shorrock, the Ulmer Museum, the Worshipful Society of Apothecaries, Chester Zoo, Russ Allen, Chris Townend, Save the Rhino and Dr. Naoko Irie and Springer Japan.

Figures: T. Milliken, J. Shaw and TRAFFIC, INTERPOL and the Rhino Impact Investment Project.

Maps: These were especially created for the book by Maxwell Boardman.

In all cases, the copyright remains with the donors.

Case studies: Many of these were based on investigations carried out by the Environmental Investigation Agency. They very kindly gave me their permission to include them.

The 'technical' side: Maxwell Boardman, of Lancashire Law School, who was a star and LIS. I couldn't have managed without them!

Professor Sir Patrick Bateson, although he was ill, helped me understand the complexities of genetics in small populations and checked my chapter for errors.

Doctors Richard Thomas and Daniel Stiles, who, although I was a stranger, always gave freely of their valuable time to answer questions, suggest material that might be useful and gently point out when I perhaps needed to think again. Dr. Stiles also allowed me to use some of his beautiful and informative photographs.

My friends and family, who never wavered in their support.

My friends and colleagues at Lancashire Law School and Harris Hub including Rachel Cooper-Green. Without their encouragement and help, this book would not have been possible.

And of course, Janet Slobodien and Rivka Kantor at Springer Nature, who are making this project come to life.

Thank you, all of you, so very much!

# Contents

# Abbreviations

| | |
|---|---|
| ACB | (Malawi's) Anti-Corruption Bureau |
| AfESG | African Elephant Specialist Group |
| ARC | The Alliance of Religions and Conservation |
| BBC | British Broadcasting Corporation |
| CBNRM | Community-Based Natural Resource Management |
| CoP | Conference of the Parties |
| DRC | Congo (Democratic People's Republic) |
| DEA | South Africa's Department of Environmental Affairs |
| Defra | Department for Environment, Food and Rural Affairs (UK) |
| DNPW | (Malawi's) Department of National Parks and Wildlife |
| DSWT | David Sheldrick Wildlife Trust |
| DWNP | (Malaysia's) Department of Wildlife and National Parks |
| EIA | Environmental Investigation Agency |
| ENV | Education for Nature Vietnam |
| EU | European Union |
| EU-TWIX | The European Union Trade in Wildlife Information eXchange |
| FARDC | Congolese Armed Forces |
| GAWPT | George Adamson Wildlife Preservation Trust |
| GEC | Great Elephant Census |
| GEF | Global Environment Facility |
| GP | Global Programme |
| ICCWC | International Consortium on Combating Wildlife Crime |
| IFAW | International Fund for Animal Welfare |
| IWT | Illegal Wildlife Trade |

| IUCN | International Union for Conservation of Nature |
| KWS | Kenyan Wildlife Service |
| LATF | The Lusaka Agreement Task Force on Cooperation Enforcement Operations Directed at Illegal Trade in Wild Fauna and Flora |
| MOU | Memorandum of Understanding |
| NEMA | National Environmental Management Act 107 of 1998 |
| NEMBA | National Environmental Management Biodiversity Act [No. 10 of 2004] |
| NESTs | National Environmental Security Task Forces |
| NGOs | Non-governmental Organizations |
| NWCU | National Wildlife Crime Unit |
| NWEN | Network of the Wildlife Enforcement Networks |
| PHASA | The Professional Hunter's Association of South Africa |
| Protect RAPID | Real-time Anti-Poaching Intelligence Device |
| RATZ | A private anti-poaching unit in Zambia |
| RhODIS | Rhino DNA Indexing System |
| RIE | Raw Ivory Equivalent Weight |
| RUSI | Royal United Services Institute |
| RWENs | Regional Wildlife Enforcement Networks |
| SADC | Southern African Development Community |
| SANParks | South African National Parks |
| SAPA | The Social Assessment of Protected Areas |
| SAWEN | The South Asian Wildlife Enforcement Network |
| SC | Standing Committee |
| SLU | Sustainable Livelihoods Unit |
| SSC | Species Survival Commission |
| TFCAs | Transfrontier Conservation Areas |
| UK | United Kingdom |
| UN | United Nations |
| UNEP | United Nations Environment Programme |
| UNEPWCMC (CITES Trade Database) | United Nations Environment World Conservation Monitoring Centre (CITES Trade Database) |
| UNODC | United Nations Office on Drugs and Crime |
| UNWTO | United Nations World Tourism Organization |
| USA | United States of America |
| USFWS | United States Fish and Wildlife Service |
| VCCI | Vietnam Chamber of Commerce and Industry |
| WENs | Wildlife Enforcement Networks |
| WISTs | Wildlife Incident Support Teams |

| | |
|---|---|
| WLFC | Wildlife and Forest Crime |
| WRAP | The Wilderness Rhino Awareness Programme |
| WRSA | Wildlife Ranching South Africa |
| WTA | Wildlife Translocation Association |
| WWF | World Wide Fund for Nature |
| ZAWA | Zambia Wildlife Authority |
| ZSL | Zoological Society of London |

# Maps

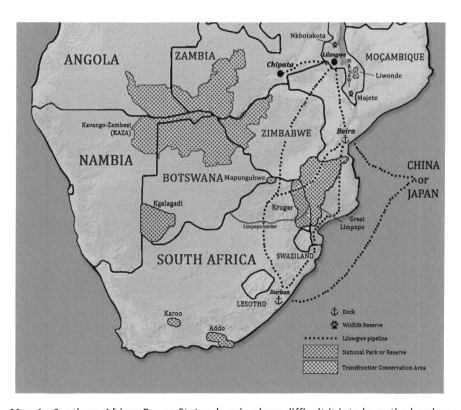

**Map 1**  Southern African Range States showing how difficult it is to keep the borders secure from poachers. Some of the national parks are marked, as is the Lilongwe Pipeline. Copyright Maxwell Boardman

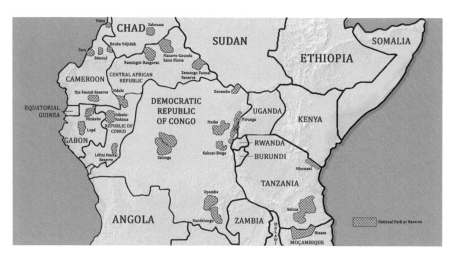

**Map 2**  Central and Eastern African countries, again showing some of the national parks and the problem of securing borders from crime gangs. Copyright Maxwell Boardman

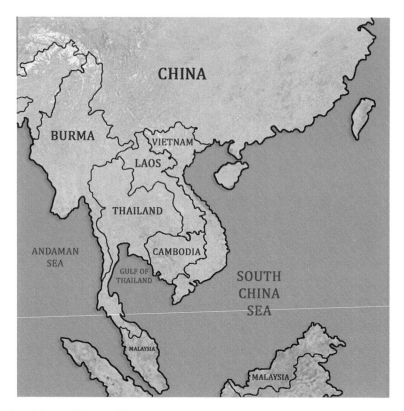

**Map 3**  China and its neighbours. Copyright Maxwell Boardman

# List of Figures

# List of MAPS

# 1

# Introduction

In April 2009, Simba, an elderly white rhinoceros, died peacefully of old age in Colchester Zoo. On 30 June that same year, Donald Allison was stopped at Manchester Airport as he was about to board a flight to China via Amsterdam. After a tip-off, officers of the UK Border Agency were looking for rhino horns, Simba's to be exact. They eventually found them in his luggage, cunningly disguised as a Vienna bronze sculpture featuring a bird on a log. Allison was successfully prosecuted and sent to prison.

They have disappeared, 'incorporated' into the Vienna bronze sculpture as the tree branches on which the little bird is perched. In the lower image, the sculpture has been partly opened to expose the hidden horns.[1]

There is a long cultural relationship between people, elephant ivory and rhino horn going back millennia. Recently, however, demand has become insatiable. The value of ivory and rhino horn is spiralling out of control, corruption is rife and organized criminal gangs are making a killing.

Elephants and rhinos are in deep trouble, subjected to an uncontrolled killing spree that shows no sign of ending. Such deep trouble that some of them are threatened with extinction. All because of a seemingly insatiable desire by some people for ivory and rhino horn, a desire other people are only too pleased to fulfil, at a price. And what a price. Animals are dying in torment, while humans grow rich beyond the dreams of avarice. Our story appears to have a heart of darkness.

But all is not as it seems. Although there are people who show no pity, who seek only to enrich themselves, there are millions of others who care desperately about the plight of the animals and are doing their utmost to save them.

---

[1] These formed part of the evidence which convicted Donald Allison.

© Springer Nature Switzerland AG 2019
B. Martin, *Survival or Extinction?*, https://doi.org/10.1007/978-3-030-13293-4_1

**Fig. 1.1** Simba's horns: Now you see them (these formed part of the evidence which convicted Donald Allison). Copyright Guy Shorrock

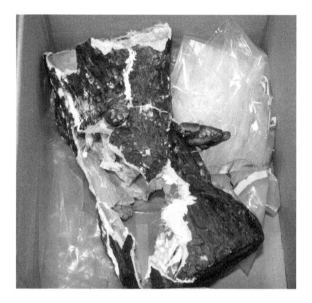

**Fig. 1.2** Now you don't! Copyright Guy Shorrock

Some, like the rangers, who are even prepared to sacrifice their own lives to this end.

Another champion is the law. Perhaps this fact is not so well known. Two treaties, CITES, the Convention on International Trade in Endangered Species of wild Fauna and Flora and the CBD, the Convention on Biological Diversity, play a critical role. Together with other legislation, they provide another key to successful conservation.

In our story, although we meet poachers, organized crime gangs and rebel militias, we shall see how people working together and with help from the law

**Fig. 1.3** The majestic Lion Man carved out of mammoth ivory by our ancestors in prehistoric times. Copyright the Ulmer Museum

are gradually bringing about their downfall. This is a global problem, but peoples of the world are united in their efforts to combat this cruel and unnecessary trade, this illegal trafficking of ivory and rhino horn.

This then is the story of elephants, rhinos and two treaties. It is the story of people, some bad but most good, and the battle between them to save these animals, these beautiful iconic creatures that have lived on Planet Earth for millions of years. And it brings a message of hope.

## 1.1 Prologue

In February 2013, the British Museum curated a thought-provoking exhibition of Ice Age art.[2] The oldest exhibit was the Lion Man.[3] Clouded in mystery, some 40,000 years ago, using a stone saw tool, he was carved from mammoth

---

[2] *Ice Age Art: Arrival of the Modern Mind*, Exhibition at the British Museum, 7 February–26 May 2013.

[3] The *Lion Man* is in the Ulmer Museum, Germany. See Martin Bailey *Age Lion Man is world's earliest figurative sculpture* The Art Newspaper, 31 January 2013. See www.theartnewspaper.com. Accessed 23/07/2013

tusk, a task that must have taken over 400 hours to complete. He was almost certainly of religious significance and had obviously been well-handled. It would seem that people's relationship with elephant ivory goes back to the dawn of time.[4]

Although some rhino horn is carved into beautiful artefacts, it is people's use of it in powdered form, as an ingredient in medicines, a use that goes back several thousand years starting in China and spreading gradually throughout east and southeast Asia, that is mainly responsible for the devastating slaughter of large numbers of rhinos today. The earliest written record of its use dates from the Han dynasty (206 BCE to 220 AD) and is found in the *Shennong Ben Cao Jing* otherwise known as the *Divine Peasants Herbal*.[5] Perhaps what is less well known is that even in the West, rhino horn was deemed to have very special properties.

---

[4] Neil McGregor *Living with the Gods*, a series on BBC Radio 4.

[5] *Species trade and conservation Rhinoceroses Assessment of Rhino Horn as a Traditional Medicine*, a report prepared for the CITES Secretariat by Kristin Nowell on behalf of TRAFFIC, April2012, SC^" Doc. 47.2, Annex (Rev. 2) – p.19.

# Part I

## Cast of 'Characters'

# 2

# The Animals: Elephants, Rhinos and People

The characters in this story are elephants, rhinos and people. Time to meet them. We will start with the animals.

## 2.1  Elephants

Found throughout much of Africa and Asia, many people consider elephants to be special animals. They have evolved large and complex brains. They have a complex social structure including highly developed empathetic behaviour, a complex system of communication and advanced learning ability.[1] They can problem-solve, and they can even use tools.[2] And yet these beautiful, intelligent creatures are being slaughtered in their thousands for their tusks.

Today, the **African elephant** wanders in small herds throughout 37 of the African countries,[3] and although two distinct species, namely, the savannah elephant (*Loxodonta africana*) and the forest elephant (*Loxodonta cyclotis*), are currently recognized, preliminary research from DNA studies suggest there may even be a third distinct species, the **West African elephant.**[4]

---

[1] Roca & O'Brien *African Elephant Fact Sheet*, Sandiego Zoo 2005, p.3. See http://library.sandiegozoo. org/factsheets/african_elephant/african_el... Accessed 14/05/2013.

[2] BBC NATURE *Tool use videos, news and facts*, 30 March 2010. See http://www.bbc.co.uk/nature/adaptations/Tool_use_by_animals Accessed 23/07/2013.

[3] WWF *African Elephant Species*. See http://worldwildlife.org/species/african-elephant Accessed 02/05/2013.

[4] See Roca & O'Brien (n. 1).

© Springer Nature Switzerland AG 2019
B. Martin, *Survival or Extinction?*, https://doi.org/10.1007/978-3-030-13293-4_2

**Savannah elephants** are the largest living land animals. Currently found in east, central and southern Africa, they live in lowland and montane forests, flood plains, all types of woodland and savannah, and they roam widely. We will only look at a few of the countries that comprise their range States. In the east, Kenya, Tanzania and Uganda. In the west, Côte D'Ivoire, Mali, Nigeria and Senegal. In all these countries, elephant populations are listed as *Vulnerable* on the *IUCN Red List of Threatened Species*. However, as we shall see later, in southern Africa the story is much more complicated. In some countries, such as South Africa, the populations are healthy, while in others, such as Angola and Mozambique, the future seems bleak.

**Forest elephants** have been distinguished from savannah elephants by DNA analysis.[5] The distinction is vital if both species are to be properly conserved, especially as numbers of these elephants have plummeted in recent years. They live mainly in the rain forests of Cameroon, the Central African Republic, the Côte d'Ivoire, the Democratic Republic of Congo, Gabon, Ghana and Liberia.

Forest elephant ivory is more highly prized than savannah elephant ivory, and unfortunately this is proving lethal to the survival of these animals. They are such a popular target for poachers that in places, there has been what can best be described as wholesale slaughter of these animals. For example, some 11,000 were killed in parts of Minkebe National Park and northern Gabon between 2004 and 2012, between 44% and 77% of the population.[6] This, added to the fact that their families tend to be smaller than savannah elephants, makes it harder for population levels to recover. Forest elephants are listed on the *IUCN Red List* as *Endangered and threatened with extinction*.

Finally we come to the **Asian elephants**. There are three subspecies, of which the **'Indian' elephant** (*Elephas maximus indicus*), living on the mainland, is the most numerous. Despite the fact that in some eastern religions elephants are considered to be deities, and in some countries they have been domesticated and used for centuries in a number of roles, Asian elephants are even closer to extinction than the African species, though the main threat is not from poaching and the illegal ivory trade. One reason for this is that many of them do not have tusks, while others merely have tushes, very small tusks that do not extend past their mouths.

---

[5] Roca et al. *Genetic evidence for two species of elephant in Africa*, 2001.
[6] WWF *Poachers kill over 11,000 elephants in Gabon* 6 February 2013. See https://www.wwf.org/updates/poachers-kill-over-1-1000-elephants-gabon Accessed 25/10/2018.

Elephants are highly social, living in family groups headed by a matriarch. The herds vary in size, with savannah elephants having most members, 4–14 animals, Asian elephants 8–12 and forest elephants 2–4, the latter being merely nuclear family groups. Although Asian elephant herds only contain the matriarch's offspring and their young, savannah elephant herds can also include the matriarch's sisters and their offspring. This may result in herds becoming too large, and if this happens, 'bond groups' split off, although these still maintain a loose association with their original group.[7] All adult members of the herd are female and all help to look after the calves.

When young males become 'teenagers', they tend to move about between families, joining up with other young males, play-fighting each other and learning how to be a male elephant. At one time, adult males were thought to be solitary, but research now shows this is not so. They form loose/transient associations, whose size depends on whether they feel safe or not, how plentiful food is and what are the prospects for mating.[8]

Savannah elephants form *complex social bonds with dozens of clan members and hundreds of acquaintances*, and in times of stress, such as drought or human interference, they have even been known to congregate in temporary herds of over 1000 individuals.[9] Forest elephants meet together, often in large numbers in forest clearings known as bais, where they bathe in the mud, drink the waters and generally enjoy each other's company.

They also range widely. Just how far they go depends on many factors. Is it the wet or the dry season? Are food and water available? Perhaps they are seeking a mate? Is there a problem with people? They migrate, covering many miles and sometimes crossing into different countries. This puts them at considerable risk, both of human-elephant conflict and from poachers.

Elephants should have a long lifespan. Asian elephants live longer than the African species, but both live nearly as long as people. The importance of this cannot be overestimated. With age comes wisdom, and the older matriarchs play an essential role caring for their families, passing on invaluable knowledge and survival skills,[10] while older males teach younger ones how to behave.

---

[7] For an interesting description, see Elephant Voices *Elephants are socially complex* https://www.elephant-voices.org/elephant-sense-a-sociality-4/elephants-are-socially-complex.html Accessed 08/11/2018.

[8] Lesley Evans Ogden *Male elephants are not the loners we once thought* BBC-Earth, 31 October 2014. See http://www.bbc.com/earth/story/20141101-male-elephants-have-a-sweet-side Accessed 11/08/2016.

[9] WWF *African elephants*. See http://wwf.panda.org/what_we_do/endangered_species/elephants?afr... Accessed 30/04/2013.

[10] BBC-Earth News *Older female elephants are wiser matriarchs*. See http://news.bbc.co.uk/earth/hi/earth_news/newsid_9425000/9425590.stm Accessed 30/04/2013.

**Mammoths** lived in the Ice Age and became extinct thousands of years ago, but rather surprisingly, they are playing a gradually increasing role in our story. And global warming is one of the reasons for this. In Siberia, the permafrost is melting, making it possible to retrieve their ancient tusks and, in the process, triggering a different kind of 'gold rush'.

## 2.2    Rhinos

These funny, prehistoric-looking creatures, beautiful in their own way, are so special yet live under such threat. Like elephants, rhinos are indigenous to both Africa and Asia, and although there are five species of rhino, these animals are much rarer than elephants just because some people value not the animals themselves, but their horns so highly.

2011 was a very bad year for rhinos. The *IUCN Red List* declared the western black rhino to be *Extinct* and the northern white rhino *Possibly Extinct in the wild*, while a report by WWF and the International Rhino Foundation declared the Javan rhino to be *Extinct in Viet Nam* after the last one was killed by a poacher shortly after it had been seen alive on a camera trap.[11]

That same year, the BBC reported the fact that rhino horn was now worth £50,000 a kilo, more than gold, diamonds and cocaine. The UK government decided to put forward measures aimed at achieving an international clampdown on the demand.[12] With little effect, as in May 2013, Aljazeera was reporting that in Viet Nam, rhino horns were *still worth more than gold,* 1 kg selling for USD 65,000.[13] Even the Chair of the IUCN Species Survival Commission was moved to comment: *'You've got to imagine an animal walking around with a gold horn; that's what you're looking at, that's the value'.*[14]

Two species of rhino are found in Africa, white rhinos (*Ceratotherium simum*) and black rhinos (*Diceros bicornis*). Like the elephants, they were once widespread, but they too are causing grave concern, particularly black rhinos, for these only hang on *in small isolated pockets.*[15]

---

[11] BBC News *Western black rhino declared extinct.* See http://www.bbc.co.uk/news/science-environment-15663982 Acc 21/05/2013; and BBC News *Javan rhino now extinct in Viet Nam.* See http://www.bbc.co.uk/news/science-environment-15430787 Accessed 23/06/2013.

[12] BBC News *UK to lead international rhino horn clampdown* 20 August 2011. See www.bbc.co.uk/news/science-environment-14603905 Accessed 29/07/2013.

[13] *Rhino horns 'worth more than gold' in Viet Nam* Aljazeera, 22 May 2013. See www.aljazeera.com/indepth/features/2013/.../2013515141111445265 Accessed 29/07/2013.

[14] BBC News *Western black rhino declared extinct.* See http://www.bbc.co.uk/news/science-environment-15663982 Accessed 21/05/2013.

[15] Emslie and Brooks, 1999 *Black African Fact Sheet* p.1. See http://library.sandiegozoo.org/factsheets/Black_rhino/black_rhino.htm Accessed 21/05/2013; and WWF *Black rhinoceros.* See http://wwf.org/what_we_do/endangered_species/rhinoceros/a... Accessed 21/05/2013.

**Black Rhino** numbers are small, and their distribution is patchy. Their range States include Ethiopia, Kenya, Namibia, Rwanda, Swaziland, South Africa, Tanzania and Zimbabwe although they have been reintroduced into Zambia and Botswana.[16] About 98% of them are found in South Africa, Namibia, Zimbabwe and Kenya, with the largest population in South Africa.

There used to be four subspecies, but in 2011, one of them was declared to be extinct. The remaining three are all listed as Critically Endangered on the IUCN Red List. The most numerous is the **southern-central black rhino** whose stronghold is in South Africa, while the **eastern african black** rhino is mainly found in Kenya.

**White Rhinos** Today, there are two genetically distinct subspecies of white rhinos, the **southern white rhino**, whose numbers have dramatically increased over the years, and the **northern white rhino**, whose numbers are now down to two, both females, who are unable to breed. The solitary male, Sudan, died of old age in March 2018. (More of their story later.)

**Southern White Rhinos** have made an astonishing comeback. Almost hunted to extinction in the late nineteenth century, in 1895 a tiny population of about 50 individuals was discovered in KwaZulu-Natal, South Africa, a population which, because of excellent management techniques, had reached an astonishing 18,000 by 2012. Listed as '*Near Threatened*' on the IUCN Red List, they are now enduring a frenzy of poaching.

These rhinos are mainly found in South Africa, Namibia and Zimbabwe, although there are a few in Kenya.

**Asian rhinos** at one time ranged widely throughout Asia, but they are now only found in small pockets of land, and two of the three species are listed as *Critically Endangered* on the IUCN Red List.

**The Greater One-Horned or Indian Rhino** is the most numerous of the three. The rarest rhino is the **Javan rhino**, and the **Sumatran rhino** is genetically quite distinct from the other two.

Greater one-horned rhinos also came close to extinction, numbering only 600 in 1975, after which valiant conservation efforts managed to raise them to a more viable 2913 animals by 2011. Once widespread, now they are mainly confined to Kaziranga National Park, in India, and the Royal Chitwan National Park in Nepal, and they are listed as *Vulnerable* on the IUCN Red List.

---

[16] WWF *Black rhinoceros* See http://wwf.panda.org/what_we_do/endangered_species/rhinoceros/a... Accessed 21/05/2013.

As their name suggests, the greater one-horned rhino only has one horn, as does the Javan rhino. All other rhinos have two horns.

**Javan Rhinos** are the rarest of the Asian rhinos[17] despite the fact that they too ranged widely throughout much of southeast Asia. Now, of the 3 subspecies, 2 are extinct, and fewer than 50 animals still exist, only found on the Indonesian island of Java, in the Ujung Kulon National Park and in range of the infamous volcano Anak Krakatau. Should this erupt, the rhinos could well be wiped out. Listed as *Critically Endangered* on the IUCN Red List, these rhinos seem doomed to a very precarious future.

**Sumatran or Hairy Rhinos** really are hairy. Their ears have a fringe of longer hairs, the ends of their tails have a tuft of thicker hairs, while the shaggy fur on their bodies gets rubbed down to short bristles by the vegetation they pass through. There are two surviving subspecies.

**Western Rhinos** are listed as *Critically Endangered* and only found in three National Parks in Sumatra.

**Bornean Rhinos** are the smallest and hairiest species of rhino, but a tiny population still exists in Sabah, Malaysia. They too are classed as *Critically Endangered*.

Unlike elephants, most species of rhino are relatively solitary animals although they do come together to mate. So what do we know of their social behaviour? That calves stay with their mothers for years. That adult and young male greater one-horned rhinos may gather at wallows or to graze. That black rhinos are perhaps the exception, for they were filmed by a BBC television crew, meeting, in the dead of night, at a particular water hole. Mothers with their calves, lone males looking for a mate, communicating in the complete darkness by smell, touch and sound. A secret gathering of about a fifth of the total population of these rhinos.[18] A wonder of the natural world.

---

[17] Grubb *Rhinoceros sondaicus* IUCN Red List of Threatened Species. See http://www.iucnredlist.org/details/19495/0 Accessed 23/06/2013.

[18] Michael Bright *BBC Africa Eye to Eye with the Unknown* Quercus, 2012.

## 2.3   People

In this story, we meet people who are devoting their lives to protecting 'their' animals, prepared to, and sometimes giving their lives to saving elephants and rhinos. Other people too. These people see tusks and horns not animals, and money, fabulous amounts of money. They are prepared, quite literally, to 'make a killing', indeed, to make many killings, slaughtering these beautiful animals almost to the point of extinction just to satisfy their own, selfish greed.

And **local communities**, who play a key role. Ultimately, the future of the animals rests in their hands, and this is a heavy burden to bear.

# Part II

## The Law

# 3

# CITES and the CBD

*Unsustainable trade in wildlife is one of the central threats to biodiversity as it con-*
*cerns thousands of plant and animal species, and can push them close to*
*extinction.*[1]

Having looked at the animals and the devastating effect poaching has on them, we must start to consider how the law can protect them. Two treaties in particular play a crucial role in determining any successful control of the illegal trade in elephant ivory and rhino horn. They are CITES, the Convention on International Trade in Endangered Species of wild Fauna and Flora, and the later Convention on Biological Diversity.

## 3.1 CITES: The Convention on International Trade in Endangered Species

CITES came into existence on 3 March 1973, in Washington, D.C., USA, as the result of an IUCN initiative taken because of a sudden and rapid increase in wildlife trade during the 1960s and 1970s.[2] Recognizing the fact that wildlife has always been traded by people, its objective was to regulate wildlife trade in those species most likely to become extinct were such trade to carry on as before. Thus, it takes a pragmatic approach to the problem.

---

[1] IUCN *About CITES*.
[2] See Bowman M., Davies P. and Redgwell C. *Lyster's International Wildlife Law* Cambridge University Press, 2nd edition, 2010, pp.483-4.

© Springer Nature Switzerland AG 2019
B. Martin, *Survival or Extinction?*, https://doi.org/10.1007/978-3-030-13293-4_3

By understanding that there will always be trade in such valuable commodities, valuable from the very fact of their rarity regardless of any other factors, CITES is able to prescribe a set of regulations that at least minimize the harm done, and it also sets out an administrative system to enforce them. To ensure maximum flexibility, the endangered animals and plants are listed on Appendices, a listing that can be altered if their protection needs to be strengthened or, more rarely, reduced. And the numbers listed continue to increase. At the start of 2016, there were some 5,600 animals and 30,000 plants included in the 3 Appendices.[3] Despite this treaty, the illegal trade in some species, including both elephants and rhinos, continues and has even increased.

The most endangered species of animals and plants, that is, those threatened with extinction, includes, for example, big cats such as tigers and jaguars, the great apes, sea turtles, many parrots and some orchids and cacti. All rhinos and most elephants also fall into this *threatened with extinction* category. These species are listed in Appendix I and the strictest regulations apply. Trade is only permitted in exceptional circumstances. The rules are more relaxed for Appendix II species because although these are not currently threatened with extinction, '*there should be some indication that they might become so if they presently are subject to trade or are likely to become subject to trade*'.[4] Therefore, the controls are still strict.[5]

Because of the implications to trade and the fact that any restrictions may be controversial, the listing procedure is rigorous. Let us see how it is done. Party P (a country signed up to CITES) wishes to list animal A in Appendix I. It produces evidence on trends in population and trading, which it submits, together with a proposal that A be listed in Appendix I, to the Conference of the Parties (CoP), for discussion and debate. The Conference will decide whether the proposal will be adopted. All Parties present vote, and acceptance requires a two-thirds majority. This applies whether the Appendix is I or II. We shall see later in our story that listing proposals play an important role.

To ensure maximum flexibility, the definition of *species* is wide, so that it also includes any *subspecies or geographically separate population*.[6] This distinc-

---

[3] Earth Negotiations Bulletin (ENB) *Summary of the 66th Meeting of the Standing Committee of the Convention on International Trade in Endangered Species of Wild Fauna and Flora, 11 – 15 January 2016/ Geneva, Switzerland* Volume 21 Number 87/Monday, 18 January 2016.

[4] CITES *Article II(2)(a)*.

[5] There is also an Appendix III. Where a country is concerned about the population of a particular animal or plant within its own borders/jurisdiction and needs the cooperation of other Parties to control any trade in it, the country can request the Conference of the Parties that the species be listed in Appendix III. If this is agreed, the species is then treated for trading purposes as though it was listed in Appendix II.

[6] CITES *Article 1(a)*.

tion enables different populations of Savannah elephants to be treated differently and thus has enabled the controversial one-off sales of elephant ivory to take place. Furthermore, because in many instances it is not the entire animal (or plant) that is being traded, the Convention applies to *specimens of species* which may be living or dead and includes *any readily recognisable part or derivative thereof.*[7] This wide definition plays a crucial role in our story as it is immediately obvious that both elephant ivory and rhino horn fall into the category of *specimens of species.* They are also *readily recognizable parts* of the animals, while much of the rhino horn sold is in powdered form, ready for medicinal use or already in medicines and thus would be classed as a *derivative thereof.*

**How Does CITES Work?**  It works by using a permit system. Each signatory country must appoint its own Scientific Authority and Management Authority.[8] The main function of the Scientific Authority is to provide whatever advice the Management Authority requires to carry out its role. In the UK, the Scientific Authority, which advises Defra (the Department for the Environment, Food and Rural Affairs), is split into two, Kew Gardens advising on plants and the Joint Nature Conservation Committee on animals. Similarly, the UK Management Authority has two *arms, the policymakers (Defra) who influence things like the listings and also stricter domestic measures and the delivery arm (Defra's agency, which used to be AHVLA, the Animal Health and Veterinary Laboratories, but is now APHA (Animal and Plant Health Agency)) the part which actually issues the permits and certificates allowing trade to take place in carefully controlled circumstances.*[9]

Any legal international trade in Appendix I species will only be authorized in exceptional circumstances.[10] However, where it is permitted, both export (and where necessary re-export permits) and import permits will be required, and these will not be granted without the consent of the Scientific Authorities of both exporting and importing countries. While the Scientific Authority of the exporting country must determine that the *export will not be detrimental to the survival of the species*, perhaps of even greater importance is the requirement that the Scientific Authority of the importing State is also satisfied, among other things, that the import is for purposes not detrimental to the

---

[7] CITES *Article 1(b).*
[8] CITES *Article IX(1).*
[9] E-mail to author from Chris.Auger@ahvla.gsi.gov.uk
[10] CITES *Article II(1).*

survival of the species.[11] In fact, the Management Authority of the State of export must be sure the import permit has been issued for the specimen *before* it grants the required export permit.

Similar export-re-export provisions apply to those species listed on Appendix II, in that the Scientific Authority must have determined that the *export will not be detrimental to the survival of the species.* However, because these species are not so vulnerable, the requirement for an import permit is dispensed with and the goods can be used for commercial purposes. In all transactions though, whether the species are in Appendix I or II, the Management Authority of the exporting State must be satisfied, among other things, that the specimen was acquired legally.

As 'working together' is a golden thread that runs throughout our story, let us look at an early example. We have just seen some of the functions of both Management and Scientific Authorities. Another function crucial to the effective operation of CITES is cooperation. For CITES to work, these Authorities must be prepared to work with not only customs and the police but also any other concerned agencies.

CITES, because it is an international treaty, can only set out policies. It is up to each Party to pass their own national legislation to implement the policies. As we shall see later, in some cases, the legislation is weak, occasionally bordering on unsatisfactory, while in others, the legislation may be even stronger than that required by CITES.

In theory therefore, all lawful trade should be sustainable, but in practice problems arise, particularly when deciding whether or not the export is detrimental to the survival of the species. Furthermore, where species are so endangered as to be listed in Appendix I, their very rarity can add to their desirability, making them even more vulnerable, for although in most circumstances commercial trade is banned, the demand for the products does not necessarily reduce and can even increase. Where that occurs, prices will rise making it even more of a temptation to carry on the illicit trade.

So has this affected our animals? Unfortunately, yes. In the case of rhinos, prices have risen to the extent that in Europe even rhino horns in museums have been 'pillaged', live rhinos in zoos have been threatened with attacks,[12] and now one has been cruelly slaughtered.[13] Needless to say, organized crime

---

[11] CITES *Article III*.

[12] Michael McCarthy *British zoos put on alert over rising threat of rhino rustlers* The Independent, 30/01/2013.

[13] Kim Willsher Rhino shot dead by poachers at French zoo The Guardian 7 March 2017. See https://www.theguardian///.com/world/2017/mar/07/rhino-shot-dead-by-poachers-at-french-zoo   Accessed 06/11/2018.

gangs are deeply implicated. Until 2011/2012, elephants had not suffered such sustained attack; then everything changed, with illegal killing and trading increasing so dramatically that there are now fears that some species, particularly the forest elephants, might become extinct in the not-too-distant future.

On the whole though, CITES works well, benefitting both producer and consumer countries and regulating trade in goods *estimated to be worth billions of dollars.*[14] Most of the consumer countries appear to take their obligations very seriously, passing legislation and setting up enforcement agencies in an attempt to control illegal trade. Perhaps the best evidence for its success is the fact that after a very slow and reluctant start, now there are very few countries not signed up to the treaty.

## 3.2 TRAFFIC

*TRAFFIC's role as an NGO combating illegal wildlife trade is to plant the seeds of change and to form partnerships that can effect the needed change in consumer behaviour in relation to wildlife products.*[15]

TRAFFIC, *the Wildlife Trade Monitoring Network*, is the result of an alliance between the World Wildlife Fund (WWF), with whom it has an active and productive partnership, and the International Union for Conservation of Nature (IUCN). Established in 1976, soon after CITES became international law, its aim is:

'*To ensure that trade in wild plants and animals is not a threat to the conservation of nature*', in other words, to make sure that trade is sustainable.

TRAFFIC has a vision and a 2020 goal. In its vision, wildlife trade is managed so that healthy wildlife populations and ecosystems are maintained, while at the same time contributing to meeting human needs and supporting local and national economies. And people are motivated to conserve wild species and habitats. Its 2020 goal is pragmatic; it is '*to help reduce the pressure of illegal and unsustainable wildlife trade on biodiversity, and enhance the benefits to wildlife conservation and human well-being that derive from trade at sustainable levels*'.[16]

---

[14] CITES Secretariat *What is CITES?* See https://www.cites.org/eng/disc/what.php Accessed 06/011/2018.
[15] James Compton in Symposium on Combating Wildlife Crime: Securing Enforcement, Ensuring Justice, and Upholding the Rule of Law. The Proceedings. Editors Kala K. Mulqueeny and Francesse Joy J. Cordon, 2014, pp.39-40.
[16] TRAFFIC, 2008.

An international organization, its headquarters are in Cambridge in the UK, and it works out of offices throughout the world. It is a treasure trove of knowledge regarding wildlife trade, knowledge that helps governments to develop policy, produce legislation, strengthen enforcement and generally work together and with other NGOs to better protect their wild animals and plants.

It is immediately obvious just how close are its links to CITES, and as our story progresses, we shall see what an important role it has played and continues to play, not only in the evolution of the treaty but also in the fate of our two animals.

**Curbing Demand** is essential if elephants and rhinos are to survive. The demand for ivory and rhino horn needs to be severely curtailed, and this is one of TRAFFIC's most important functions.

## 3.3    The CBD: Convention on Biological Diversity

It too plays a fundamental role in the protection of endangered species though its function is very different. Unlike CITES, which sets out a system of regulations whereby permissible trade in endangered species may be undertaken, this treaty merely sets out a list of principles. These are intended as a framework to guide the Parties as to how they can implement the treaty; hence, this kind of treaty is known as a Framework Convention. The CBD attempts to fill in gaps in existing legislation concerning biodiversity, and it is the first international treaty to address every aspect of biodiversity.

Most countries are Parties to this treaty because they are '*conscious of the intrinsic value of biological diversity and of the ecological, genetic, social, economic, scientific, educational, cultural, recreational and aesthetic values of biological diversity and its components*'.[17]

The following provisions are of particular concern to our story.

**Article 1** sets out the objectives, which are:

- The conservation of biological diversity
- The sustainable use of its components
- The fair and equitable sharing of the benefits arising out of the use of genetic resources

---

[17] Convention on Biological Diversity *Preamble* 1992.

We shall only consider the first and second objectives.

The conservation of biological diversity is concerned with ecosystems and genetic diversity. We shall be looking at this later in our story.

And the principle of sustainable use is considered to be so important that it has been written into the Constitutions of some countries, one being South Africa, home to both elephants and rhinos. Sustainable use implies there will be some use of 'commodities' such as elephant ivory and rhino horn but with the very important proviso that this use be sustainable.

**Article 3** recognizes the sovereign right of States to exploit their own resources provided these activities do not cause damage to other States, while **Article 5** recognizes the fact that there must be cooperation if the treaty is to succeed. **Articles 6 and 7** set out general measures for achieving conservation and sustainable use.

**Articles 8 and 9** are concerned with ecosystems.

Research and training (**Article 12**), access to and transfer of technology (**Article 16**), exchange of information (**Article 17**) and technical and scientific cooperation (**Article 18**) are all incorporated into the treaty, and provision is made for public education and raising awareness.

All this is going to be costly, so **Article 20**, which deals with financial resources, makes express provision for aid to be given to poorer countries. In effect, while all contracting Parties undertake to spend what they can afford on achieving the objects of this Convention, '*the developed country Parties shall provide new and additional financial resources to enable developing country Parties to meet the agreed full incremental costs to them of implementing measures which fulfil the obligations of this Convention …*'. In other words, developed countries will be expected to help out financially, if called upon to do so. We shall be returning to this very important provision.

The administration of the Convention is, in many respects, similar to that of CITES, having not only a Secretariat but also a Conference of the Parties (CoP). It has, in addition, a Subsidiary Body on Scientific, Technical and Technological Advice to provide advice (evidence) to the CoP. Further 'evidence' of how the treaty is being implemented can be found in the reports which each contracting Party must, from time to time, submit to the CoP. These reports set out what the Party has done to implement the Convention and how effective this has been.

We've not yet mentioned **Article 10**. This is concerned with sustainable use which plays a key role in our story and we will return to it later.

## 3.4   The Precautionary Principle

This important principle, which is not legally binding, was first defined in the 1992 Rio Declaration on Environment and Development. It is concerned with the prediction of environmental effects, and because there can be no certainty as to how certain events will work out in the future, it is always best to err on the side of caution.

> *In order to protect the environment, the precautionary approach shall be widely applied by states according to their capabilities. Where there are threats of serious or irreversible damage, lack of full scientific certainty shall not be used as a reason for postponing cost-effective measures to prevent environmental degradation* (**Principle 15**).

The precautionary principle is one of the key principles of the European Union (EU) where it must be taken into account whenever new policy and legislation are considered.

It is also included in the Preamble to the CBD, although, perhaps because its implications could be controversial, it never made it into the main text.

And it reappears in the *Addis Ababa Principles and Guidelines for the Sustainable Use of Biodiversity*, agreed at the CBD's seventh Conference of the Parties (CoP7), as one of the underlying conditions of sustainable use, and in the operational guidelines accompanying **Practical Principle 5**.

In other words, it is very important. So, in the context of our story, what does the principle mean? Suppose we were to take a worse-case scenario, a perceived threat of the extinction of some species of elephants and rhinos and their ecosystems. Applying the principle would allow us to take immediate remedial action without being delayed by the lack of 'full scientific certainty'. It could even be argued that had it been strictly applied to the one-off sales of elephant ivory, they may never have taken place. But, as we shall see, matters are never that simple.

## 3.5   CITES and the CBD

The close link between CITES and the CBD was examined in a workshop in 2004, *Promoting CITES-CBD Co-operation and Synergy*, which was organized and financed by a number of governmental and non-governmental bodies (NGOs).[18] Biological diversity requires animals such as elephants and rhinos to survive, but they are under threat, with an uncertain future because some

---

[18] *Proceedings of the Workshop, 20–24 April 2004*, held at the International Academy for Nature Conservation of the Federal Agency for Nature Conservation, p.5.

people value their tusks and their horns, parts of their bodies, more highly than they do the living beasts. So cooperation and synergy are fundamental to the protection of these animals.

## 3.6    What Is International Law?

Both CITES and the CBD are examples of international law, and there is a fundamental distinction between international and national law, a distinction that needs to be understood as it plays a crucial role in the next stage of our story. So let us see what it is.

If, in the UK, Oliver steals the eggs of an osprey or kills a badger by digging it from its sett and then baiting it, he will have committed not just a criminal offence but a number of them, set out in legislation such as the Wildlife and Countryside Act 1981 and the Protection of Badgers Act 1992. Assuming that Oliver is caught and sufficient evidence is found to show that he did indeed commit the offences, he will be arrested by the police and then prosecuted. The legislation has been passed by the government and will be enforced. In other words, any breach of national law has repercussions. International provisions, however, are different. The CITES Secretariat cannot prosecute poachers or the criminal gangs involved in illegal trade in rhino horn. It cannot even take action against erring States as we will see later.

In the UK, legislation starts life as a Bill, passes through Parliament, becomes law and then takes effect on a specified date, normally soon afterwards. However, a treaty does not come into force until a certain number of countries have decided to sign up to (ratify) its provisions. Thus although CITES was concluded in 1973, it did not take effect until 1975, after it had been ratified by ten countries. The final text of the CBD was negotiated and adopted at the Earth Summit at Rio in June 1992, taking effect as early as December 1993 because the importance of maintaining biodiversity had, by then, been recognized by so many countries. However, signatory States must pass their own legislation to implement the provisions of a treaty, and although this should not present a problem, not all legislation proves to be satisfactory.

**The CITES National Legislation Project** classifies the national legislation of each Party into one of three categories, according to *how fit for purpose* it is. The legislation in **Category 1** *is believed generally to meet the requirements for the implementation of CITES*, while that in **Category 2** *is believed generally not to meet all requirements for the implementation of CITES*, and **Category 3** legislation *is believed generally not to meet the requirements for the implementation of CITES*.

## 3.7    Brief Examples of the Implementation of CITES by Some of the Players in our Story

Here we take a quick look at the legislation of two countries that have been allowed to sell their legally held stockpiles of elephant ivory in the one-off sales: Botswana and South Africa. Then, by way of contrast, at the Regulations of the European Union (EU). We are focussing particularly on provisions relating to elephants and rhinos.

**Botswana** has been a Party to CITES since 1977 and has a number of pieces of legislation in place to protect its wildlife. The Wildlife Conservation and National Parks Act 1992, section 62, sets out explicit provisions relating to the export and import of animals in compliance with CITES. The requisite permits will only be issued where the Management Authority is satisfied that a number of conditions have been fulfilled and anyone who fails to comply is guilty of a criminal offence. The punishments are harsh. A fine of P(ula)10,000 and to up to 7 years in prison. Where the contravention is in respect of an elephant or any elephant trophy, the penalties are even higher, the fine being increased to P50,000 and the prison term to 10 years. And for a rhino or a rhino trophy, the fine is higher still, sometimes P100,000 although the prison term remains at 10 years.

**South Africa** was one of the first countries to sign up to CITES, in 1975, although for many years, indeed until Nelson Mandela became the President, there was a reservation on elephants. This meant that South Africa could treat elephants as though they were not protected by the treaty.

The current law dealing specifically with matters relating to CITES is GNR.173 of 5 March 2010: Convention on International Trade in Endangered Species (CITES) Regulations (Government Gazette No. 33002). Regulation 10 sets out the provisions relating to permits and certificates, and Regulation 16 lays down a series of offences and penalties if an accused person is found guilty of committing an offence. Punishment can be severe. A first-time offender can expect either a fine not exceeding R(and)5,000,000 or a term of imprisonment not exceeding 5 years, whereas for a repeat offender the punishment is doubled, the maximum fine being R10,000,000 and the custodial sentence not exceeding 10 years.

### 3.7.1    EU Legislation

The EU has passed a series of regulations, initially implementing CITES and then updating them when required. The first of these Wildlife Trade Regulations was Council Regulation (EC) No. 3626/82 which came into effect in 1984 and which, after amendment, lasted until 1997. It was replaced by Council Regulation (EC) No. 338/97 which has been amended a number of times but is still in force. As with the first regulation, its object is to protect species and guarantee the conservation of wild animals and plants. There are four Annexes, which offer better protection than CITES to some species. Detailed provisions require each Member State to designate its own Management and Scientific Authority, and levels of trade must be monitored. The four Annexes together offer a higher level of protection to some species.

Import, export and re-export of specimens of species, as well as specimens in transit through the Union, are all regulated, and, in order to ensure compliance, Member States must impose sanctions for infringements, monitor compliance, investigate infringements and inform the Commission of any steps taken in relation to significant infringements including seizures and confiscations.

Because these are regulations, any country joining the EU, immediately has to comply whether or not it has ratified CITES.

In 2015, the EU itself became a Party to CITES, thus becoming the first regional economic integration organization to sign up to the treaty and bringing a new power base to fight illegal wildlife trafficking. Of course, each Member State will still be able to act in its own right as well.

# 4

# Sustainable Use: Golden Thread or Fool's Gold?

Sustainable use is another key component in our story and we need to know its history.

## 4.1 The Stockholm Conference

In 1972, the Stockholm Conference on the Human Environment produced the Stockholm Declaration. This recognized people's responsibility to protect both the environment and the earth's natural resources, and, most importantly, it linked environmental protection for the benefit of future generations. This has now become the concept known as intergenerational equity and has been written into a number of treaties, including the CBD. It is concerned with the avoidance of irreversible harm.

The Stockholm Conference also noted the difference in priorities of the developed, mainly northern, countries and the developing countries mostly found in the southern hemisphere. In many of the latter, poverty and poor economic development are key factors in poaching and the illegal trade in elephant ivory and rhino horn.

## 4.2 Our Common Future

The early 1980s marked the beginning of global economic growth and accelerating ecological degradation. The dangers were becoming more apparent.

© Springer Nature Switzerland AG 2019
B. Martin, *Survival or Extinction?*, https://doi.org/10.1007/978-3-030-13293-4_4

**The World Charter for Nature** set up in 1982, among other things, wanted special protection for rare or endangered species, and where living resources such as elephant ivory and rhino horn were to be used, the use should take account the animals' capacity to reproduce. There should be no irreversible damage to nature.

**The World Commission on Environment and Development (the Brundtland Commission)** This was established in 1983, its brief to focus on environmental and developmental policy. The result was *Our Common Future*, a significant and influential report.

This stressed the importance of sustainable development, which it defined as:

> *development that meets the needs of the present without compromising the ability of future generations to meet their own needs.*

Furthermore, it recognized that both poverty reduction and wealth redistribution were intimately linked with environmental conservation. From now on, the role of economic development could no longer be ignored. Sustainable development had to be on the agenda of world governments.

## 4.3   The Rio Declaration 1992

The **1992 Earth Summit, the UN Conference on Environment and Development**, was held in Rio de Janeiro, Brazil. Its underlying scheme was sustainable development, and one result was **the Rio Declaration**, which was adopted by consensus at the Summit and later endorsed by the UN General Assembly. Although it laid down 27 principles, we are only concerned with those relevant to our story.

**Principle 1** places people '*at the centre of concerns for sustainable development*'. They are entitled to '*a healthy and productive life in harmony with nature*'.

**Principles 3–9** are concerned with redressing the imbalance between developed and developing countries by prioritizing the needs of the poor while, at the same time, not disadvantaging later generations. Poverty is to be eradicated and differences in living standards between the developed and developing world are to be reduced. There are to be transfers of scientific knowledge and technology. We will see later just how important a role this is playing in the protection of both elephants and rhinos.

**Principle 4** is concerned with the integration of economic development and environmental protection. Sustainable development, it is clear, can only be implemented through political action. This implies that States will be prepared to carry this through, which, in some cases, will not be easy for them.

**Principle 8** restates the requirement that there should be fair use of resources and requires States to '*reduce and eliminate unsustainable patterns of production and consumption*'.

**Principle 15** is the precautionary principle.

The Rio Declaration makes it quite clear that, provided the needs of present and future generations are met, countries have a right to exploit their own resources in accordance with their own environmental and development policies.

## 4.4   The CBD

So how does this treaty add to our understanding of sustainable use? Because it is concerned with conserving biological diversity and the sustainable use of biological resources is a key component.

The **Preamble** begins by recognizing '*the intrinsic value of biodiversity*' and, at a later stage, '*a resource*', and it also makes clear that conservation and sustainable use of biodiversity are critical for meeting the food, health and other needs of a growing world population. **Article 1** sets out the second objective of the Convention, which is '*the sustainable use of its components*', while **Article 10** requires Parties to adopt measures relating to the use of biodiversity to avoid or minimize impacts on biological diversity.

The Convention therefore provides a framework whereby there is a trade-off between conservation and economic equity. Although it still remains to be seen how effective this will prove to be, the importance of the principle of sustainable use continues to grow.

## 4.5    The Addis Ababa Principles and Guidelines for the Sustainable Use of Biological Diversity

In 2002, the **Plan of Implementation of the World Summit on Sustainable Development** restated how sustainable use is an effective tool to combat poverty and achieve sustainable development. Two years later, in his **Foreword to the *Addis Ababa Principles and Guidelines for the Sustainable Use of Biodiversity***, Hamdallah Zedan wrote that '*it is also an effective tool for achieving the Millennium Development Goals, eradicating extreme poverty and hunger and ensuring environmental sustainability*' and '*it can generate incentives for the conservation and restoration of biodiversity because of the social, cultural and economic benefits that people derive from it*', provided of course that the use is sustainable and not excessive.[1]

The 14 interdependent principles and accompanying guidelines have been designed to provide a framework to help governments, indigenous and local communities, resource managers, the private sector and other stakeholders. They provide the tools that, hopefully, will make sure that the use of biological diversity will indeed be sustainable and will not lead to its long-term decline.

They have been endorsed by the IUCN[2] and also by CITES, initially at CoP13 and again, in amended form, at CoP14 in Conf. 13.2 (Rev. CoP14).

## 4.6    The Strategic Vision of CITES

Finally we return to CITES, to see how it intends to support biodiversity and sustainable use.

In 2000, CITES CoP11 adopted the **Strategic Vision through 2005**, together with an **Action Plan**. At CoP13, both of these were extended until 2007 and then extended again, this time to 2020, at CoP16. The Resolution (Conf. 16.3) passed at CoP16 also included amendments which will contribute to achieving the **Strategic Plan for Biodiversity 2011–2020**, relevant **Aichi Biodiversity Targets** and relevant outcomes of the **UN Conference on Sustainable Development**. These will be strengthened by ensuring that inter-

---

[1] Secretariat of the Convention on Biological Diversity *Addis Ababa Principles and Guidelines for the Sustainable Use of Biodiversity* 2004.

[2] IUCN (the World Conservation Union) Recommendations February 2004 to the Seventh Meeting of the Conference of the Parties to the Convention on Biological Diversity *Sustainable Use: Adoption of draft Addis Ababa Principles and Guidelines* (Agenda item 19.5).

national trade in wild animals and plants is conducted at sustainable levels. Of course, this applies to both elephants and rhinos.

The **2008–2020 Vision** sets three strategic goals. This is the current CITES Vision Statement:

> *Conserve biodiversity and contribute to its sustainable use by ensuring that no species of wild fauna or flora becomes or remains subject to unsustainable exploitation through international trade, thereby contributing to the significant reduction of the rate of biodiversity loss and making a significant contribution towards achieving the relevant Aichi Biodiversity Targets.*

# Part III

## The Commodities

There is one thing more to consider before our story can start. We must look at the commodities themselves.

# 5

# Elephant Ivory and Rhino Horn

## 5.1 Understanding the Problem

We must recognize the fact that elephant ivory and rhino horn are 'commodities' that can be and are bought and sold. Goods whose value continues to skyrocket up into the stratosphere. Goods that can only further increase in value as their 'origins' continue to almost free-fall into extinction.

**Elephant Ivory:** Here it is important to realize that tusks are not the only part of an elephant that can be sold. Elephant hide can be fashioned into objects tourists will buy, their flesh can be eaten as bushmeat and in Myanmar, for example, wild elephants are captured then smuggled into Thailand where they are used to give rides to tourists. Our story, however, is about the illegal trade in ivory.

People's love affair with ivory and carving it goes back to the dawn of time. There are so many beautiful artefacts made of ivory, some of them very important, some simply objects to delight the eye and some of deep religious significance. Some are modern, some are antique. All of it is very valuable not just in financial terms.

Ivory was high on the list of 'must have' items favoured by the rising middle class in China. This should have come as no surprise, since the love affair of the Chinese people with ivory is deeply embedded in their culture and many of them were unaware of its origin. That most modern ivory came from elephants that had been cruelly poached. That elephant numbers might be approaching a tipping point towards extinction. But ivory is bought by people from all over the world. This is a truly global trade.

© Springer Nature Switzerland AG 2019
B. Martin, *Survival or Extinction?*, https://doi.org/10.1007/978-3-030-13293-4_5

**Rhino Horn:** So is the trade in rhino horn. We shall see how battles rage between the 'good guys', which is most people in the world, and the 'bad guys', those who care nothing for the animals except in so far as they can be exploited to make fortunes for them, despite the best efforts of governments and the rest of us.

Like elephant ivory, rhino horn has been prized for thousands of years. However, there is a tendency to believe that only the peoples of southeast Asia are interested in its supposed medicinal properties. Not so. Our story now moves to Europe, to ancient Greece and to Iran or to be precise to ancient Persia. Rhino horn was used by the ancient Greeks to purify water. Even more important was the widespread belief in its power to detect poison. Originating at some time in 500 BC, the Persians used to pour suspected liquids into rhino horn drinking cups, to see whether they would start to bubble. Bubbling indicated the presence of poison. In fact, this belief may be well founded, for modern-day chemists have concluded that such a reaction is possible provided that the poison is an alkaloid.[1] Hindus, Muslims Buddhists and Christians have all believed in the power, though only from time to time. Indeed, until comparatively recently (namely, up to and through the eighteenth and nineteenth centuries), it was even believed to be true by the crowned heads of Europe.

Now to a mystery. What is the representation of Durer's rhinoceros doing on top of the coat of arms of the Worshipful Society of Apothecaries, one of the craft guilds or livery companies of the City of London, in the UK? The shield is supported by two unicorns, one on either side of it and on top is the crest, in which is the rhino. Hubert Chesshyre, Clarenceux King of Arms, who researched the matter, concluded in an excellent article that all three beasts were probably there because of the medicinal properties attributed to their horns, mediaeval druggists deeming powdered unicorn horn to have prophylactic powers as well.[2]

But rhino horn has another use. Like ivory, it can be carved. And it makes beautiful objects which are found in museum collections all over the world. Although most of them are of high value, such as the libation cups that, for centuries, Chinese noblemen gave their emperors as birthday gifts, there have also been fripperies. This happened in Europe, particularly during the 1920s, when rhino horn could be found as door handles, pistol grips and sometimes walking sticks. It was even used as decoration in some expensive cars. In Yemen, it was traditional to present adolescent boys with special curved daggers with rhino horn handles, *as a sign of their manhood and devotion to Islam.*[3]

---

[1] US Fish and Wildlife Service *Facts about Rhino Horn* Office of Law Enforcement. See https://www.fws.gov/le/pdf/rhino-horn-factsheet.pdf Accessed 22/09/2015.

[2] Hubert Chesshyre Chester Herald of Arms *The Worshipful Society of Apothecaries* British Heritage Dec/Jan 1981/82.

[3] See US Fish and Wildlife Service (n.45).

**Fig. 5.1** The Coat of Arms of the Worshipful Society of Apothecaries. Copyright the Worshipful Society of Apothecaries

In other words, like elephant ivory, rhino horn has been treasured for centuries by the peoples of the world.

## 5.2   What Is Ivory?

*Ivory on sale in Vietnam is commonly mixed in with pig teeth and carved bone, perhaps in an attempt to dupe government inspectors.*[4]

So how do we know which artefacts are made out of or contain elephant ivory when there are so many kinds and they can be difficult to identify? There is a book an *Identification Guide for Ivory and Ivory Substitutes*, a handbook for enforcers and for the wildlife law enforcement officers, scientists and managers who are involved in stamping out the illegal trade. However, the book is for guidance only, to provide a first step, sufficient information to

---

[4] Daniel Stiles (2009). *The elephant and ivory trade in Thailand* TRAFFIC Southeast Asia, Petaling Jaya, Selangor, Malaysia.

judge whether or not the ivory comes from an elephant. Then, if there is doubt about its legality, it can be seized. A trained scientist must positively identify the source of the species.[5]

Other people concerned to ensure their ivory is legal are antique dealers and dealers in collectibles, consumers as opposed to producers. In this case, in addition to publishing literature to help with identification of different ivories, there are also books full of pictures of artefacts which provide an important insight into potential end products.[6]

## 5.3   Some Different Ivories

Ivory comes in many forms and from a number of sources. Although the most common source is teeth or tusks, one ivory surprisingly is vegetable, and others are purely synthetic. Ivory artefacts must also be distinguished from those made of bone or even shell, so it is essential to be able to distinguish both elephant and mammoth ivory.

**Elephant Ivory:**  Coming from its tusks or upper incisors, it is a dense substance, chemically similar to bone and antler. It has a fine, even grain and lines of Retzius which make it unique and give it its great desirability as a substance to be carved. These are growth rings and they are only found in teeth. They are the tooth equivalent of tree rings and they give the age of the tooth's owner. When the tusk is carved, they produce an *engine-turned effect (intersecting lines with a diamond shape between them)*.[7] For identification purposes, this is the equivalent of seeing a hallmark on gold or silver. But elephant ivory can also be identified by the presence of Schreger lines, which show up as a *chequer board* or *diamond* pattern when the tusk is seen in cross-section.[8]

As we saw earlier, although mammoths have been extinct for over 10,000 years, the bodies and tusks of one species, the woolly mammoth (*Mammuthus primigenius*), continue to be found, often in such good condition

---

[5] Edgard O. Espinoza & Mary-Jacque Mann *Identification Guide for Ivory and Ivory Substitutes* WWF TRAFFIC in cooperation with the CITES Secretariat. This online version was published in 1999. See http://www.cites.org/sites/default/files/eng/resources/pub/E-ivory-guides.pdf Accessed 9/06/2014.

[6] *Identifying Different Types of Ivory* Part of the Uniclectica Antiques and Collectibles Online Series *Caring for your Antiques and Collectibles*. See http://www.uniclectica.com/conserva/ivory1.html Accessed 9/06/2014.

[7] Ibid.

[8] Jacqueline Codron *Annals of Ivory: Perspectives on African Elephant Loxodonta africana* (Blumenbach 1797) *Feeding Ecology from a Multi-Decadal Record* 2008 (Ph.D. Thesis). See http://uctscholar.uct.ac.za/PDF/91348_Codron,J.pdf Accessed 17/06/2014.

that the tusks provide high-quality carvable ivory. Sometimes they can be over 13 ft. long and thick as tree trunks. And they can fetch large sums of money as the ivory is highly desirable.[9]

**Mammoth Ivory:**  While the outer layer of the tusk may be coloured by minerals found in the soil, inside, it is a light cream colour, and it polishes to a bright shine.[10]

Other ivories include:

- **Hippopotamus ivory**, the second most commonly used ivory. It is found in inlays or as small items like buttons.
- **Walrus ivory**, which comes in three forms. Its teeth are normally carved into small objects like jewellery, inlay or even small scrimshaw, while Oosik is used for knife handles.[11]
- **Sperm whale and orca ivories**, which are hard to distinguish from walrus ivory. They also have a long history of carving, particularly into scrimshaws.
- **Narwhal ivory**, which comes from its single tusk or horn which has ancient historical significance. It was thought to possess mystical powers and is associated with unicorns, and the Vikings believed that *a cup made from the tusk of a narwhal would save the owner from poisoning by his enemies.*[12] Rather like rhino horn cups.
- **Warthog ivory**.
- **Hornbill ivory** (from the bird, the helmeted hornbill which is now under threat)
- **Vegetable ivory** (from the Tagua palm tree nut)
- **Synthetic ivory**, of which there are a number. Some are so realistic that chemical tests are needed to spot the fakes. Netsukes, or Japanese perfume bottles, are normally carved from elephant ivory and are very decorative, are highly collectible and fetch high prices, but during the latter part of the twentieth century, there were so many excellent fakes, most made of synthetic ivory and so hard to detect that the bottom fell out of the market.

---

[9] Brook Larmer *Of Mammoths and Men* National Geographic, April 2013. See http://ngm.nationalgeographic.com/2013/04/125-mammoth-tusks/larmertext Accessed 17/06/2014.

[10] Canadian Ivory Inc. See http://www.canadianivory.com/ Accessed 12/06/2014.

[11] Ibid.

[12] Ibid.

## 5.4   The CITES Ivory and Elephant Seizure Data Collection Form

Elephant ivory has been divided into categories for evidential purposes. In the form there are five:

**Raw Ivory:** includes tusks from various sources. They may be whole, mounted or polished or broken or cut into pieces, but not otherwise shaped or processed.

**Semi-Worked Ivory:** has only undergone some processing. A good example is ivory blocks made into name seal blanks.

**Worked Ivory:** has been made into finished products such as piano keys, chopsticks, game pieces, carvings and composite sculptures.

**Fresh Ivory:** is ivory that has been recently poached. It can be identified in a number of ways. Blood or particles of flesh may still be attached to it, and sometimes there are signs of an axe or machete used to hack it from the elephant. It will not be discoloured or cracked from ageing.

**Old Ivory:** looks weathered and discoloured. It is also cracked and brittle.[13]

Unfortunately, as yet, there is no on-the-spot test to determine whether or not the ivory is old or comes from a lawful source so can be legally sold, although some antiques come with provenance and some artefacts have 'a passport'.

## 5.5   Mammoth Ivory, Other Ivory, Bones and Synthetic Substitutes

Because mammoth ivory in any form can be legally exported, imported and sold and there are no restrictions on other ivories or bones, as the price of ivory has risen, the temptation to use substitutes for elephant ivory has increased and with it the temptation to pass them off as elephant ivory.

---

[13] Milliken, T., Pole, A., and Huongo, A. (2006) *No Peace for Elephants: Unregulated Domestic Ivory Markets in Angola and Mozambique.* TRAFFIC International, Cambridge, UK, p.6. See also Conf. 10.10 (Rev. CoP14), *Regarding marking.*

It is mainly Chinese craftsmen who carve these substitutes, and although some artefacts were produced before the total ban on international sales of elephant ivory, it was the ban that really caused this industry to take off. Particularly items carved from mammoth ivory. And to such an extent that the State Forestry Administration in China produced a leaflet setting out the differences between elephant and mammoth ivory, explaining how to distinguish the two.

Siberia exports mammoth ivory, and China imports it, often via Hong Kong where there is no import tax. Some factory owners even go to Russia to buy mammoth tusks. All the ivory is graded and it fetches good prices. Furthermore, it can be exported legally.

So what is happening in China? Researchers found that mammoth tusks were being carved in factories in Guangzhou and Fuzhou, where elephant ivory is also carved. In one of the three factories in Guangzhou, almost half of the items were carved from mammoth ivory, while in the other, most were carved from cow bone and consisted of elaborate landscape scenes, magic balls and some composite pieces similar to those carved from elephant ivory. In 2007, this factory bought a supply of mammoth ivory and, in 2010, hippo teeth. Although the researchers didn't see any mammoth ivory in the third factory, the manager told them that it was used.

Mammoth tusks were also being carved in the three factories in Fuzhou. All three factories exported this ivory, and although it only amounted to about 10% of the sales in one factory and 40% in another, the third factory exported most of its output. Its ivory carvings, of human and animal figures, were particularly fine. Some of the tusks were merely polished.[14]

The use of ivory substitutes seems to be increasing in Thailand as well, although no substantive data is available at present. Worked mammoth ivory is being imported from China, and mammoth ivory, bone and resin pieces are being painted to resemble ivory, which can then be offered for sale mixed in with authentic elephant ivory.

## 5.6   What Is Rhino Horn?

The most important fact we need to recognize about rhino horn is that it is composed of hard keratin proteins, alpha keratins, which are common to most mammals. Indeed, our hair and finger nails are made of these same proteins.

---

[14] Esmond Martin and Lucy Vigne (2011) *The Ivory Dynasty: A report on the soaring demand for elephant and mammoth ivory in southern China* Published by Elephant Family, the Aspinall Foundation and Columbus Zoo and Aquarium 2011.

There are no magic ingredients. Yet down the ages, it has been imbued with more and more magical properties, the latest being its ability to cure cancer. And it is these allegedly healing properties that are mainly responsible for the near extinction of these beautiful animals.

## 5.7    Distinguishing Rhino Horn from Fakes

Unlike elephant ivory, distinguishing between fake and real rhino horn can be dealt with relatively briefly. The most realistic false rhino horns are apparently cow or buffalo horns, which are a deep black in colour. At least these are harmless if they are used in medication, unlike other fake horns made from hair and plastic pressed together or with plastic playing some part in their making. And with the high demand for rhino horn in Viet Nam, there is no way it can be ever be met with the real thing, so a 'cottage industry' making fake horn is beginning to develop in villages in some parts of Viet Nam.[15] When researchers visited a factory where fake horns were manufactured, they were able to film the production process with a hidden camera.

The level of ignorance about a rhinoceros is staggering. Some traditional medicine practitioners appear to believe that rhinos *have two kinds of horn: one on the top of the head and one under the jaw*, and one market seller claimed to his prospective buyer that the horn he was trying to sell him was *the one on the top of his head*.[16] Perhaps this is one reason why, according to Milliken and Shaw, fraud *characterises a major part of the trade itself.*[17]

## 5.8    Rhino Horn as a Medicine

Now we must consider rhino horn as a medicine, and this must be carefully done for it is always important not to dismiss potential medicinal benefits out of hand. The people who use these remedies deserve respect. Take bear bile, for example, another ingredient of traditional Chinese medicine. When this was investigated, it was found to contain ursodeoxycholic acid, an active ingredient that can help some liver conditions. In other words, it did have some medicinal use. Even more important was the fact that, once this

---

[15] Milliken, T. and Shaw, J. (2012) The South Africa – Vietnam Rhino Trade Nexus: *A deadly combination of institutional lapses, corrupt wildlife industry professionals and Asian crime syndicates.* TRAFFIC, Johannesburg, South Africa; see Anon, 2011c p.129.

[16] Ibid, p.129.

[17] Ibid.

capacity was known about, work was done and now this substance can be produced synthetically. No need for any more bears to be killed (or farmed) to get this product!

Rhino horn is normally used medicinally in its powdered form, and in fact very little research has been carried out as to exactly what, if any, healing properties the horn might have. However, it is supposed to be able to chill and detoxify. These properties, first described in the Han dynasty's *Divine Peasants Herbal*, were re-empathized as recently as 2012 by none other than the President of the UK Association of Traditional Chinese Medicine. In other words, they deserve serious consideration.

By 2012, with the illegal killing of rhinos reaching ever more alarming proportions, and with seemingly wilder and wilder claims being made for their horns' apparently magical healing powers, it was obvious something had to be done. A report was produced by TRAFFIC for the CITES Secretariat. It was an overview of the current uses of the horn in five countries/territories, namely, China, Taiwan (Province of China), Japan, the Republic of Korea and Viet Nam. All of them had a strong tradition of using the horn as a form of medicine, and all had historic consumer markets. Even more important was the fact that all the countries had banned international trade in the horn by 1994 and all domestic trade by 1995. Despite this, the trade continued, the belief in its healing powers seemingly as strong as ever.

But we already know that rhino horn is made up entirely of alpha keratins, just like our hair or finger nails. Furthermore, there are no significant differences in composition between it and water buffalo horn.[18] So does it have a medicinal use? Unlike bear bile and its ursodeoxycholic acid, few attempts have been made either to isolate or identify any active components the horn may contain, and those that have been undertaken have not yielded any results.[19]

Fortunately, the only trial involving humans did apply best practice, being randomized and double-blind. The results showed *a short-lived significant effect on fever in children*, but its use was not recommended because another drug, a common non-steroidal anti-inflammatory drug, performed better.[20]

In order to identify viable substitutes, China allowed research to be carried out using actual rhino horn. The tests, which used in vitro techniques as well as laboratory animals, produced some interesting results. All of them found *statistically significant pharmacological effects* when the horn was used, among

---

[18] Kristin Nowell *Species Trade and Conservation, Rhinoceroses: Assessment of Rhino Horn as a Traditional Medicine* A report prepared for the CITES Secretariat on behalf of TRAFFIC, April 2012, SC62 Doc.47.2, see Lee and Kim, 1974, pp. 6–9.

[19] Ibid, p. 2.

[20] Ibid, p. 1.

other things, to reduce fever, as an anti-inflammatory, an analgesic and a procoagulant. Far more important was the fact that these results were replicated when using animal horn substitutes.

So has there been any research outside southeast Asia? In fact, two studies have been carried out, one in the UK and one in South Africa, and their results were rather different. They *found no pharmacological effects* for any of the animal horns, while *some traditional medicinal plants that were tested as possible substitutes for the rhino horn did produce some significant results.*[21]

The most recent and most extravagant claims that rhino horn offers a potential treatment for both cancer and strokes remain to be substantiated.

Japan included rhino horn in its traditional medicine system 300 years ago during the Edo period, and, more recently, in its national pharmacopoeia between 1962 and 1980. It was used in two forms. As rhino horn slices, its main use was for treating children with colds, fevers and measles, but it was also included in well-known manufactured household medicines, one of many ingredients. Parents used these medicines to sedate their children, while adults used them as a cardiotonic. When the sale of rhino horn was banned in 1980, the Japanese started using saiga antelope horn instead and found it was just as efficacious.[22] As a result, now these animals are becoming threatened as well.

Perhaps we can derive a kernel of hope from the fact that China has carried out a series of properly controlled experiments and found that both rhino horn and animal horn substitutes produce the same results. And Japan has managed to remove virtually all rhino horn from its traditional medicines.

---

[21] Ibid, pp. 1–2 and 9–18.
[22] Kitade and Toko 2016.

# 6

# The Products, Their Retail Outlets, Their Sellers and Their Buyers

So 'who will buy my bright white ivory, who will buy my rhino horn?' We need to see not only what ivory and rhino products are for sale but also who sells and, more importantly, who buys them. For the key to stopping this illegal trafficking is to reduce, and if possible, eliminate all demand.

We will start with ivory, reminding ourselves once again that people have been carving ivory for generations. That carving runs in families and that in many countries still producing ivory artefacts, it is embedded in their culture. Our information/evidence comes from reports.

## 6.1　Ivory Products

Perhaps the most important thing to note is the great variety of objects for sale, ranging from simple bangles to complex landscapes carved from whole tusks. So the objects themselves can provide clues, and experts on the illegal trade use these, together with other clues, to work out the nationalities of the people buying this ivory.

The first thing to note is that items varied considerably in size, some large, some tiny, some over 30 cm in length or height, some under 10 cm and all sizes in between. Many of the small items could be classed as jewellery and

© Springer Nature Switzerland AG 2019
B. Martin, *Survival or Extinction?*, https://doi.org/10.1007/978-3-030-13293-4_6

trinkets. They included necklaces, bracelets, bangles, rings, combs, and hair-clips, ivory toothpicks, key-rings and, in China, charms[1] and brush holders.[2]

Ivory animals were popular. In one market in Mozambique, the researchers saw a tiny sable antelope and a turtle, and one hotel shop was selling six small ivory swans, while a small curio shop had dolphins, giraffes, zebras, okapis and kangaroos for sale. Carved fruit, chopsticks, unfinished name seal blocks and sometimes even dragons could be bought.

Busts and heads were medium to large items. Geisha girls jostled with figures of the Buddha.

In Egypt, researchers found figures of humans, Gods and animals, as well as pieces of jewellery, particularly pendants. Lots of scarab beetles, as might be expected. But also some antique ivory objects as well as some newer items clearly produced in other countries especially China, Japan and France, walking sticks and rosary beads.[3]

Sometimes there were ivory/wood composite pieces. An obvious example of these would be boxes. In Mozambique, researchers found one market selling, amongst other items, four wooden guinea fowls with ivory beaks and three ivory/turtle shell necklaces. A wooden mask was inlayed with ivory and turtle shell, and there were even Maasai figures, all carved from ebony and inlaid with ivory and tortoiseshell. The researchers saw 42 of these, mainly small (24) or medium (14) or life-size (4).[4] Games were sometimes offered for sale.

However these were not the only composite pieces. For example, in both Thailand and China, researchers saw ivory belt buckles, knife and sword handles.[5]

Intricate landscapes carved out of whole tusks were usually antique. Similarly intricate were the magic balls once so beloved by the Chinese, but no longer so popular. These are a speciality of the Chinese city of Guangzhou and consist of as many as 52 intricately carved hollow, concentric balls, one inside the other.[6]

[1] Esmund Martin and Lucy Vigne (2011) *The Ivory Dynasty: A report on the soaring demand for elephant and mammoth ivory in southern China* Published by Elephant Family, the Aspinall Foundation and Columbus Zoo and Aquarium 2011

[2] Grace G. Gabriel, Ning Hua, Juan Wang *Making a Killing: A 2011 Survey of ivory markets in China* IFAW 2011

[3] See Martin and Vigne (n.1).

[4] Milliken, T., Pole, A., and Huango, A. (2006). *No Peace for Elephants: Unregulated Domestic Ivory Markets in Angola and Mozambique.* TRAFFIC International, Cambridge, UK

[5] Daniel Stiles (2009). *The elephant and ivory trade in Thailand* TRAFFIC Southeast Asia, Petaling Jaya, Selangor, Malaysia

[6] Martin 2006

Occasionally, actual tusks or segments of tusks unworked but polished could be bought, though these were rare. However, in one Angolan market, researchers saw as many as 10 polished tusks, whose sizes ranged from 25 to 90 cm long and whose total weight the researchers estimated to be 105 kg.[7]

## 6.2    Ivory Vendors and Their Retail Outlets

So a great variety of goods were for sale. Who was selling them?

In the countries surveyed by the researchers, most people tended to buy ivory from markets. This is not an entirely unexpected result. After all, as well as catering for the needs of local people, markets are something of a magnet for the tourist industry. And if you, the vendor, are aware that selling ivory to international tourists is illegal, it is much easier to cover your tracks in a market. By their very nature, markets are hard to police.

Ivory was also sold in shops of various kinds. While some were in streets, many of them were located in hotels, a few even in airports. For example, in 2005 there were four shops in Mozambique's Maputo Airport selling ivory items, two of them only accessible after passengers had passed into the departure lounge and these were not selling exempted ivory. In fact, there were 588 pieces of ivory on offer for sale, which the purchasers could only take out of the country. In other words, any sales would be blatantly illegal, being precisely the type of sales banned under CITES.[8] Whoever the vendors were, it would seem that the authorities were turning a blind eye.

Under Chinese legislation, all ivory should be sold in registered shops, but not all of it is. We will be looking at this in more detail later, as well as the ever-present problems with enforcement.

**Is This Sale Legal?**  An important question that demands answers urgently is: 'Do you, the ivory seller, realize that by selling this particular ivory product to a foreign tourist, you are breaking the law?'

The surveys showed, without a doubt, that some sellers knew these transactions were illegal.

Researchers described an encounter in one tourist shop in Angola, a shop which was actually being run by two Europeans. There were an estimated 7.8 kg of ivory goods openly on display, which they were quite blasé about

---

[7] See Milliken, Pole and Huango (n.4), pp.13–21.
[8] Ibid.

selling to tourists for illegal export, even suggesting that *small items could easily be put in a suitcase or a pocket and that Angolan authorities would not bother with small amounts of ivory.*[9]

In Egypt, researchers recorded the slick operation of the vendors, noting that *many Egyptian sales personnel are very savvy operators on the international tourist scene and would often quote prices for ivory products in US dollars (USD) or euros once they ascertained the nationality of prospective buyers.*[10] Nor did the buyers need to worry about getting their purchases out of the country as apparently some sellers were prepared, indeed offered *to write a receipt indicating that a piece was an antique or made of camel bone.*[11]

As we shall see later, in some retail outlets in China, ivory laundering has become something of a fine art. However, there is 'being aware' and 'being aware'. Undoubtedly some of the sellers are fully aware of what they are doing, realizing they get away with it perhaps because of poor enforcement, but there will be many others who, having no appreciation of the devastating consequences of their ancient trade, merely see its regulation as more government aggravation.

## 6.3   The Customers

The first point to note is although the markets and many of the shops were open to the public, some of the items being sold were unlikely to be bought by other than international tourists. For example, Europeans preferred carved animals and sculptures of African figures, while Asians were attracted by chopsticks, blank name seals, Geisha girls, Buddhas and dragons. A further distinction is that ethnic Thais preferred amulets and good luck charms, while Chinese Thais bought Chinese-style figures.

Some of the stallholders were quite prepared to talk about the nationalities of their customers. From them we learnt that buyers at the Mercado do Artesanato in Angola included Americans, Italians, Portuguese, English, Spanish and Chinese, as well as Angolans of course, while two women running the shop in the Alvalade, a new hotel, again in Angola, said that most of their customers were Europeans and that some were diplomats.[12] It is also clear that different countries had different customers, so that in 2005 in

[9] Ibid.

[10] Martin, E. and Milliken T., 2005 *No Oasis: the Egyptian Ivory Trade in 2005* TRAFFIC International, Cambridge, UK.

[11] Martin E. and Vigne L. *Illegal Ivory Sales in Egypt* TRAFFIC Bulletin, 30 January 2012.

[12] See Milliken, Pole and Huango (n.4).

Egypt, unlike Angola, they were mainly Europeans from Italy, Spain and France. Others were from Latin America and a few from the Gulf States.

The second point to note is that there were very few buyers from China, which was perhaps why the artefacts on sale were not of the kind that would have appealed to Chinese consumers. And no Egyptians, as apparently they didn't buy ivory.[13]

Customers were different again in the West African countries, so that in Senegal, although both the Europeans and Chinese were in evidence, there were also Koreans and the local Marabout traders, while in Côte d'Ivoire, the European buyers were mainly French, and Americans were buying far more ivory. There was also the curious fact that some of the local African business men buying ivory were traders from Senegal. In Nigeria, once again there were American, Chinese and European (this time Italian) buyers, as well as Koreans and some Lebanese.[14]

But of course customers do not stay the same. We have already remarked that during the 2005 survey in Egypt, very few Chinese and no Egyptians were buying ivory products. By 2011, however, this had all changed. The Chinese had become the principal buyers, and they were prepared to spend big money. One seller told the researchers that *Chinese buyers would sometimes spend USD50,000 on ivory during one bargaining session.*[15] And the Egyptians themselves were buying it.

There was a similar switch in Thailand. After 2008/2009, a significant change occurred in that the Thais started to buy ivory,[16] whereas before that, the buyers were mainly Europeans, Americans, Japanese and ethnic Chinese from Malaysia, Singapore, Taiwan and mainland China.[17]

Some customers bought in bulk, others for resale, like a few of the European traders in Senegal. And some were diplomats, an important and disturbing discovery. Let us look at two examples. The two women running the shop in the Alvalade hotel in Angola said that most of their customers were European and many were diplomats,[18] while traders in Senegal claimed Chinese diplomats to be frequent customers.[19] In Côte d'Ivoire, there were allegations that some Asian diplomats were buying ivory products for resale abroad.[20] We shall return to this.

---

[13] Martin and Milliken 2005.

[14] Couroble, Hirst and Milliken 2003.

[15] See Martin and Vigne (n.11).

[16] See Martin and Vigne (n.1).

[17] Luxmore 1989; Martin and Stiles 2002

[18] See Milliken, Pole and Huango (n.4).

[19] See Martin and Vigne (n.1).

[20] Ibid.

## 6.4   Rhino Products

In Viet Nam, rhino horn products tend to fall into four categories:

- A 'wonder drug', a curative medicine for people suffering from serious, possibly terminal illness. It is, for example, touted as a cure for cancer although there is no evidence to support this claim.
- A detoxifier. The powdered horn is normally mixed with water and used as a cure for a hangover and a general health, body-rejuvenating tonic. This has now been taken a step further. Instead of water, the horn is added to rice wine, the concoction then being drunk in elaborate social or business settings.
- A product for the family medicine chest, the powdered horn to be used as part of a medicine that will reduce a high fever.
- A high-value gift. This may well take the form of part of a horn, as they are so expensive to purchase and it has recently led to the extraordinary and occasional use of horn as currency.[21] *'The man with the rhino horn was a car salesman and had acquired it as a client as part payment for a new car in lieu of [US] $5,000'*.[22] luxury payment for luxury items.

There is also a market in hunting trophies and not just in Viet Nam. Some of these are old, even antique, some very recent, like last week, and they all bear heavily on the fate of rhinos. Then there are the antique artefacts, many of which are very beautiful. High-value articles in their own right and not just because they are carved from rhino horn. Both hunting trophies and antiques play an important role in our story.

## 6.5   The Customers/Consumers

As we have already seen, the products themselves give a clear indication of who might be buying and/or using them, and the trend is worrying.

Rhino horn is bought mainly by the young members of a swiftly growing middle class not by the rather more elderly people who tend to be far more frugal in their attitudes. And in modern Viet Nam, 65% of the population is

---

[21] Milliken, T. and Shaw, J. (2012) *The South Africa – Vietnam Rhino Trade Nexus: A deadly combination of institutional lapses, corrupt wildlife industry professionals and Asian crime syndicates.* TRAFFIC, Johannesburg, South Africa.

[22] See Smith 2012b in Milliken and Shaw (n.21).

under 30 years old[23] and growing at a phenomenal rate. The young and wealthy want to spend some of their money on ultimate luxury items like rhino horn. Which is very bad news for all of us, especially the rhinos! Even worse is the fact that *in recent years, rhino horn consumption has seemingly grown exponentially and (now) includes a broad spectrum of Vietnamese society.*[24]

In 2011, Gwin researched rhino horn use for an article later published in National Geographic. This extract is illuminating:

> *Most of the users we met belonged to Viet Nam's fast growing middle class and included Western-trained doctors, a bank executive, a mathematician, a real estate salesman, an engineer, and a high school teacher, among others. Often families would pool money to buy a piece of horn and share it. Some donated it to gravely-ill friends who couldn't afford it. Mothers gave it to children with measles. Old people swore it cured poor circulation and prevented strokes. Many considered it a sort of super-vitamin.*[25]

The most pathetic are the seriously ill, those who put all their faith in rhino horn as a possible cure. Not only are they being traduced by false hope, in the process, they are also being defrauded of large sums of money.

Another group using rhino horn for its supposed medicinal properties are affluent, middle to upper income young mothers. They use it to reduce high fevers especially in their children, and although they only keep small quantities of the horn, it is enough to be significant.[26]

Then there are consumers who buy rhino horn just to give it away as *an elite gift*. Once again, this practice has its roots in the past. An *ancient attribute of Vietnamese social and political intercourse* is the giving of precious presents *to curry favour and gain influence.*[27] There is no way of knowing how many horns are involved or how much money is exchanged. But this is a form of bribery.

However, by far the largest and most important group of rhino horn consumers are the hedonists, who are normally well-to-do men living in cities. Again, a quotation vividly illustrates just how difficult it will be to curb such frivolous use of rhino horn:

> *Nguyen Huong Giang loves to party but loathes hangovers, so she ends her whiskey benders by tossing back shots of rhino horn ground with water on a special ceramic*

---

[23] See Ray et al. 2007 in Milliken and Shaw (n.21).

[24] Milliken and Shaw (n.21) p.134.

[25] Ibid.

[26] See Milliken and Shaw (n.21), p.136.

[27] Ibid, p.137.

*plate' 'says she's unfazed by doctors' assessments of the substance's efficacy and doesn't care to know how her father acquired the horn.* (Ibid)[28]

Milliken and Shaw describe them as an *obsessed group of buyers*, so it is most unfortunate that there are so many of them. The local conservation group Education for Nature-Vietnam suggests there is a *Ferrari factor* in operation, their term for it. That *rising income and a strong desire to flaunt new money and success amongst friends and business partners may be a more important factor driving the trade with Vietnam.* They are referring to the growing prevalence of expensive sports cars on the streets of Hanoi. According to Douglas Hendrie, one of their technical advisors *the new rich want luxury goods that are rare, exotic, and expensive as indicators of their success.* It is *these values in addition to the fact that rhino horn is also supposed to be good for you* that *may be driving the surge in demand here in Vietnam.*[29]

Even more alarming is the fact that the group *also includes mainland and overseas Chinese, Korean, and Japanese expatriate entrepreneurs living in Viet Nam, especially those forging business relationships through active participation in rhino wine associations with rich Vietnamese colleagues and associates.*[30] And it's all tied up with 'face consumption', a potent force in both Vietnam and China.[31]

The concept of face consumption plays an important role in collectivist societies. In order to enhance, maintain or save face, consumers must engage in acts of conspicuous consumption. Rhino horn falls within the category of conspicuous, so the concept encourages ever more frivolous use of the substance by each member of their particular social group. Their efforts must show they are complying with group values.

> *That same night I went out to a Turkish restaurant … with a friend and his date, a Vietnamese woman who owned a business in Taipei. Diem was obviously well-off, so … I casually asked her about rhino horn. She (not knowing why I was asking) told me enthusiastically that she had tried it recently for the first time at a private Chinese Lunar New Year (Tet) party at her mother's house in Go Vap District near the airport. She shared a situation akin to any sharing of illicit substances at a party, the woman's friend had invited family and close friends aside to share some rhino horn that he had received in a business deal. In a private room, he produced a section of horn that she described as round as a beer can and about 2 cm high. He also produced*

---

[28] Ibid, see Ives 2012.
[29] Ibid, p.135.
[30] Ibid, p.135.
[31] Cimigo 2010 in ibid.

*a small grinding stone, ground it, mixed it with water and served the milky liquid in tiny cups… She partook and said everybody was impressed and curious to take part in the impromptu ceremony.*[32]

Another aspect of face consumption is the use of rhino horn as a cure for impotence and as a sex aid. *If key individuals in a group use rhino horn for sexual purposes, others will unquestioningly follow suit and continue the narrative irrespective of whether there is any discernible improvement in their sexual prowess or not.* This is of course, encouraging even more users, and although once again the majority are men, women also seem prepared to go along with it, one even claiming it was *good for women's complexions.*[33]

Face consumption is all about acceptance as the following quotation illustrates. It is a conversation between Phung, the Vietnamese owner of a video production company and a Western journalist:

*Phung read the site then translated. The article said the whole body of the rhino is a "miracle medicine". Even its shit is a miracle - a pain killer – when drank with alcohol. In an article called "Why is rhino horn more expensive than gold" the writer claimed to have drank some of this rhino-dung alcohol made by a man named Viet from Cat Tien National Park… He claimed it had reinvigorated him after an exhausting day of trekking in the forest. I told Phung cynically that it would be a miracle feeling better after a strong drink mixed with rhino dung and added that the account of drinking shit only showed how desperate some Vietnamese were to experience the famed rhino elixir. Phung totally gobsmacked me by saying that he believed it. "Why would there be so much talk if it wasn't true," was his reasoning.*[34]

And this is why the last Javan rhino in Vietnam was killed by a poacher!

## 6.6 The Vendors

In most circumstances, it is illegal to sell rhino horn, so how do the Vietnamese consumers get their supplies? Some dealers make a point of targeting the seriously ill. They or their 'agents' seek out and then home-in on individuals, deliberately marketing the supposed curative qualities of their commodity. And the government does not stop them. Mothers get their horn from various sources, which probably includes both the internet and the practitioners they

---

[32] Smith 2012b in ibid.
[33] Ibid.
[34] Ibid.

consult, because they are using it in traditional medication. While rhino horn given as a special gift comes from very shadowy sources, the hedonists get theirs from more informal channels of trade, personal connections through social networks which sometimes include government officials and internet distributers.[35] Neither consult doctors.

Fortunately, as we shall see, after a great deal of hard work and dedication, things are beginning to change.

---

[35] Ibid, p.136.

# Part IV

## Poaching and Smuggling

# 7

# Poaching and Its Consequences

We start with horror.

## 7.1 Poaching

Satao, *Kenya's king of elephants*, was killed by a poisoned arrow and then had his face hacked off, just so that his killers could remove his magnificent tusks. And all this happened in Tsavo East National Park, where tourists flock, where everyone hoped he would be safe. Satao had wonderful tusks. They were so long, they swept the floor, and they weighed more than 45 kg each. They were the tusks of an elephant king, and they were his undoing. They had made him an irresistible temptation, a target for the hunters. Perhaps it was amazing that in these greedy times he had survived so long, for he was more than 45 years old. He was *one of the world's last great tuskers,*[1] and his ending was cruel in the extreme. He was mourned and will be missed by his many admirers. Rest in peace, old elephant.

> *Elephants are now at dire risk due to a dramatic rise in poaching for their ivory. Reports have reached CITES and the media on mass and gruesome killings of elephants, with their heads and tusks removed, from near every corner of their range in Africa.*

This was John E. Scanlon, the Secretary-General of CITES, writing in 2013, in the **Preface** to a report *Elephants in the Dust, the African Elephant*

---

[1] Jerome *Starkey Ivory poachers kill Kenya's king of elephants* The Times, 14 June 2014.

© Springer Nature Switzerland AG 2019
B. Martin, *Survival or Extinction?*, https://doi.org/10.1007/978-3-030-13293-4_7

*Crisis: a Rapid Response Assessment* carried out to determine just how threatened the elephants of Africa were by this possibly catastrophic, epidemic of poaching and how they might be better protected.[2]

However, this was not the first time African elephants had been subjected to such an excess of poaching. During the 1970–1980s, so many of the animals were unlawfully killed for their tusks that Kenya took the simple yet drastic and dramatic step of burning its ivory stockpile. Thousands of pounds of money in the form of ivory simply went up in smoke so that the world would sit up and take notice. Which it did. In 1989, most international trade in elephant ivory was banned, a ban which, with the occasional exemption such as antique ivory (dated as pre-1947 ivory) continues to this day. Central and Eastern Africa bore the brunt of the killings and continues to do so, although the populations in Southern Africa are now under attack. Elephant numbers continue to plummet. Forest elephants are threatened with extinction. All range States struggle to care for their remaining animals.

Rhinos have also been subjected to episodes of mass killings, and now some rhino horn stockpiles have been burnt as well. At the present time, rhino poaching is being carried out at such a rate that the very existence of these iconic beasts is in jeopardy.

The three case studies that follow later speak for themselves. They are a dramatic demonstration of what has been happening and also how hazardous it is to carry out undercover investigations.

### 7.1.1   The 'Killing Game'

Now for the poachers and their grisly work.

Death came silently to Satao, winging its way through the air in the form of a poisoned arrow, but death by poachers can take many forms when it comes to killing elephants. Methods range from the primitive to the high tech. Anything goes, provided the elephant dies and they can hack out its tusks.

Elephants are killed singly or in families. Families are more productive of course! Although they are normally shot, some are poisoned. Satao was killed by a poisoned arrow, but Klao, another royal elephant, ate poisoned bananas. Klao lived in Thailand. He was 50 years old when he died. During his lifetime, he had taken part in parades celebrating the Thai royal family and had appeared with Angelina Jolie in Oliver Stone's film *Alexander*. He should have

---

[2] UNEP, CITES, IUCN, TRAFFIC (2013), *Elephants in the Dust – The African Elephant Crisis, a Rapid Response Assessment*, United Nations Environment Programme, GRID-Arendal.

been safe in his home the Ayutthaya Elephant Palace, an elephant sanctuary, but he wasn't. The poachers used an electric power saw to cut through his metre-long tusks and botched it.[3]

Sometimes water holes are poisoned, with potentially devastating repercussions on all the wildlife in the area. In Zimbabwe, 2013, at least 90 and possibly as many as 300 elephants were poisoned with cyanide, which is readily available on the black market there. Cyanide is used by artisanal gold miners to leach out gold from its ore.

More dead elephants were found in autumn 2015. They too had suffered an agonizing death, killed by cyanide. The first grim discovery was of 14 dead animals. The rangers found them in three different places, two in Hwange National Park and one in the north of the country. A fortnight later, 26 more poisoned elephants were also found in the park, again in 2 different locations. The poachers, who had poisoned salt licks as well as watering holes, had got away taking 7 tusks with them, but wildlife officials recovered 14 others. According to the Zimbabwe Parks Authority, the rest of the elephants were too young to have any ivory. Park officials made sure the damage was contained.[4] Finally, in 2017 and after 4 years on the run, Tony Maphosa was arrested in Hwange National Park after a tip-off, on suspicion of killing more than 100 elephants with cyanide.[5]

And in one truly dreadful episode in 2012, a gang of poachers from Sudan killed over 200 elephants in northern Cameroon, yet another catastrophe for the beleaguered forest elephants.[6]

Sometimes the poachers come in armed gangs, like the 20 who slaughtered 12 elephants in Tsavo East, a national park in Kenya. The elephants died within 500 m of each other, probably shot with automatic weapons.[7] But the maximum number of poachers is normally 12. This would include the gunmen/killers, a local guide, porters, and an axeman to hack off the elephant's face and retrieve its tusks. A backup team is also needed, to bring in food and

---

[3] Darren Boyle *Thai ivory poachers kill 50-year-old elephant that featured in royal processions and Oliver Stone movie Alexander sanctuary* Mail Online 11 July 2014. See http://www.dailymail.co.uk/news/article-2688770/Thai-poachers-kil... Accessed 27/01/2015.

[4] *Another 22 elephants poisoned with cyanide in Zimbabwe reserve* The Guardian, 26 October 2015. See https://www.theguardian.com/world/2015/oct/26/22-more-elephants-poisoned-cyanide-zimbabwe-reserve Accessed 13/11/2018.

[5] Caroline Mortimer *Zimbabwean poacher arrested after 4 years on the run for 'killing more than 100 elephants'* Independent, 25 July 2017. See https://www.independent.co.uk/news/world/africa/poacher-zimbabwe-kill-100-elephants-arrest-four-years-run-tony-maphosa-ivory-tusks-tshotsho-poison-a7859421.html Accessed 13/11/2018.

[6] UNEP Report, September 2013.

[7] The Times, 9 January 2013; and Jerome Starkey *Albino calf is the latest victim of elephant slaughter* The Times, 12 January 2013.

carry out the tusks. These personnel usually ride motor bikes. Mobile phones are essential for communication, so everyone in the gang has one. As a general rule, these are new smartphones that have been misregistered and only contain a few numbers.[8] These people mean business. They are very dangerous, and this is the reason so many rangers lose their lives.

Unlike elephants, rhinos are normally killed with automatic rifles, AK 47 s, Kalashnikovs. Like elephants, most of them are killed in national parks and game reserves. Even 24 h protection by dedicated and devoted rangers doesn't always protect them. One pregnant female was shot within earshot of the headquarters of the Kenya Wildlife Service.[9]

A 2002 report by WWF on the killing of Sumatran rhinos, found a variety of methods being used. They make chilling reading and included overhead electric cables cut and draped over hundreds of yards of forest to electrocute the poor creatures and pits dug into the ground to trap the hapless animals who could then be shot.[10] Now less than 100 of these rhinos remain.

In South Africa, poachers used to be recruited from local communities who had always carried out subsistence hunting for rhinos. Theirs was, in effect, a traditional occupation. The next poachers came from the ranks of those who had been given specialized training, so had developed their skills in tracking or shooting. These included former military personnel, police officers or game scouts, and these poachers overlapped with the earlier ones. They *joined the ranks of those profiting from the illegal killing of wildlife*, and they did so for various reasons. They coexist with the newest type of poacher. Frequently coming from inside the wildlife game industry, these are skilled marksmen who use heavy-calibre weapons that instantly kill the rhinos. But there are others who use crossbows, silent and deadly. Yet another method used by these 'state-of-the-art' poachers is to dart the rhinos with immobilization drugs, then move in and hack off their horns leaving the creatures to die a dreadful, long-drawn-out death, because they do not administer a reversal agent. '*In rare instances, rhinos have survived some period of time … often suffering horrific facial injuries*'.[11] Darting takes place on the ground or from helicopters, which often don't display identification numbers. Using helicopters makes it easier to find a rhino and track the animal once it has been darted.

---

[8] Jerome Starkey *Rangers outwit armed poachers with gun sensors* The Times, 23 October 2015.

[9] *Pregnant white rhino is butchered* The Times, 14 August 2013.

[10] WWF *Wanted Alive: Asian Rhinos in the Wild* 2002.

[11] Milliken, T. and Shaw, J. (2012) *The South Africa – Vietnam Rhino Trade Nexus: A deadly combination of institutional lapses, corrupt wildlife industry professionals and Asian crime syndicates.* TRAFFIC, Johannesburg, South Africa, pp.74–75.

And they are invaluable when men and rhino horns need to be swiftly removed from the area.[12]

Even more sinister is the use of immobilizing drugs. How are such restricted veterinary medicines and equipment obtained? And special training is needed to use them. Milliken and Shaw are in no doubt as to who is involved:

*Since 2008, a small proportion of the wildlife industry, including game ranch owners, professional hunters, game capture operators, pilots and wildlife veterinarians have become active players in the rhino poaching crisis ... (it) is a significant factor not only behind the record levels of rhino losses since 2008, but also the insidious spread of rhino poaching across the country. Exactly how extensive this phenomenon is within the game industry is difficult to quantify, but it remains a serious, corrupting force that undermines rhino conservation and stains the image of a community that should be on the forefront of wildlife conservation.*[13]

Although this situation is unique to South Africa, this is very cold comfort since most of the world's population of both white and black rhinos is located in that country. We shall see later whether the situation has changed.

But the killing game can be a two-way process. In Assam, India, there is the Kaziranga National Park, home to some 2300 one-horned rhinos, more than in any other reserve in the world. It is so important that the forest guards operate a shoot-to-kill policy when dealing with poachers. In January 2015, rangers received a tip-off that a gang of poachers was about to enter the park. They were waiting for them. A gun battle followed, and three of the four suspected poachers were killed.

So much pain and suffering and, sadly, the cycle continues.

## 7.2 Devastating Consequences

Elephants have been more researched than rhinos. These highly social animals, a cognitively advanced species that lives in close-knit family groups, are more obvious in the bush, more easily found and observed than rhinos who are mainly solitary. There are also far more elephants than rhinos. NGOs such as Save the Elephant have been studying their animals for years, steadily building up an important data-base of information, a treasure trove of facts.

The Center for Conservation Biology has examined long-term impacts on elephant populations in the Mikumi-Selous Ecosystem in Tanzania, an area containing *one of the largest and most heavily poached elephant populations on*

---

[12] Ibid.
[13] Ibid.

*the continent prior to the 1989 ivory ban.* In other words, before the current poaching crisis. Concentrating their research in the northern part, the Mikumi National Park, they were able to distinguish herds that had experienced poaching because of *their peculiar group structure based on age and relatedness of their adult females.*

Their findings were distressing. Despite increased protection since the ivory ban, 30% of all adult females were solitary, which was very similar to what had been observed back in 1989. As well as this, *a high percentage of single adult female families had an atypically small family size, with an average of only 2.2 adult females.* The researchers had predicted that such a heavily poached population would contain some families that consisted entirely of non-kin, and although a most unusual situation in elephant families, this was what they found. Furthermore *small families tended to be less closely related, and some exhibited very fluid, uncohesive social behaviour, whereas on average larger families, which are more consistently closely related, tended to form stronger bonds and tighter social units.*

There were also long-term effects on both stress levels and reproduction. Once again, the most stressed groups were those who had no matriarch and few closely related adult females, while *groups still ranging in areas where poaching was historically high* had particularly raised levels of stress. These were areas that were closest to villages and furthest from park headquarters, areas where poachers had little difficulty finding their prey. So this was where much of the killing took place. And, all these years down the line, it was still affecting the elephants' reproduction. Young females in their reproductive prime produce calves, the next generation of elephants, except that those from a genetically disrupted group were not producing as many calves.[14]

Elephants can live to be as old as 70, and their years make them wise. Matriarchs play a vital role, not only as a repository of knowledge but also by passing on their wisdom to the younger members of their families. Indeed, their social hierarchy is primarily based upon age, yet we have seen that it is the older elephants, the ones bearing the largest tusks who get targeted by the poachers. When these are killed, what is the effect on their families? Thanks to some excellent recent research, we now have a far better idea.

Because elephants are so social, they have developed a considerable number of different calls, a form of 'elephant language' if you will. This has enabled a group of eminent researchers to carry out a series of playback experiments, playing appropriate calls to determine the effects of social disruption in the

[14] Center for Conservation Biology *Effects of poaching on African elephants.* See http://conservationbiology.uw.edu/research-programs/effects-of-poac... Accessed 27/01/2015.

herd and whether these persisted for any length of time. The disturbing results provided *the first systematic evidence that fundamental social skills may be significantly impaired by anthropogenic disruption*, where anthropogenic disruption includes *processes such as illegal and legal hunting/culling, translocation and habitat fragmentation.*[15] In other words, poaching is in this frame.

They looked at two populations of African elephants, one in South Africa, which had been subjected to culling and translocations, and the other in Kenya, which carried out its normal free-ranging life in Amboseli National Park. They were investigating two specific effects:

- The initial trauma
- *'The subsequent loss of opportunities for interacting with older group members that could act as appropriate role models or repositories of knowledge'* (the loss of role models)

Losing members of your family to poachers is a highly traumatic event, and whether you are an elephant or a human suffering from post-traumatic stress syndrome, the result can be one or more abnormal behaviours during later life. These may include *persistent fear, hyper-aggression and infant abandonment*. And such behaviour has been observed in the South African elephants, the elephants who might be traumatized. One dramatic observation was that over a 10-year period, 107 rhinos had been killed by orphaned male elephants who were exhibiting *abnormal hyper-aggressive behaviour*.[16]

The second specific effect, the loss of role models, is particularly important in species like elephants. The older animals *play a key leadership role and coordinate decision-making in the context of social and ecological threats*. Without them, not only are the younger group members given fewer opportunities *to learn the most appropriate response in dangerous situations* but also, and perhaps even more worrying, *any abnormal behaviour patterns that may have arisen from socially disruptive events have the potential to be passed between the generations and may persist in the long term.*

*Cognition encompasses the mechanisms by which animals acquire, process, store and act on information from the environment, including perception, learning, memory and decision-making* and elephants are *cognitively advanced mammals.*

---

[15] Graeme Shannon, Rob Slotow, Sarah M Durant, Katito N Sayialel, Joyce Poole, Cynthia Moss and Karen McComb *Effects of social disruption in elephants persist decades after culling* Frontiers in Zoology, 23 October 2013.

[16] Slotow R., Dyk G. *Role of delinquent young 'orphan' male elephants in high mortality of white rhinoceros in Pilanesberg National Park, South Africa* Koedoe 2001, 44:85–94. See http://www.frontiersinzoology.com/content/10/1/62 Accessed 14/04/2015.

The researchers' findings suggested that this disruption to their family relationships was affecting their decision-making abilities, with potentially very bad results for such long-living animals. They also provided '*the first direct evidence that abilities to process information on social identity and age-related dominance are severely compromised among African elephants that had experienced separation from family members … decades previously*'.

More research has shown just how very intelligent and discriminating elephants are. The participating elephants were 47 family groups in the Amboseli National Park in Kenya. They listened to the recorded voices of Maasai men, who kill elephants from time to time when conflicts arise over water or land for cattle-grazing, and Kamba men, who farm crops and rarely get into conflicts with elephants. The playback words were '*Look, look over there, a group of elephants is coming*'. The results were remarkable. There was a marked contrast in the elephants' reaction to the two recordings. While it was already known that the animals could distinguish between the Maasai and the Kamba by their clothes and smells, their reactions made it quite clear that they could also distinguish their voices.

The researchers decided to take it one step further. Would the elephants flee from any Maasai person, or was it only from the men, their attackers? The elephants reacted to the women's and boys' voices as they had to the Kamba's voices. But there was something else. *Most intriguingly, the researchers noted that elephant families led by matriarchs more than 42 years old never retreated when they heard the voices of boys, but those led by younger matriarchs retreated roughly 40% of the time.* One of the researchers, Karen McComb, suspected that the knowledge was cultural, *younger elephants following the lead of their matriarchs who remember spearings from long ago.*[17]

While the devastating consequences losing a matriarch can have on her family are well documented, we are only now finding out that losing older male elephants can be equally important to their kin.

And now elephants are growing shorter tusks. A physical consequence that makes very grim reading.

Although we now have some understanding of the damage poaching is inflicting on elephants, unfortunately similar information about rhinos is proving very difficult to obtain, mainly because rhinos are solitary animals.

Chester Zoo provides some insight into the pain caused by a mother's loss of her baby. In early 2015, Kitani, one of its eastern black rhinos, gave birth to Fara in 2015, but sadly, the calf had a hole in her heart, a congenital heart defect. And her protective mother made it impossible to detect this fault, so that, a few weeks later, she died in the arms of a keeper.

---

[17] See Shannon et al (n.15).

A rhino pregnancy lasts a long time, and a strong emotional bond develops between mother and calf, so when Kitani lost her baby after only 6 weeks, she became very distressed and quite depressed. She kept looking for her, seemed to be *calling* for her. She also became aggressive to her surroundings, pushing at gates and fences with her horn until eventually it became so loose it fell off. Kitani was lucky. She had kind and caring keepers who tried to understand her grief, working out the best way to help her. She now has another calf, this time a lively youngster, growing up fast—as well as a regrown horn.[18]

With solitary animals, scent plays a key role in information gathering, so the Zoo funded one of its Conservation Scholars to research how urine spraying was helping their eastern black rhinos to communicate. Essentially, it was a means of marking their territory, checking out who might have come into their territory and whether, or not, there might possibly be some interest in mating.[19]

Of course, the most devastating and final consequence of poaching can be extinction, of which the killing of the last rhino in Vietnam is a dreadful example.

The three case studies that follow speak for themselves. They are a dramatic demonstration of what has been happening and also how dangerous it is to carry out undercover investigations. All three were carried out by the Environmental Investigation Agency (EIA).

Tanzania and Zambia are elephant range States where poaching has been rife for several years, yet both countries have been trying their hardest to persuade the CITES Secretariat and the CoP to allow them to sell off their legally held stocks of elephant ivory. These are very valuable and would bring in much needed revenue which could be spent on elephant conservation, for example. The argument goes: if South Africa, Namibia, Botswana and Zimbabwe can do it, why not us. It is also sustainable use of a valuable commodity. Or is it?

### Case Study 1: Elephants in Tanzania

This particular investigation was triggered by two important happenings. In early 2009, there had been major seizures of ivory, over 11 tonnes in total, in the Philippines and Viet Nam. All the ivory originated in Tanzania, coming from the port of Dar es Salaam. That was the first event. The second was the request by both Tanzania and Zambia to sell their legal ivory stockpiles despite the fact that reported incidents of elephant poaching in their countries were

---

[18] Channel 4 *The Secret Life of the Zoo;* and help from Chester Zoo 2018.
[19] Article in Chester Zoo's *Z Magazine* Summer 2018, p.33.

showing a disturbing increase, up by more than threefold, from 18 to 53, in one part of the Selous Game Reserve. This persuaded the EIA to return once again to both Tanzania and Zambia, countries they had been investigating since 2002.

The Selous Game Reserve, situated in southern Tanzania is a World Heritage Site and Africa's largest protected reserve. During the blizzard of poaching that took place in the 1980s, as many as 70,000 elephants could have been slaughtered. The problem is, 50,000 km$^2$ will always be difficult to police.

The investigators discovered a continual flow of ivory out of the Selous. This was how it was achieved. '*Poachers enter the reserve for periods of around two weeks and kill an average of ten elephants each trip. The poached ivory is then hidden, buried at remote locations on the edge of the reserve until it is sold to traders, usually from Dar es Salaam. The transactions take place in villages that have become known hotspots for ivory trading … local villagers describe a thriving trade in ivory, with buyers from Dar staying in local guest houses when summoned by local traders, with deals being done for around Tsh 45,000 (USD34) per kilogramme of ivory, and the ivory transported to Dar es Salaam by bus, and even on government vehicles*'. This, they found, was happening right across the Selous.[20]

Enforcement was poor, with ivory seizures only occasionally occurring. However, when they did take place, the amount of ivory found was considerable. The report included figures for November 2009, when the police made 3 seizures, 2 in Liwale on the southern side of the Selous, where 20 tusks were found hidden in a guest house and 3 businessmen were arrested transporting 28 tusks. The men were from Dar es Salaam and that was where the third seizure took place, 33 tusks, from a house on the outskirts of the city. As a result of these operations, 7 suspects were arrested and over 100 kg of ivory recovered, all of which came from the Selous.[21] Although some of the ivory poached from the Selous was sold in the domestic markets, most was transported to Dar es Salaam then smuggled in containers to markets in the Far East.[22] In 2010, the investigators teamed up with the Journalists Environmental Association of Tanzania, with the result that local villagers were prepared to talk to them and explain how the ivory trade worked.

The report showed just how difficult it was going to be to stop this cruel killing spree. For example, *one villager claimed that poachers usually come from outside Mloka* (a village and gateway to the northern Selous), *but would some-*

---

[20] *Open Season: The Burgeoning Illegal Ivory Trade in Tanzania and Zambia* Environmental Investigation Agency (EIA), 2010, p.3.

[21] Ibid, p.13.

[22] Ibid, p.4.

*times employ villagers to act as guides inside the Selous, paying them (USD15) per trip.* According to the villagers, rangers worked with the poachers, and the buyers were often *retired army, police and government officials, who also help provide firearms to the poachers.* Poaching *peaks in the rainy season when large areas of the Selous are difficult to access and tourist operations close.*

The ivory was taken to Dar es Salaam in small consignments, by bus and/ or hidden in government vehicles *which are never stopped at checkpoints.* Or carried by motorbikes with modified silencers.

The investigators vividly described one encounter. After receiving intelligence that there was fresh ivory for sale in one remote village, they decided to pose as timber and ivory dealers and see what was going on.

'*(They) were introduced to a local shopkeeper who also poaches elephants in the Selous. The poacher said he had five kilogrammes of ivory for sale, with a further 100 kilogrammes available within two days. He claimed that around 500 kilogrammes of tusks are traded in Chumo (the village) every month. When the investigators asked to see the ivory, they were led to the dense bush surrounding the village, where the hidden ivory had been buried prior to sale. Under intense pressure from increasingly aggressive poachers to buy the ivory, the investigators had to flee the scene'.* They were chased by the poachers who were riding on motorbikes with modified silencers.[23]

Investigating poachers can be a dangerous business!

## Case Study 2: Zambia

The EIA also investigated in Zambia in 2009–2010, because the elephants in the Lower Zambesi National Park were being targeted by poachers yet again. Their findings were disturbing. One poacher was prepared to kill 4 elephants and deliver the tusks to the EIA investigators, an unwelcome offer. Not only that, he was employed as a village scout by the Zambia Wildlife Authority (ZAWA), and only too pleased to show them the ZAWA 'Warrior of Wildlife' T-shirt he was wearing. Another poacher, who additionally carved tusks, would *poach to order*, selecting his elephants from the nearby Sioma Ngwezi National Park. National parks are not necessarily safe places for elephants.

The investigators also heard accounts of poachers *being kitted out in expensive outdoor clothing to enable them to operate despite the elements* and even of poachers with satellite phones, though these were very rare.[24]

---

[23] Ibid, p.7.
[24] Ibid.

Much of this poached ivory would have been fed into domestic markets to be bought mainly by tourists, the remainder smuggled by various routes out of Africa to southeast Asia, particularly China.

The third case study involves rhino horn. Parts of this investigation were recorded, and some of it was shown on prime-time television. Harrowing at times, and disturbing to watch, it was a very good way of raising awareness of the issues surrounding illegal trafficking in rhino horn. The events, which were filmed using hidden cameras, took place during the frenzy of poaching that decimated the rhino population in the late 1980s and early 1990s.

## Case Study 3: Rhino Poaching

Rumours were rife in the late 1980s that a syndicate in China had started to stockpile rhino horns as a very unusual form of speculation. A futures market with a difference. After all, rhino horn was more valuable than gold. Were these rumours true? If so, were these horns coming from the rhinos that were being so cruelly killed in Africa? It was these questions the team set out to answer.

The investigators looked at a number of poaching incidents, one of which was particularly distressing. Two heavily pregnant females had been gunned down with Kalashnikovs. Their horns sawn off, they were abandoned to a lingeringly painful death. So four rhinos had been killed in all, and rhinos take a long time to reproduce. They interviewed a poacher, and they followed the trail of the rhino horn from South Africa to Taiwan then on to mainland China where it ended.

On their journey, they visited pharmacists who were selling rhino horn undercover and customs officials at one of Taiwan's major ports who seemed unaware that they were supposed to be seizing any rhino horn coming into their country, perhaps because it wasn't on the itemized list they used to check for CITES prohibited goods. The trail led to China.

They visited a market full of wildlife products. Was there any rhino horn for sale? They met a trader who told them that although none was openly on sale, traders were buying up any horn available because it was so valuable. Everything done undercover of course. But this horn was nothing to do with the mythical stockpile. Finally, they made contact with an agent for the syndicate. The rumours were true!

The owner of the stockpile was a government official in a small town situated in a remote part of China. He had originally amassed 5 tonnes of horn and had already sold some 4 tonnes of it. The team arranged, through an intermediary, to buy the rest. They travelled by train to the town where they finally saw for themselves enormous numbers of rhino horns in a warehouse. *More rhino horn than on top of the heads of all the rhinos in Zimbabwe.* None of it fake, they carefully tested to check.

Although everything about the deal was supposed to be top secret, the owner was only too ready to boast about his good connections in local government *who would oil the deal.* He even suggested the local police would help transport the horn over the border into Hong Kong. The horn was seized.

Shocking revelations, so now it is time to look at the poachers themselves.

## 7.3   Who Are the Poachers and Why Are They Doing It?

There are three men, sitting, waiting, waiting patiently, high up in the mountains. Waiting and watching. Watching the men in the wild lands below. The men who are on guard. Constantly on guard. Protecting their precious animals, the rhinos. The rangers hope they are safe from prying eyes, but know they probably aren't. But they continue on guard, continue to hope. Continue to protect.

Darkness falls. The men climb down from their mountain fastness. Down, down, down to the lands where the rhinos are, where the men who guard the rhinos are. They have located their target. One man takes careful aim and fires. A direct hit. The rhino falls. They are indifferent to whether or not she is dead, because it is a 'she'. And not only a 'she', a mother rhino with her baby beside her. She falls and the other men move in, one with a chain saw which he puts to immediate dreadful use. Her horns are gone. They must be quick. No time to get back before dawn, before the treacherous light comes which will betray them to the rangers. They hide and return to their homes under cover of darkness the following night. The little orphan alone with her dying mother, squeals piteously.

There is a war going on. A war between poachers and rangers. Shoot to kill, except many rangers are unarmed. Who will win? If the poachers are caught poaching, they might well be shot, injured rather than killed. But not always. And rangers are killed as well. It is a dangerous existence for men and beasts. Terrible prices to pay or be paid. And the rhinos are becoming extinct.

Why do the poachers carry on with their missions of death? Money. Always for money.

Men have hunted animals for millennia. In the dawn days their targets were woolly mammoths and woolly rhinos. And they hunted for food and to protect themselves and their 'families'. Poachers are hunters, but hunters with a difference. The Oxford English Dictionary describes *poaching* as *the taking of game illegally* and *to poach* as *to trespass for this purpose*. There is a fine line between poaching and hunting.

Until recently, most of the poachers came out of the local communities, poaching to get themselves some food, bushmeat or some extra money because so many of them were so desperately poor. This is subsistence poaching, and in some circumstances, it may be driven by the presence of refugees.[25] In the UK, it is the occasional pheasant poached from the local landowner. They were merely carrying on the same activities as their forebears. A traditional activity. Our three poachers could have come from Mozambique, still a very poor country. They might even have been offered bribes 'to help them decide' to take on such a potentially hazardous enterprise. Poverty and bribery. Two important determinants in whether to become a poacher. And they apply to poachers in Africa and Asia.

However, everything changed once the prices of ivory and rhino horn began to rise. Enter the crime syndicates and the criminal gangs. The poachers now tend to be 'the poor bloody infantry' in an all-out war between those who desperately want to protect the elephants and rhinos and secure their place in the future of the planet and those who merely wish to exploit. Exploit man and beast for their own selfish benefit. And, of course, those who poach from helicopters tend to remain unscathed.

And then there is war. Wars need lots of money, and in Africa there continue to be warring factions destabilizing many areas. Money from ivory helps fight these wars. Blood ivory. Death to elephants used to finance death to people. Way back in the 1970s, the South African Defence Forces entered into an unholy alliance with poachers, using money from selling this ivory to help fund their battle to support apartheid. It was Nelson Mandela who finally lifted the CITES reservation South Africa had taken out on elephants and from then on, South Africa has protected its elephants. More recently, the notorious Janjawid regularly swept on their horses out of Sudan and into Zakouma National Park in neighbouring Chad, there to slaughter elephants, dozens every month. The money they kept getting from this frenzy of killing enabled them to carry out other frenzies of killing, in neighbouring Darfur, except that these victims were humans.[26] Sometimes it helps terrorists too.

---

[25] Harrison et al. 2015 in Rossi, A. (2018). *Uganda Wildlife Trafficking Assessment*. TRAFFIC International, Cambridge, United Kingdom.

[26] Joshua Hammer *The Race to Stop Africa's Elephant Poachers* Smithsonian Magazine, July 2014. See www.smithsonianmag.com/science-nature/race-stop-africas-elephant-poachers Accessed 05/04/2015.

## 7.3.1   Poaching in India's Jaldapara National Park

There are obvious parallels with Africa. Initially local people were used. They were accustomed to killing rhinos, as it only became illegal to sell rhino horn in Assam in 1980, and they hunted the traditional way, with bows and arrows and spears. But everything changed when Viet Nam suddenly decided that rhino horn was their 'must-have' commodity. The crime gangs arrived with their AK 47 s and M16, automatic rifles, by 2018 fitted with silencers.

More than 100,000 people live either within or on the periphery of Jaldapara National Park. They are small, forest-dwelling tribal communities. Both poverty and corruption are involved. In fact, one recent arrest was so unexpected, the villager had *even been involved in conservation initiatives with the forest department and would deliver good speeches before the villagers on the importance of rhino and forest protection*, that nine other people, with the same name, were picked up before he was arrested. The forests are managed by the Forest Department and local Joint Forest Management Committees.

An article in Mongabay describes the process. The traders (members of the crime gang) send a 'linkman' into the village to select a person, normally naive and/or unsuspecting to organize the poaching. Field men, two to three villagers, carry supplies and act as guides, while others act as lookouts. After the rhino has been killed, a field man removes the horn and leaves the forest with the shooter. They boil the horn in water to clean it, and then the shooter and another villager go to collect payment from the trader. The amount of money varies according to the weight of the horn, but traders frequently only pay part of what is due, *promising to pay in full after future kills*.[27]

China is the suspected recipient of these rhino horns, which are smuggled out of India through the porous border with Myanmar.[28]

## 7.3.2   A Shocking Revelation

Although poachers are normally creatures of the shadows, some have high profiles. After a former vice president of the Dallas Safari Club, a trophy hunter, was convicted of poaching in a national park in Zimbabwe, US officials carried out a sting operation which netted a South African hunting

---

[27] Moushumi Basu *In unsuspecting Indian villages, the international rhino horn trade takes a toll* Mongabay, 22 May 2018. See https://news.mongabay.com/2018/05/in-unsuspecting-indian-villages-the-international... Accessed 29/05/2018.

[28] See https://cites.org/sites/default/files/eng/cop/17/WorkingDocs/E-CoP17-68-A5.pdf A report by IUCN-Species Survival Commission and TRAFFIC.

guide, who was caught offering to organize a trip to poach elephants. According to the indictment before the court, in his correspondence to the supposed client, he had alleged corruption was rife, so bad, that with the right money he could buy the necessary permission and make things right if necessary.[29] The year was 2018.

---

[29] Jane Flanagan *Big game hunter faces arrest after offering poaching trip* The Times, 23 May 2018, see later in '*Prosecutions*'.

# 8

# The Survivors

We know that poaching has harrowing consequences, but what happens when the animal is not quite killed?

## 8.1 Some Can Be Saved

Vets regularly attend elephants with spear wounds and they often survive. Here is the story of two of them.

In June 2014, two bull elephants were found injured on the northern edge of the Maasai Mara National Reserve in Kenya. They had been attacked, probably by poachers using spears, and one had sustained a serious head wound. Fortunately his brain was uninjured, but the metal tip of the spear was still embedded in his temple, and he had severely cut his trunk trying to pull it out.

Vets from the David Sheldrick Wildlife Trust's Mara Mobile Veterinary Unit and the Kenya Wildlife Service (KWS) darted the elephant and then managed to carry out the risky procedure of removing the spear tip without the wound bleeding excessively. There had been a danger it could have bled to death. The operation was a success, as was the one on the other elephant, a younger animal who had been standing guard over his friend. He too was darted and his leg injury treated. Both their wounds were then packed with green clay which is a natural antibiotic and given further medication. Job done, the animals got to their feet and ambled away.

© Springer Nature Switzerland AG 2019
B. Martin, *Survival or Extinction?*, https://doi.org/10.1007/978-3-030-13293-4_8

In Mozambique, the International Anti-Poaching Foundation are involved in saving injured elephants, like the young bull who had been shot in one of his hind legs. He was very lucky, the veterinarians were able to operate on him in the bush and he was later seen to be fully recovered.[1] Normally the prognosis is poor where the injury is a gunshot wound.

While operating on rhinos who have lost their faces is rather more difficult, even the worst injuries never appear to daunt the vets. But only they can judge when the battle cannot be won, and it is time to euthanize their poor patient.

**Thandi and Themba:**  Early in March 2012, three white rhinos on the Kariega Game Reserve were attacked by poachers. Their horns were hacked off and they were left for dead. When a ranger found them, one, the largest male in the Reserve, was already dead, but the other two were still alive, though only just.

Themba, a young bull, was lying under a tree, minus his horns and half his face. All hacked off. And Thandi was writhing in agony. There was a pool of blood close by.

Dr. William Fowlds, the Medivet Rhino Project coordinator for the Wilderness Foundation, and his team of helpers then attempted to work miracles.

The poachers had darted both rhinos with a strong opiate anaesthetic, tranquillizing them to enable the criminals to carry out their horrific work. So the first thing to do was inject an antidote to waken the animals and then get them back onto their feet. When a rhino lies for very long on its side, this is potentially life-threatening as it interferes with their ability to breathe, which, in turn, cuts off blood supply to vital muscles, causing tissue death. These two rhinos, it was estimated, had been lying like that for almost 8 h.

Themba was the first to struggle to his feet, and everyone hoped he would survive. He appeared to be recovering well although he was obviously very weak, and then 3 weeks later, he drowned while drinking water. An autopsy revealed that death had only been a matter of time. Eight hours had been too long for the poor rhino. He had developed an infection in his leg joints, and his muscle tissues were beginning to disintegrate.

Although Thandi's injuries were even more severe, they managed to save her. She had lost a great deal of blood, and when the poachers had hacked off her horns, they had done a very thorough job, exposing her nasal passages and her sinuses. Like Themba, Thandi managed to struggle to her feet, but she was fighting to breathe. However, over the next few weeks, she began her slow recovery. Both physical and psychological.

---

[1] International Anti-Poaching Foundation (IAPF) Newsletter 2016.

Then, a year later, there came an unexpected setback. A new male rhino was introduced onto the Reserve, to replace the two who had died at the hands of the poachers. Although he had been dehorned, a few weeks after his arrival, he had a brush with Thandi, which damaged the precious new skin on her nose.

Fortunately during that same year, medicine had advanced. Veterinary consultants had been developing new methods of skin grafting, and they agreed to try them out on Thandi. Four different techniques were tried to find out which would be the most effective, and although Thandi was not the best of patients because she would keep rubbing off the fragile grafted tissue, eventually the grafts took, and the skin grew back.

The ultimate miracle came nearly 3 years later, when this remarkable rhino gave birth to a baby, a beautiful calf to an incredibly brave mother, a rhino who, with rather more than a little help from her friends, survived literally, against all the odds!

Thandi was not the first white rhino to receive a skin graft. This honour goes to two other white rhinos with the colourful names of Lions Den and Dingle Dell. They were the very first of their species in the world to have skin transplants, true pioneers. The benefit of this essential technique is that it gives the injured part, normally the rhino's nose, a chance to grow back normal skin, thus replacing possibly all of the scar tissue.

**Hope:** Only 4 years old, Hope was in a parlous state when she was found, missing two horns which had been savagely hacked off and with the nasal bone badly fractured and partly removed. This had exposed the sinus cavities and the nasal passage. Maggots had got in and she had a severe infestation.

Once again, human heroes took over. They moved her to holding facilities in her home, the Shamwari Private Game Reserve, and after cleaning up her wounds, they fitted her with an artificial cast/shield to cover her terrible injuries. Shields are essential, giving the damaged area a chance to recover. In Hope's case, not only would this protect her from further damage as she foraged for food, it would also prevent further infestation by maggots and keep out infection.

Normally shields are attached to undamaged bone, but Hope's face had been so severely damaged that this was not possible, and she managed to loosen the first shield. So the bottom part of the shield had to be sutured, with the UK's Prince Harry (we meet him again), as always 'hands-on', helping during the operation.

It's a good thing she ended up in such loving and capable hands, for Hope's medical team expected she would take at least 12–18 months before she recovered. In the end she needed 16 surgical procedures, including skin grafts from hippo and elephant hide.[2] However, the team was upbeat. *'What Hope is teaching us now is invaluable. We can draw from her treatment when treating other victims of poaching with severe facial wounds. We must be patient and give her body time to heal. We are providing her with the best medical care we have and making her as comfortable as possible'.*[3] The vets were right. Hope lived up to her name and made a speedy recovery.[4] Sadly though, she died a few months later from an unrelated infection.

Saving Hope was a brilliant piece of team work, for her medical team was composed of experts from a number of organizations: Dr. Johan Marais from the Faculty of Veterinary Science, Onderstepoort; Dr. Gerhard Steenkamp from the University of Pretoria and a co-founder of Saving the Survivors; and Dr. Will Fowlds together with a team of people from the Wilderness Foundation, Medivet, UK, and Dr. Johan Joubert, with his team on Shamwari Private Game Reserve, and his ground crew (led by Bruce Main). Everyone working together.

Injuries caused by snares can be equally fatal if the animal is not rescued in time. Snares are uninhibited in their choice of victim. Set to catch one animal, they catch another. Deaths are long, lingering, cruel, so much pain, so much suffering. Rangers are constantly on the lookout for snares. Both elephants and rhinos can be killed by these horrible devices, but sometimes, as Baraka's story shows, there can be a happy ending.

**Baraka:** This rhino lives in Ruma National Park, in Kenya. In October 2015, KWS and the David Sheldrick Wildlife Trust (DSWT) received a report that a bull rhino had been seen dragging a log attached to a cable snare which was around its neck. A crude, but effective method of killing animals. Two vets flew out, located the animal, and then went in on foot because the vegetation was so dense and successfully darted him. It was Baraka. They knew him well.

Baraka fled, and although he ran some distance before he collapsed, the backup aerial support enabled the veterinary team to quickly find him. His breathing stabilized; they used wire cutters to remove the braided winch snare that was strangling him. Fortunately, the wound around his neck where the wire was cutting into him was relatively new so only needed to be cleaned and

---

[2] See http://www.nhm.ac.uk/visit/wpy/gallery/z Accessed 08/03/2018.
[3] Marais 2015.
[4] Valentine Low *Harry's despair in killing fields of Africa* The Times, 3 December 2015.

treated. The antidote for the immobilizer drug was then administered, and Baraka was back up on his feet, all ready to go!

However, while it is sometimes possible to save rhinos with the most horrific facial injuries, elephants die. Their faces are completely hacked off to get out the tusks.

## 8.2   Orphans and Orphanages

It is perhaps more difficult to treat the orphans, the calves who have seen their mothers, and perhaps other relatives savagely slaughtered. Many die in the bush. Baby rhinos will not leave their mother's body. For the lucky ones though, there are orphanages and skilled people devoted to helping them recover.

More than 50 years ago, the David Sheldrick Wildlife Trust (DSWT) was learning how to save and then return these young animals to the wild, and now their elephant orphans project is not only one of the most famous elephant orphanages in Africa but is arguably the most successful orphanage for baby elephants there is. It both rescues and rehabilitates its small charges.

Orphaned by poachers, or perhaps as a result of human-wildlife conflict, if the babies make it to this sanctuary, help is at hand, although even at the orphanage surviving can be a challenge. This is partly age-related, because baby elephants are completely dependent on their mother's milk until they are 2 years old and partly dependent on it for the next 2 years as well, so, to raise a baby elephant, as with raising other baby mammals, it was imperative to find and feed only the correct milk formulation.

Dr. Dame Daphne Sheldrick perfected the first suitable formula, but problems remained, and 40% of the little orphans continued to die. This was because:

- Calves in the wild are suckled for about 3 years
- During that time, the composition of the milk alters making it difficult to replicate.

And it had been virtually impossible to get samples of milk from lactating mothers who only produce milk when the calf is suckling and, understandably, are aggressive should any human try to interfere in this process. That was until Shorty came along. She is a tame Zimbabwean elephant who lives at a safari lodge near Harare, the capital city. She gave birth and then very kindly allowed some of her milk to be harvested, which meant that further research could at last be carried out. Now it can be seen how the milk changes over a 2-year period, and there are plans to develop at least four formulas. While vets from AWARE Trust, based in Zimbabwe, are taking care of Shorty, in South Africa, Professor Osthoff, a food biochemist, will develop the new formulas.[5]

Only once an orphan is feeding can other matters be addressed, problems such as post-traumatic stress syndrome, which can be just as deadly as eating the wrong food. Although it took time and a lot of research by dedicated people to recognize as a fact that this mental condition could affect these young animals, now even this is treatable.

So how do humans such as Dr. Dame Daphne Sheldrick and her team of dedicated keepers care for and rehabilitate their young charges? The first step is to recognize their state of mind. Dr. Dame Sheldrick describes elephants as *'very human animals'* whose *'emotions are exactly the same as ours'*. The orphans have not only lost their families but have also often seen their mothers killed. They arrive at the Nursery *'filled with aggression—devastated, broken, and grieving. They suffer from nightmares and sleeplessness'*. Some are physically injured as well, such as *'Murka, rescued near Tsavo National Park with a spear lodged deep between her eyes and gaping spear and axe wounds along her back and sides. The spear had penetrated ten inches, rupturing her sinuses, which prevented her from using her trunk to drink. Her deep wounds were filled with maggots'*. She had probably been orphaned by poachers and then attacked by local Maasai, angry about losing their traditional grazing land to the Park.[6] Under the DSWT's gentle care, she made a full recovery, both physically and mentally.

The orphans' keepers are devoted as well as dedicated to their charges. Every day, the young elephants are led to the bush to browse. They drink their midday bottle of milk at a designated mud bath and return to the Nursery for their 6 o'clock feed. At night, each orphan sleeps with a keeper, though never with the same keeper two nights running in case either orphan or keeper becomes too attached to the other. And because of their ability to empathize, the orphans are also able to absorb new arrivals into their herds, even when a young calf has been reared by humans from a very young age so is becoming imprinted on its human 'mother'. Moving footage of this has been captured on camera, and it is essential, for the youngsters are deliberately raised in herds that will, eventually, return in their own time to the wild.

The infants are hand-raised in Nairobi until they are 3 years old, and then they are moved to one of the DSWT's three Reintegration Units in the Tsavo Conservation Area. At these locations there are wild elephants all around, and

---

[5] Aislinn Laing *Elephant orphans to taste milk of human kindness* The Times, 9 May 2017.
[6] Charles Siebert *Orphans No More* National Geographic, September 2011.

they have no hesitation in inviting themselves in. They like to drink from the always-full water trough. Siebert describes *'a dreamlike scene'* he was part of while visiting the DWST's Ithumba Reintegration Unit in Tsavo East National Park. Between 25 and 30 wild elephants, *'massive, long-tusked bulls and matriarchs, adolescent males and females, some ex-orphans, and several newborn calves all came to drink. The orphans were already there, collected for the night, staring over at their wild counterparts, who, between sips, stared back'.* The Ithumba head keeper, Benjamin Kyalo, explained. *'They have let the wild ones know it is OK. The word is clearly spread around Tsavo: Good humans. Good water. Let's go'.* All this time, the humans were a mere 30 yards from the wild elephants. Rehabilitation is a natural process here.

Very small rhinos are much more difficult and often fail to thrive even with the best possible attention. Of course, they have no chance of survival unless they are found quickly and taken to a place of safety. Both foot patrols and helicopters are used, but their results are often far from satisfactory, and helicopters are very expensive, so the Rhino Conservation Project decided to train dogs to see whether they were any better at detecting and tracking these precious youngsters.[7]

Once again, Dr. Dame Daphne Sheldrick worked out the correct composition for the milk formula that gave these and other rhino babies a chance of survival. Coping with their trauma can be much more problematic, and at the Rhino Orphanage in Limpopo, South Africa, women rather than men look after the orphans who can react badly to men.[8]

While there has not been the same amount of long-term research carried out into rhinos that there has into elephants, we do know some things. There is evidence that rhinos can become very stressed and that a strong bond exists between mother rhinos and their babies, and we have already seen how distressed Kitane, one of Chester Zoo's black rhinos, became after her baby died.

Returning to Limpopo, Peter, who with his team guards rhinos, bears witness to the fact that *'during a poaching, the calf will do anything to stay next to the mother, even if poachers start hacking away at them. The calves will squeal and cry throughout the entire poaching until killed or knocked out'.*[9] Peter describes how *'this bewildered young rhino calf (was) bellowing mournfully while nudging its dead mother, imploring her to get up. The mother of course had bled to death after the front of her face was laid open when her horn*

---

[7] Endangered Wildlife Trust *Rhino Conservation Project The use of dogs.* See https://www.ewt.org.za/RHINO/dog.html Accessed 07/01/2016.

[8] Amanda Levine *PSTD* in Rhinos Youth 4, African Wildlife.

[9] Ibid.

*was severed by a machete. Apparently a rescued rhino calf will call for its mother for months, especially at sunrise and sunset'.* Peter says that *'like people, these young ones can be traumatized for life'.*[10]

The DSWT provides guidelines for *raising rhino orphans*. There are seven things that are *important to do*. Briefly, they include:

- Using several keepers to replace *the mother figure*. Babies follow their keepers, who need time off.
- Keeping each calf in a small stable at night and providing a familiar scent with an article of a keeper's clothing.
- Instructions on bottle feeding and weaning.
- *Walk a calf around dung piles and urinals of the established wild community on a daily basis from dawn to dusk giving it time to investigate scent trails and contribute to the dung piles. Allow it to eat the dung of the other animals (which all rhinos do) and which establish the stomach bacteria needed.*
- Make sure it has mud wallows on hot days, and plaster its body in soft mud, as it gives protection from the sun and biting insects. *Mud is part of good skin care in all animals that do not have fur.*
- Growing calves should spend their nights in a spacious stockade, to prevent claustrophobia. *Be sure to take its dung and establish a dung pile within the new quarters and also have a Keeper handy to calm the animal.* This is because *'any break in the daily routine or happening is traumatic'* to these animals of habit.
- When the calf is no longer dependent on its keepers, at about 3–4 years old, the stockade doors can be left open at night to allow the young animal to *make physical contact with the other rhinos*, if and when it wants to. *But it is important that it can return to something familiar ('home base') whenever it feels the need to.* It will always respond to its keepers. When a former orphan, Scud, was injured as an adult, she *returned home* but would only let her old keepers handle her.

These orphan rhinos continue to be monitored once they have been reintroduced to the wild. Sometimes they return, but normally this is to seek help, and they go to the keepers that they know.[11]

---

[10] *Save the Last Rhinos* See http://www.nikela.org/projects/stop-rhino-poaching/ Accessed 27/01/2015.

[11] If you go to the DWST website, you can read some of their stories, which are very moving.

# 9

# Smugglers and Smuggling Routes

Once the animals have been poached, these are the key components of the next stage of the illegal trade.

## 9.1 Smugglers

Who are the smugglers? The answer might surprise you. It could be you! Many of us like to visit a local market when we are on holiday, perhaps treat ourselves to a bangle or buy a tiny figurine as a present for a friend not thinking for one moment it is made of ivory. That it might be carved from the tusk of a poached elephant. Wearing the bangle on your return home. Taking it through customs, never thinking, never even realizing you should have declared it. Done in all innocence. Yet this makes you too a smuggler of elephant ivory and part of the consumer boom that is driving the illegal trade.

Professional smugglers come from many countries, may have specialized knowledge of the links in the trafficking chain and work with various people from poachers to middlemen and even corrupt officials, to carry out their deadly trade. From the market surveys, we can see that anyone involved in an international sale of illegal ivory or rhino horn is a potential smuggler. In the case of ivory, we know the nationalities of many of the customers. We can surmise that many of them would be shocked if they realized what harm they were doing. But not all would care. In any case, the results are the same, more animals cruelly slaughtered by poachers to satisfy the never-ending demand.

© Springer Nature Switzerland AG 2019
B. Martin, *Survival or Extinction?*, https://doi.org/10.1007/978-3-030-13293-4_9

Corruption, of course, also plays a key role. Corrupt officials at sea and airports enable smugglers to ship their illicit cargoes. And the sellers themselves speak of diplomats who play a very active part in smuggling operations, making full use of the protected status of their diplomatic bags.

And then there are the organized crime gangs who are amassing fortunes from 'white gold' and rhino horn. In fact, rhino horn smugglers from Viet Nam are more like drug smugglers, actively recruiting couriers ('mules') in Viet Nam for the specific purpose of smuggling horn from Africa. Mules come from various walks of life. Those caught have included students, a security guard and a cameraman.[1] Apparently they work to a tight schedule, turn-around missions, spending only a day or two in South Africa while they pick up the contraband, before flying back to Viet Nam. Some even come and leave without any check-in luggage. In fact researchers have found rhino horns in both check-in and carry-on luggage at South Africa's O. R. Tambo International Airport, with, allegedly, the occasional handover in the departure terminal itself.[2]

Milliken and Shaw include the following account which clearly indicates why such smuggling will be difficult to prevent:

*From a wealthy Saigonese family, Nga was a sales-rep for a foreign firm at that time. As her father moved in powerful circles I thought Nga might know something of sung te giac. I said I was looking for the people who smuggled it into the country and asked if she had heard anything about it. At first she said no, but after a long pause she decided to share a story with me. It turned out that Nga's father had been visiting a rich family friend when the woman invited him to view something special in her room under oath of secrecy. In the room she produced seven rhino horns that she claimed to have travelled to South Africa to get. Nga said she had brought them into Vietnam herself through customs.[3]*

Smugglers come in many guises, and the methods employed by, for example, rhino horn smugglers *'are infinitely versatile, limited only by imagination and opportunity'*. Furthermore, *'as new smuggling methods are uncovered … trafficking networks quickly adapt and refine their tactics, finding new methods of concealment and new weaknesses to exploit'*.[4]

[1] Milliken T. and Shaw, J. (2012) *The South Africa – Viet Nam Rhino Trade Nexus: A deadly combination of institutional lapses, corrupt wildlife industry professionals and Asian crime syndicates.* TRAFFIC, Johannesburg, South Africa, p.32.

[2] TRAFFIC research in South Africa, 2012.

[3] Smith, 2012b in Milliken and Shaw (n.1).

[4] Moneron, S., Okes, N. and Rademeyer, J. *Pendants, Powder and Pathways*, TRAFFIC, East/Southern Africa Regional Office, Hatfield, Pretoria, South Africa, p.10.

## 9.2    Smuggling Routes

From smugglers to smuggling routes and a key report (data collected between 2008 and 2014), which will enable us to work out some of the many routes smugglers use to get their ill-gotten gains to their ultimate destination.

Most illegal elephant ivory and rhino horn originates in Africa. Because it is illegal, it cannot be exported lawfully and must be smuggled out of its country of origin. It may well pass through one or more countries before reaching its final destination. To be completely successful, the illegal ivory must get from **A** (its country of origin) to **B** (its final destination) without being seized and without the smugglers getting caught, which requires a complex and well-organized operation.

So how is this done? And what about the various staging points en route, the transit countries? Let us call these **X**. Sometimes there are many **Xs** and sometimes none.

Let us begin our journey with a case study, an earlier investigation carried out by the EIA. They called it the *Lilongwe Pipeline* because it acted as a conduit through which many consignments of illegal ivory left Africa and were successfully smuggled into Asia. It will show us how some smugglers operate.

**Case Study 4: The Lilongwe Pipeline**

This was an important discovery, as the Pipeline was used for trafficking illegal ivory from Zambian elephants all the way to both Singapore and Southern China. And not just Zambian elephants either, for although most of the ivory came from Zambia, some allegedly came from other countries such as Mozambique and Zimbabwe. In other words, this was a major route.

Most of the elephants were killed in the South Luangwa National Park, although the poachers carried out their deadly activities throughout Zambia.

The collection/starting point was Chipata in Zambia (**A**), described as '*a dusty trading hub*'. From there, the '*cargo*' passed, by road, through a porous border, into neighbouring Malawi (**X**) and thence to Lilongwe, its capital. Once there, it was taken to an ivory factory where it was inspected by buyers from south-east Asia, who then paid for any ivory they bought. After purchase, an inventory was made of that ivory and it was packed and stowed in a container for shipment.

The container then passed by land through a number of countries, such as Mozambique (**X**) and Zimbabwe (**X**), down South Africa (**X**) to the port of Durban, where it was transferred to a ship and dispatched to Singapore (**X**) or

Southern China (**B**). Sometimes though, the containers went straight to the sea, to the port of Beira in Mozambique and from there to their final destination. We note that even then Mozambique was playing a pivotal role in the illegal trade[5] (see Map 1).

## 9.2.1   More Smuggling Routes

At that time, the ivory was destined both for China (**B**) and Japan (**B**), and the containers were '*often aboard vessels owned by the Japanese company Mitsui OSK Lines*'.[6] Now, thankfully, Japan is no longer a main destination for smuggled ivory, but other countries are, although the situation is fluid. And the amount of ivory being trafficked was mind-blowing, as much as 170 tons between 2009 and June 2014. Or rather, 229,729 dead elephants.[7]

Data collected between 2008 and 2014 shows some major changes occurring since the case study, although most of the illicit ivory was still passing through seaports.[8]

**Seaports:** The first change showed the points of departure had moved from Southern Africa to East Africa, with the most active ivory export hubs now found in Kenya (**X**), Tanzania (**X**) and Uganda (**X**). The Kenyan port of Mombasa had replaced Dar es Salaam as the main port of clandestine ivory exports. Zanzibar (**X**) was another key hub, while in West Africa, Abidjan in the Ivory Coast (**X**) was beginning to cause concern. Shipment by container was still the most popular method of transport and indeed the only possible one where there was a large quantity of ivory to be 'moved'. Singapore was still an **X** country but was only one in a number of other **X** countries.

As we saw earlier, ivory could travel considerable distances both before and after it had been consolidated and containerized. Its African journey would end either at an airport, for small amounts are transported by air, or, more probably, a sea port. Large ports tended to be selected, particularly the commercially important ones and those used by most of the travelling public. The evidence appeared to show that the traffickers mainly used legal systems of transport. We shall see later why this might be.

---

[5] Jo Hastie, Julian Newman and Mary Rice *Back in Business: Elephant Poaching and the Ivory Black Markets of Asia* Environmental Investigation Agency, 2002, p.3.

[6] Ibid.

[7] Varun Vira, Thomas Ewing and Jackson Miller *Out of Africa: Mapping the Global Trade in Illicit Elephant Ivory* Commissioned by Born Free and carried out by C4ADS, published 2014.

[8] Ibid.

By 2014, the number of **X**s had increased. The report listed the following places as appearing to be '*currently, the most important ivory transhipment hubs*'. They included:

- Hong Kong SAR
- In Viet Nam, Hai Phong and Ho Chi Minh City
- In Malaysia, port Klang
- In Singapore, Pasir Panjang
- In Sri Lanka, Colombo
- In the United Arab Emirates, Jebel Ali

Speedboats and river barges were taking illegal ivory to the Chinese mainland via Hong Kong and Macao. Where **X** was one of the ports in Northern Viet Nam, the ivory would be taken across the border in trucks and/or buses. Sometimes it entered China from the Philippines, travelling in fishing boats and tugs. Later in our story, we shall see that the Philippines is an important **B** country as well and Viet Nam also keeps some illicit ivory.

China was the most important **B** country, the final destination for most of the illegal ivory. But mere arrival did not signal the end of its long journey. That continued on its complex and convoluted way until it reached an ivory carving factory or other processing centres.

There were other, less used routes and plenty of unknowns. For example, some quantity of ivory, probably from Central African (**A**) and Kenyan (**A**) elephants, was leaving Africa by way of ports in war-torn Somalia (**X**) en route to the Persian Gulf (**X**). Even West African ports (**X**) were becoming smuggler-active as poaching increased in Gabon (**A**), Republic of Congo (**A**) and Cameroon (**A**).

The researchers concluded that:

- *Ultimately, both West and East Africa are likely to grow in tandem as trafficking hotspots, especially as Central Africa's last elephants are exterminated and poaching shifts closer to each coast*
- *It is possible that as few as 100 large-scale ivory consignments move annually, but account for 70–80% of the illegal ivory trade. Intercepting these containers and dismantling the networks that transport them is vital.*[9]

You may be wondering why most of the smuggling takes place through the large and most obvious ports rather than through small, more secretive places.

---

[9] Ibid, see p.20.

The answer is quite straightforward. Where large quantities of ivory are being smuggled from **A** to **B**, the best chance of success is to manipulate the legal transport system. Corruption, in the form both of bribery and forged paperwork, can work wonders, as can the rapid loading of containers onto a ship. The latter can be particularly important as most seizures are still triggered by tip-offs. And logistically, it is quite impossible to search all the containers in a major port. Only a few can be examined.

Smuggling ivory is all about making quantities of money. Wherever possible then, costs need to be kept to a minimum so as to maximize the final profit. Money is what the smugglers see, not dead elephants cruelly slaughtered, and transport is a major cost. Transport by sea involves both direct and indirect costs. The initial cost, the direct cost, is that of leasing as many containers as are required to hold the consignment of ivory. The indirect costs are concerned with corruption, with '*bribing or co-opting customs officials, freight logisticians, and other facilitators who are all needed to successfully move illicit products through the international shipping system*'. Shell companies are usually established. These enable the traffickers to keep their identity hidden.

However, although traffickers try to keep their costs as low as possible, it is always going to be very expensive to transport illegal ivory from Africa to southeast Asia, and criminal gangs prefer to use ships to transport their ivory, so what role does air transport play in smuggling illegal ivory?

**Airports:** Aircraft are unsuitable for carrying much ivory and so are rarely used except by small-scale criminals, although some of these, like a former US defence attaché to Kenya, can be quite important people. Furthermore, with so many security measures and widespread screening now in place, this ivory is much more likely to be discovered than the large quantities of ivory that is often cunningly concealed in containers.[10] But Chinese business and tourism in Africa are growing, and it can provide tempting opportunities for both the unscrupulous and the ignorant, so smuggling is likely to increase.

Let us now look at **A**, **B** and **X** in the context of air transport.

There were no direct flights from Chinese air carriers to Africa, and only a few African airways flew directly to China, although one of these, a direct flight from Nairobi to Guangzhou, was to a key location in China's ivory carving industry. Kenya, Ethiopia, South Africa, Algeria, Mauritius, Egypt and Angola all had direct flights to China and were all **X**s.

The three African airports that accounted for most of the ivory trafficked by air from the continent were Jomo Kenyatta International Airport, at Nairobi,

---

[10] Ibid.

Kenya; Bole International Airport, at Addis Ababa, Ethiopia; and O. R. Tambo International Airport, at Johannesburg, South Africa. These were categorized as '*key air transit chokepoints*'.

Because there were so few direct routes from Africa to Southeast Asia, Malaysia, Korea and Singapore, all countries whose airlines had some direct routes to China were also **X**s, while Turkish airlines and Gulf carriers were expanding their routes. Dubai featured in seizures, as did France's Charles de Gaulle International Airport (Paris).

However, this gradually changed, and by 2017, airlines being used by the traffickers to transport ivory included Emirates Airline, Etihad Airways, Qatar Airways, Kenyan Airways and Ethiopian Airways.[11]

In 2016, Uganda was identified as an important hub for the transit of elephant ivory, going to both Kenya and Malaysia and then onwards further east. There have been a number of major seizures at Entebbe International Airport.[12]

Although most of the ivory ended up in China, making it once again the main **B** country, with Thailand another important **B**, air travel ensures there are many other **B** countries. These are the home countries of the many tourists who are taking back some ivory momento from their holidays.

## 9.2.2   Other Countries

**Mozambique:** Although elephants were becoming extremely rare in Mozambique, two of its ports, Beira and Pemba, were still involved in smuggling both ivory and rhino horn, and there continue to be well-founded concerns about its current role as a trafficking hub.

**Egypt:** We've already seen that Egypt was another destination for illegal ivory, another **B** country, particularly important because most/many of the tusks probably came from the forest elephants of DRC (**A**) and the Central African Republic (**A**), in other words, those elephants most threatened with extinction. Other ivories came from Kenya (**A**) and the Cote D'Ivoire (**A**). Although most of the ivory came in small quantities by air to Cairo Airport, some came by ship and some came overland from Sudan, with tusks occasionally concealed in personal luggage coming directly from India.

---

[11] Rossi A. (2018). *Uganda Wildlife Trafficking Assessment.* TRAFFIC International, Cambridge, United Kingdom.
[12] The UNODC Wildlife Crime Report 2016 in Rossi (n.152).

**West Africa:** In this part of Africa, the situation was even more complicated.

**Côte D'Ivoire and Senegal:** We have already seen that both these countries obtained most of their illegal stocks of ivory from various parts of Central Africa (**A**). Both were **B** countries, but Côte D'Ivoire was an **A** country as well, because it was just possible that a very small amount of ivory still originated there.

Until it went bankrupt in about 2001, most of the ivory transiting in from Central Africa was allegedly carried by Air Afrique. In Senegal, the port of Dakar appeared to be a centre for the sale not only of ivory but of all manner of wildlife products which came in either by sea, in large mixed shipments or by road from Guinea (**X**), one section of which passes along the east side of the Niokolo-Koba National Park, where there are elephants.

### 9.2.3    2014–2016

Recent research has shown that both DRC and South Sudan, politically unstable countries, provide smuggling routes for ivory coming from Central and West Africa, while a seizure in Thailand of ivory destined for Viet Nam enabled the Lusaka Agreement Task Force to discover a new overland route, from Malawi to DRC, through Congo River, to Brazzaville and on to the coast. The UNODC World Wildlife Crime Report 2016 identified Uganda as a key transit country, with ivory going to Kenya and then to Malaysia, itself acting as a transit port to other destinations in East Asia. And Uganda's role as a major transit hub for ivory coming in from Central and East Africa was borne out when large seizures (over 500 kg) of ivory were made there.[13]

## 9.3    Rhino Horn

We already know that in the 1980s and early 1990s there were Chinese officials stockpiling rhino horn, some of which came to them via Taiwan (**X**), at that time one of the largest centres for trade in this commodity.[14] Then that dubious distinction passed to Viet Nam (**B**), which has now become '*arguably the world's premiere destination country and end-use market*'.[15]

---

[13] Rossi (n.11).

[14] Nowell et al. 1992 in Milliken and Shaw (n.1).

[15] Milliken and Shaw (n.1).

Most of the rhino horn came from South Africa (**A**) and was destined for Viet Nam (**B**), via, of course, the inevitable transit countries (**X**) since there were no direct flights from South Africa to Viet Nam. So at least one connecting flight was needed to get there.

Most rhino horn left by air from Johannesburg with various options available. South African Airways and Cathay Pacific flew to Hong Kong several times a week, while other airlines flew to other destinations. For example, Thai International flew to Bangkok, Malaysian Airways flew to Kuala Lumpur and Singapore Airlines flew to Singapore. As all of them connected to Hanoi or Ho Chi Minh City and there were many flights a day, policing became very difficult.[16]

The main land route into Viet Nam (**B**) ran from neighbouring Lao PDR (**X**), although there was evidence to suggest the rhino horn first passed through Thailand (**X**).[17] There was also evidence that the smugglers used specific border crossings between Lao PDR (**X**) and Viet Nam (**B**).[18]

China was also a **B** country, with rhino horns passing northwards by land through Viet Nam before crossing into China, another land route and one which turned Viet Nam into an **X** country as well.

By 2014, as a result of research carried out by the Wildlife Conservation Society (WCS), Viet Nam, the situation had become clearer.[19] They had estimates suggesting that:

- About 63% of all recorded cross-border trade passed from Viet Nam to Guangxi (China) and vice versa.
- About 95% of all wildlife trade to China PRC passed through Ha Long and Mong Cai.

As they wanted to look at the significant illegal trafficking route of, among other things, wildlife, they collaborated with Viet Nam's CITES Management Authority and the General Department of Viet Nam Customs.

They studied the transit point Quang Ninh on an international border crossing between Viet Nam and China. It has more than 60 river and sea ports as well as the Ha Long-Mong Cai expressway, which carries about 500 container trucks daily between the 2 countries.

---

[16] Ibid, p.133.

[17] TRAFFIC monitoring 2011 in Milliken and Shaw (n.1), p.133.

[18] Ibid, Khong and Quang 2008.

[19] Robertson in *Symposium on Combating Wildlife Crime: Securing Enforcement, Ensuring Justice, and Upholding the Rule of Law. The Proceedings* Editors Kala K. Mulqueeny and Francesse Joy J. Cordon, 2014.

Mong Cai was very important to the smugglers. Between 2009 and 2011, 18,500 kg of elephant ivory were reported to have been seized, and in January 2013, 14 pieces of rhino horn were seized in Dong Xing, a lot of dead rhinos.

In Mong Cai, they identified three major wildlife trade routes. By analysing the nearly 34,000 vehicles that had passed through the transit points, they found that:

- Seventy-six percent went through the smuggling points.
- Twenty-two percent through the customs clearance points.
- Two percent through the Bac Luan International border gate.

Once these routes had been analysed, the researchers turned their attention to shipments or, more precisely, to estimating the number of shipments taking place at these various points on an average day. This led to two findings:

- Everyone in the city knew where the points were.
- Office hours were always the busiest.[20]

So who was benefiting, apart from the smugglers themselves? Well the border officials certainly were, being paid '*huge sums of money in "unofficial crossing fees*"'. The losers were the States involved, who lost not only customs fees but also fines.

Why wasn't something done about it? It was, but it made no difference. Unfortunately, during the crackdown, more vehicles than normal passed through the transit points.[21]

These invaluable reports have shed light into some very dark corners, and the following case studies demonstrate just how difficult it can be to control smuggling.

### Case Studies 5 and 6

Both involved consignments of elephant ivory. They were seized by Customs officers at Safaga seaport in Egypt in October and November 2014.

The first consignment was of 300 kg of raw ivory mainly from elephants in East Africa (**A**). From there it passed to Kuwait (**X**), after which it was trucked via Saudi Arabia (**X**) to Dubai seaport (**X**) and thence to Egypt where it was seized.

---

[20] Ibid.
[21] Ibid.

The second seizure consisted of both raw and worked ivory, some 200 kg of it. It had been imported from Saudi Arabia (**X**). Often attempts are made to disguise the ivory. This had happened to some of the raw ivory in this case, which '*had been painted with a black stain to resemble coal or wood*'.[22]

## 9.4   Mapping Wildlife Trading

Analysis of ETIS, the Elephant Trade Information System (and the world's most comprehensive database of its kind as we shall see later), for CoP16 showed that, since 2009, nine countries in various combinations '*formed the principal trade chains through which an estimated three quarters of the illegal ivory trade*' moved. These were identified as:

* Source countries (**A**): Kenya, Tanzania and Uganda
* Transit countries (**X**): Malaysia, Hong Kong, the Philippines and Viet Nam
* Destination countries (**B**): China and Thailand[23]

At the Symposium on Combating Wildlife Crime: Securing Enforcement, Ensuring Justice, and Upholding the Rule of Law held in 2013, TRAFFIC's Executive Director Mr. Steven Broad explained how wildlife trade took place. That there were common trade routes and vectors at both local and international level, the commodities transiting through hubs which ranged from road and rail networks through border crossings to seaports and airports. Shipping companies, airlines and couriers were all involved. And, most disturbing of all, the entire banking system.

Most of the illegal ivory finding its way into the USA was coming from Africa via China, where it was carved into artefacts, figurines and jewellery, indeed, the usual suspect commodities, in factories in Guangdong and Fujian. Smuggled into the USA, much of it was bought by Chinese nationals (and other Asians) who then smuggled it back into China.[24]

---

[22] TRAFFIC Bulletin, April, p.28.
[23] Milliken Tom *Elephants* Report of 16th CITES Meeting, TRAFFIC Bulletin, Vol. 25 No.2, October 2013, p.60.
[24] Daniel Stiles *Ivory Trafficking in California, USA* 2014.

## 9.5     Xie Speaks

In 2017, the EIA published their report '*The Shuidong Connection: Exposing the global hub of the illegal ivory trade*', and in it the Chinese smuggler Xie explained how they operated and why they chose to do it that way.[25]

The correct route was a key factor: '*The goods can't come directly from Africa. They must first go to Singapore or elsewhere. There must be a transit point. If they came directly from Africa, they would definitely check the container*'.

Switching the Bill of Lading: '*further hides the origin of the container and conceals the identity of the sender and the recipient*'.

Complicit freight agents were involved in '*submitting documents for customs clearance along the route*'. Xie mentioned '*a Chinese freight forwarder operating in South Korea*' whose speciality was to arrange '*onward shipments of containers holding illegal wildlife products*', the rate depending on the species. It was '*$145 per kg for ivory*'.

EIA: '*Ivory smugglers who have set up trustworthy routes with accomplices at every stage are said to "own the road"*'.

## 9.6     Pendants, Powder and Pathways[26]

The data used in this '*rapid assessment of smuggling routes and techniques used in the illicit trade in African rhino horn*', came from TRAFFIC's global database of wildlife seizures and covered 2010 to June 2017. The report, published in September 2017, not only identified current key hotspots and smuggling techniques but also common smuggling routes.

Ever more ingenious methods were being used to evade detection, such as wrapping pieces of horn '*in foil*' and '*coating it in toothpaste or shampoo to defeat x-ray machines and mask the stench of decay*', in addition to hiding it in '*box wine cartons and in consignments of timber and cashew nuts*'.[27] And unfortunately there were still some Customs officials, airline staff and airport police prepared to take bribes.

As expected, most of the horn was being shipped from South Africa, Mozambique, Zimbabwe and Namibia, with export or transit points including Cambodia, Ethiopia and Kenya (both key transit countries with direct

---

[25] Environmental Investigation Agency *The Shuidong Connection: Exposing the global hub of the illegal ivory trade* 2017.

[26] Moneron, S., Okes, N. and Rademeyer, J. (2017), *Pendants, Powder and Pathways*. TRAFFIC, East/Southern Africa Regional Office, Hatfield, Pretoria, South Africa.

[27] Ibid, p.2.

links to Asian countries), the EU, Hong Kong SAR, Indonesia, Lao PDR, Malaysia (still a regional hub for wildlife traffickers), Qatar, Singapore (an important transit State for elephant ivory), Thailand and the United Arab Emirates. Seizures have implicated Lao PDR as a destination point for rhino horn for the Chinese market. Viet Nam was acting as both a transit and a consumer country.[28]

According to the data, air rather than sea transport was still the most commonly used method of smuggling rhino horn from source to consumer countries, with vehicles also continuing to play a key role.

---

[28] Ibid, p.7.

# Part V

## The One-Off Sales of Elephant Ivory

# 10

# The One-Off Sales of Elephant Ivory and Their Aftermath

## 10.1 The Early History of Elephant Population Listing on CITES

The history of listing different populations of elephants on the CITES Appendices shows how difficult it is to protect these iconic animals. At the first ever meeting of the Conference of the Parties, CoP1, in 1976, Asian elephants, always more endangered than the African species, were placed in Appendix I to prevent international trade in their ivory, and African elephants in Appendix II, so that there could be controlled trade in their ivory.

But as countries in southeast Asia grew prosperous during the 1970s and 1980s, so did their demand for ivory also grow. As did poaching, to fulfil their demand. The result was *'the serious decline of elephant populations in many Asian and African range countries'.*[1] The trade was not properly controlled, and elephant populations in many of the African range States fell so dramatically that in July 1989, the Kenyan Government announced that it would no longer sell any ivory it had confiscated from poachers. Three months later, in October 1989, the controversial decision was taken at CoP7 to reclassify all populations of African elephants into Appendix I. In effect, there was now a ban on all commercial international trade in elephant ivory, a ban that, with exceptions, continues to this day.

Overnight, the bottom fell out of the market. Indeed, the impact was so severe that *'the day before the meeting, a pound of ivory sold for more than one hundred dollars: the day after, a seller would have been lucky to get five dollars'.*[2]

---

[1] Daniel Stiles (2009). *The elephant and ivory trade in Thailand* TRAFFIC Southeast Asia, Petaling Jaya, Selangor, Malaysia, p.3.
[2] Ibid, Leakey and Morell 2001: 118.

© Springer Nature Switzerland AG 2019
B. Martin, *Survival or Extinction?*, https://doi.org/10.1007/978-3-030-13293-4_10

The apparently draconian measure was not universally popular, particularly among some of the African range States previously benefitting from the trade. The demand for ivory significantly decreased in Japan, the USA and Europe. And poaching declined in many parts of Africa, thus allowing some elephant populations to recover.[3]

Unfortunately though, it was not all good news. Soon evidence became available indicating that, in other countries, Asian ivory traders were attempting to replace African ivory with Asian ivory, which was very bad news for the rarer Asian elephants.[4] The traders could do this because states such as Cambodia, Vietnam and the Peoples' Democratic Republic of Lao had very poor law enforcement. In fact wild elephant numbers in these three countries *dropped from an estimated total of 6250 in the late 1980s to 1510 in 2000*.[5] The demand for local ivory came mainly from China, where the standard of living was rising for the rapidly growing middle and upper classes, and Thailand, where tourism was the main driver.[6]

It was at this point in time that the first downgrading of some populations of African elephants from CITES Appendix I to Appendix II took place, and it was done solely to enable the first of the one-off sales of elephant.

## 10.2    The Downgrading of Certain Elephant Populations and the One-Off Sales of Elephant Ivory

By 1997, elephant numbers had increased to such an extent in eastern and southern Africa that, in some places, they were beginning to cause problems. The numbers of human-elephant conflicts were growing, and some ecosystems were starting to degrade. It was therefore decided at CoP10 that the elephant populations in Namibia, Botswana and Zimbabwe should be downgraded to CITES Appendix II in preparation for the first of the one-off sales to Japan. All the ivory came from legal sources and had been stockpiled by the three governments. The sale took place in 1999 and was an exception to the ban on international trade in elephant ivory.

---

[3] Ibid, Martin and Stiles 2009, 2003, 2005, 2008; Dublin et al., 1995; Nash 1997.

[4] Ibid, Stiles, 2004a, p.3.

[5] Ibid, Santiapillai and Jackson, 1990: Kemf and Santiapillai, 2000; Martin and Stiles, 2002; Stiles, 2004a.

[6] Ibid.

The South African elephant population was the next to be downgraded, in 2000 at CoP11. Since then, despite considerable pressure from both Tanzania and Zambia to downlist their elephant populations, no other elephant populations have been downgraded.

## 10.3   Laws to Protect Wildlife in Botswana and Namibia

Before we proceed further, we will take time out to look briefly at the legislation Botswana and Namibia had in place to regulate and enforce the legal possession of both elephant ivory and rhino horn, to see whether it was strong enough to protect elephants and rhinos.

We have already seen that Botswana's Wildlife Conservation and National Parks Act 1992 incorporates CITES into its legislation. Schedule 5 includes both elephants and rhinos on the list of species requiring regulation. Anyone possessing ivory trophies or jewellery, or a whole elephant's tusk, or a rhino horn, must have a certificate of ownership or, failing that, evidence that they are the legal owner of the article. Additionally, the ivory (or other commodities) must also be registered with the Department of Wildlife and National Parks.

Failure to establish lawful possession could result in a charge of unlawful possession, the penalty for which is severe, namely, a heavy fine and 10 years in prison. Elephant hunting is only permitted for personal use, for example, to acquire a trophy, which is marketed according to CITES regulations. Its owner would be issued with an ownership certificate.

Namibia presented a problem in that its main piece of legislation for the protection of wildlife, the Nature Conservation Ordinance (No. 4 OG. No. 3469), was out of date. It was passed in 1975 and only slightly amended in 1990.

Elephants and rhinos were classed as *specially protected* species, so that hunting, capturing, transporting and being in possession could only be undertaken with the correct permits, to which conditions could be attached. Trophy hunting, possession and export of trophies were covered in Chap. 36. Only permit holders could possess elephant tusks, rhino horns or parts of them. But lawful hunters or importers could own them, and anyone could own objects manufactured out of them, provided it was done legally. Any tusks or horns merely found in the territory were state property, and the state could decide what to do with them.

Penalties were strengthened in the Nature Conservation General Amendment Act No. 31 of 1990 and were high. People convicted of illegally hunting elephants and rhinos could expect a fine of up to R200,000 or imprisonment for a maximum of 20 years, or both.

Three of the countries involved in the one-off sales, Botswana, Namibia and South Africa, were considered to have inadequate legislation under the CITES national legislation project, and although Zimbabwe's was good, since then it has become a troubled State.

**The One-Off Sales:** There have only been two one-off sales of elephant ivory. The first took place in 1999, when government stocks of legally sourced ivory, that is, ivory obtained from elephants that had died of natural causes or had been lawfully killed, were sold to Japan. The ivory came from Namibia, Botswana and Zimbabwe, and the sale raised some USD5,000,000. All the proceeds went to elephant conservation. It was a good example of sustainable use.

In 2002, a further sale of ivory was conditionally approved, this time from South Africa as well as Namibia and Botswana, although it was not until 2006 that the CITES Standing Committee finally approved Japan as the trading partner.

That same year, the Accra Declaration on the conservation of African elephants was signed by 19 African elephant range States. It called for a ban on the international trade in elephant ivory. Perhaps it was prescient. Perhaps the leaders of those countries which included Benin, Burkina Faso, Cameroon, the Central African Republic, Chad, Côte d'Ivoire, the Democratic Republic of the Congo, Ethiopia, Ghana, Kenya, Liberia, Malawi, Mali, Nigeria, the Republic of the Congo, Senegal, Southern Sudan, Togo and Uganda realized that a partial lifting of the ban would put too much temptation in the way of some of its 'weaker' citizens.

The final decision was deferred in 2007 to CoP14 where it was decided that China should also be included in the sale. A major, possibly the biggest consumer of ivory in the world, China, had recently strengthened its legislation to ensure that no imported ivory would be re-exported and that domestic manufacturing and trade requirements had been met. One of the main reasons the Standing Committee supported the sale was that it *'would flood the market with legal ivory, thereby ceasing the illegal trade by undercutting the inflated illegal ivory market prices (up to US$ 1500/kg)*[7]*) with cheap ivory'.*[8]

---

[7] Daniel Stiles (2008). *An assessment of the illegal ivory trade in Viet Nam* TRAFFIC Southeast Asia, Petaling Jaya, Selangor, Malaysia.

[8] CITES Secretariat Press Release *Ivory Auctions raise 15 million USD for elephant conservation* Geneva, 7 November 2008. See http://www.cites.org/eng/news/press/2008/081107-ivory.shtml Accessed 12

The conditions for selling the ivory and the generation of resources for conservation in the countries involved were set out in a number of decisions. Certain procedures had to be completed including declaring the stocks to the CITES Secretariat before the elephant populations could be transferred to CITES Appendix II. The one-off purchase was to be for non-commercial purposes.

The presale procedure was this. First the ivory had to be correctly marked. Then an independent audit had to be carried out under the auspices of TRAFFIC International in cooperation with the CITES Secretariat. Only after all that could the ivory finally be consolidated in predetermined locations ready for sale.[9] All the money from the sale would be deposited in and managed by conservation trust funds, which had to have *a positive rather than a harmful influence on elephant conservation*.[10] This should have been an excellent example of sustainable use.

**What Happened Next?** A series of decisions were then adopted to strengthen elephant conservation, which included the *Action plan for the control of trade in elephant ivory* (Decision 13.26(Rev. CoP14))[11] and an *African elephant action plan* to improve elephant management (Decision 14.75).[12] The Standing Committee not only had to prepare a proposal for a decision-making process for future trading in ivory, to be ready for CoP16 at the latest (Decision 14.77),[13] but also *conduct ongoing comprehensive reviews of the status of the elephant, trade in its specimens and the impact of the legal trade*, using data from:

* MIKE (Monitoring the Illegal Killing of Elephants)
* ETIS (Elephant Trade Information System)—more of these later

Because the sale was going ahead, both action plans had to be implemented (Decision 14.78),[14] and everything had to be done to ensure that it did not lead to a downturn in elephant protection and an increase in illegal killing. In

---

December 2008. See also EIA Briefing Document for the 61st Meeting of the CITES Standing Committee, *Elephants*, p.2.

[9] Decisions of the Conference of the Parties to CITES in effect after the 14th meeting, *Interpretation and implementation of the Convention, Species trade and conservation, Fauna*, 10.2 c (Rev. CoP 11), p.2. (Full details in this document).

[10] Ibid, 10.2 b)ii (Rev. CoP 11), p.1.

[11] Ibid, 13.26(Rev. CoP14), p.2.

[12] Ibid, 14.75, p.2.

[13] Ibid, 14.77, p.3.

[14] Ibid, 14.78.

other words, they adopted a precautionary approach, correctly anticipating that there would be pressure for further sales of ivory, even though the opinion was that *'this sale would flood the two markets thereby satisfying the demand and reducing poaching'*.[15]

On 16 July 2008 in Geneva, the CITES Standing Committee gave their permission for the sale to take place. This time four countries, Namibia, Botswana, Zimbabwe and South Africa, were authorized to sell a total of 108 tonnes of government-owned ivory to both Japan and China, with all proceeds from the sale to be used exclusively for both elephant conservation and the local communities living side by side with elephants.[16] Although this was supposed to be the last of these sales for at least 9 years, a deal having been struck at CoP14 and at CoP15, both Tanzania and Zambia sought permission to sell their legally held stockpiles of elephant ivory. This was refused.

## 10.4 Resolution Conf. 10.10 (Rev. CoP12 and CoP14) Trade in Elephant Specimens

We turn now to a key piece of legislation regulating trade in elephant specimens. Applying to both Asian and African elephants, it recognizes the fact that although *'elephant range States are the best protectors of their elephants …  the majority of them lack adequate enforcement capacity to ensure the security of their elephant populations'* (Conf. 10.10 (Rev. CoP14)). Besides making a number of recommendations, it stresses the important role that monitoring systems such as MIKE can play and the need for the range States to work together.

After defining the different forms of ivory and recommending a system of marking, it sets out (again by recommendation) a series of measures to control the internal ivory trade which apply both to Parties with an ivory carving industry that is not already structured, organized or controlled and to Parties designated as ivory-importing countries. Because this is international legislation with all its concomitant problems of enforcement, the Secretariat is urged to assist Parties wherever possible, and the Standing Committee is directed to review regularly what actions consumer states have taken to improve legislation and enforcement measures.

---

[15] *Open Season: The Burgeoning Illegal Ivory Trade in Tanzania and Zambia* Environmental Investigation Agency 2010, p.1.

[16] CITES Home Page Press Release. See www.cites.org/eng/news/press/2008/080716-ivory.shtml Accessed 1 November 2011.

The Secretariat will use evidence taken from MIKE and ETIS, the systems that monitor illegal hunting and illegal trade in elephants to identify Parties that continue to trade illegally and then, if financially possible, help them develop practical measures to regulate their internal ivory trade. MIKE and ETIS are to be expanded. Range States are to be helped manage and conserve their elephant populations *through improved law enforcement, surveys and monitoring of wild populations*. Detailed quotas for trade in raw ivory are laid down.

It is interesting to see what measures the Resolution recommended for Parties with an unregulated domestic ivory market:

(a) Register or license all importers, manufacturers, wholesalers and retailers dealing in raw, semi-worked or worked ivory products;
(b) Establish a nationwide procedure, particularly in retail outlets, informing tourists and other non-nationals that they should not purchase ivory in cases where it is illegal for them to import it into their own home countries; and
(c) Introduce recording and inspection procedures to enable the Management Authority and other appropriate government agencies to monitor the flow of ivory within the State, particularly by means of:

- Compulsory trade controls over raw ivory; and
- A comprehensive and demonstratively effective reporting and enforcement system for worked ivory.[17]

## 10.5 The 'Action Plan for the Control of Trade in Elephant Ivory'[18]

This sets out a comprehensive set of measures to regulate and thus control any trade in ivory.

It starts with Article 1 which is wide-ranging in its scope and applies to many countries including all African range States, other Parties (to CITES) and even non-Parties (there are very few of these now), if they either have an ivory carving industry or an unregulated (i.e. not controlled by legislation) domestic trade in ivory. The required measures are to be carried out as a

---

[17] See Stiles (n.1), pp. 3–4.
[18] See http://www.cites.org/eng/dec/valid144/annex2.shtml Accessed 15 September 2011.

matter of some urgency, and only Parties authorized to trade in worked ivory are exempt.

Article 1 (a) seeks to ban unregulated domestic sales of ivory (as opposed to international sales of ivory which are unlawful anyway). Countries should pass legislation which should include the situation where, if a person were to be found with ivory in his possession, which he was probably going to sell illegally, he would have to prove that he legally possessed this ivory.

Article 1 (b) and (c) set out basic instructions on enforcement and public awareness/education, both of which, as we shall see, are essential components in combating illegal trade. Other measures follow, and there are detailed procedures to ensure the plan is complied with.

Governments, international organizations and NGOs in both Africa and Asia should help the Secretariat eradicate illegal ivory exports and unregulated domestic markets. And the Secretariat should provide technical assistance to countries who ask for help implementing the action plan and with any other Parties with an ivory carving industry or internal trade in ivory (Article 4). It should also work with *national, regional and international law enforcement organisations and networks (such as the ASEAN Wildlife Enforcement Network, ICPO-INTERPOL, Lusaka Agreement Task Force and the World Customs Organisation) to assist in combating illicit trade in ivory* (more of these later). While Article 4 is all about working together, Article 5 is about priority. Countries identified by ETIS as being *significantly affected by illicit trade*, like Cameroon, the Democratic Republic of the Congo (DRC), Nigeria and Thailand, should be given particular priority to help them to comply.

However, the document does not confine itself to tackling Africa's unregulated domestic ivory sales; *all domestic markets outside Africa* should be continually monitored with priority given to China, Japan and Thailand. In other words, southeast Asia is also in the frame.

## 10.6    MIKE: Monitoring the Illegal Killing of Elephants

The one-off sales made it essential that all elephant populations in Africa and Asia should be intensively monitored to see what effects, if any, they might be having, but no system was in place to provide the detailed assessment that was needed. So CoP10 passed a Resolution to set up two systems, MIKE, Monitoring the Illegal Killing of Elephants, and ETIS, Elephant Trade Information System. Both are mandated to share the same objectives, which are set out in Resolution Conference 10.10 (Rev. CoP12) on *Trade in elephant specimens*. When the information from MIKE is combined with that from ETIS, it provides as complete a picture as possible of what is actually happening to elephants.

MIKE is a site-based system stretching across the entire range States of African and Asian elephants, monitoring and measuring trends in their illegal killing. As a reliable, standardized monitoring system, it provides essential evidence and has become part of the decision-making process at the CoPs on matters regarding the protected status of elephants. Set up in 1997, in 1999 it was endorsed at Standing Committee 41. A year later CoP11 was held in Kenya, always a fierce defender of its elephants, and Resolution Conf. 10.10 was revised to broaden MIKE's objectives to include:

- Establishing an information base to support decision-making on appropriate management, protection and enforcement needs
- Capacity building in the range States

Because funding has always been (still is) a problem, CoP12 made provision to ensure MIKE would continue to run after external funding came to an end.

**What is MIKE for?** It collects data, which enables it to provide information to both African and Asian range States to help them manage their herds and decide what enforcement is needed to provide their elephants with good protection.

When various factors are taken into account, it should be possible to:

- Measure levels and trends in the illegal hunting of elephants
- Determine changes in these trends over time
- Determine the factors causing or associated with such changes
- Try and assess in particular to what extent observed trends are the result of any (such) decisions taken[19]

Its main benefit is to increase *knowledge of elephant numbers and movements and a better understanding of the threats to their survival*, thus making sure that the animals and their ecosystems are managed sustainably. It also provides data on other species including rhinos, and range States are enabled to develop their own capacity to gather information which is particularly important.[20]

---

[19] MIKE See http://www.cites.org/eng/prog/mike/intro/index.shtml Accessed 12/09 2013.
[20] Ibid.

States can appoint their own site managers and use scarce resources to their best advantage. Real cross-border collaboration and the harmonization of different monitoring systems can be developed.[21]

In fact MIKE has been essential in helping understand the trends in illegal killing and what is driving them.

**How Does Mike Work?**  Because MIKE provides important information, it must be as reliable as possible. These are the factors it takes into consideration when monitoring dead elephants and determining which have been killed illegally:

Ecosystem type/habitat; elephant population levels; elephant/human conflict levels; adjacent land use; human access; human population pressure; availability of water; land tenure systems; development activities; tourism activities; history of illegal killing; proximity to international borders; cross-border incursions; civil strife or military conflict; law enforcement effort levels; judicial severity; corruption; illegal drug/arms trafficking; ivory trade patterns; CITES decisions.[22]

The list is comprehensive and informative, including both the obvious (such as information on ecosystems) and the more obscure (such as land tenure systems). Unfortunately, much of the illegal killing takes place in protected areas such as designated national parks and reserves where, of course, larger populations of elephants are to be found.

Not all sites are selected for monitoring and many are in African range States. A number of factors are included in the choice:

Forest vs. savannah; relative size of elephant populations; protection status of site; historical incidence of illegal killing; ivory trade situation; incidence of civil strife and military conflict; level of law enforcement; and CITES Appendix status.[23]

All of this is essential if MIKE is to identify national and subregional trends, as well as patterns in elephant populations and the illegal killing of the animals.

At present, the data is collected from 60 sites throughout 31 range States in all 4 subregions of Africa, with the encouraging result that some subregions no longer feel deprived of much needed information.[24]

---

[21] Ibid.

[22] Ibid.

[23] Ibid.

[24] UNEP, CITES, IUCN, TRAFFIC (2013). *Elephants in the Dust – The African Elephant Crisis*. A Rapid Response Assessment. United Nations Environment Programme, GRID – Arendal. www.grida.no

MIKE's 'Carcass Form': When the rangers in designated MIKE sites find a dead elephant, they determine what caused its death. It could be natural. Perhaps the animal died of old age. Or unnatural. Killed by poachers, an unlawful death which is usually established by the presence of bullet holes and the absence of tusks. The rangers complete a standardized *carcass* form, which records:

* Its cause of death
* Its age and sex
* The place the body was found, including the GPS coordinate
* Its state at the time of discovery, so that an estimate can be made of when it was killed

The form is then submitted to the CITES-MIKE Programme.

**MIKE and PIKE (Proportion of Illegally Killed Elephants):** The data from MIKE can be used to calculate PIKE, Proportion of Illegally Killed Elephants, which corrects any enforcement and monitoring effect. This enables elephants that have been killed legally or may have died naturally in the same area to be taken into account. The PIKE from any particular site is determined by taking the total number of illegally killed elephants found and dividing it by the total number of dead elephants on that site. Where all the elephants have died of natural causes, the PIKE value would be calculated as 0.0, but if they had all been illegally killed, the value would be 1.0, or 0.5 if half the elephants had been unlawfully killed.[25]

It is hard to overestimate just how important the MIKE/PIKE combination is in terms of evidence gathering. The PIKE data shows that poaching has been steadily increasing since 2006.

One final point to make before we move on. Some of the range States would also like to extend MIKE into an early warning system indicating increases in poaching.

# 10.7   ETIS: Elephant Trade Information System

Like MIKE, ETIS, Elephant Trade Information System, was mandated at CoP10. It too is a monitoring programme, this one tracking illegal trade in elephants. The system, supervised by the Standing Committee and managed

---

[25] Ibid, p.32.

by TRAFFIC, records data of elephant product seizures and analyses this to show levels and trends in the illegal trade.

The database is complex, consisting, among other things, of a central component and subsidiary databases to iron out bias. There is *'a law enforcement effort measure based on the CITES National Legislation Project'* and *'a law enforcement efficiency measure based on the Corruption Perceptions Index of Transparency International'*.[26]

The system has four main objectives:

1. *'measuring and recording current levels and trends, and changes in levels and trends in ... trade in elephant range States, and in trade entrepots;*
2. *assessing whether and to what extent observed trends are related to changes in the listing of elephant populations in the CITES Appendices and/or the resumption of legal international trade in ivory;*
3. *establishing an information base to support the making of decisions on appropriate management, protection and enforcement needs; and*
4. *building capacity in range States'*.[27]

Soon there were records of over 10,394 cases representing 77 countries, and ETIS has continued to grow steadily with ever more Parties regularly contributing data. The first ETIS analysis was produced in 2002 for CoP. It not only showed that illegal trade was closely linked to poorly regulated large-scale domestic ivory markets, a situation that had not changed by 2007 when the second one-off sale was being considered, but it also identified China, not Japan, as the major ivory market.[28]

ETIS has been able to identify key countries trading in illegal ivory and to document this trend (information produced for CoP13). It has provided TRAFFIC with facts which has enabled it to link the growing presence of organized crime gangs with an increase in the ivory trade and to identify those countries most implicated *'in terms of trade flows from source ... to transit points to end markets'*.[29] They can then be classed either as priority countries or countries of secondary concern, and appropriate action can be taken. Perhaps even more exciting is its potential to counter illegal trade by providing intelligence

---

[26] See Stiles (n.1), pp.1–2.

[27] Ibid, p.3.

[28] Ibid, p.5.

[29] Broad in *Symposium on Combating Wildlife Crime: Securing Enforcement, Ensuring Justice, and Upholding the Rule of Law. The Proceedings.* Editors Kala K. Mulqueeny and Francesse Joy J. Cordon, p.27.

**Fig. 10.1** Tusks to be killed for. A magnificent African elephant, peaceful in the bush. Copyright Chester Zoo

reports that not only *'help focus enforcement effort where it could be most effective'* but also contribute to better law enforcement.[30]

But even now, more remains to be done as not all countries have fully implemented the system. A recent TRAFFIC report set out the deficiencies and steps needed to be taken to remedy the situation in Central African countries.[31] The governments/CITES Management Authorities needed to become more involved in basics like collecting seizure information and correctly filling out ETIS forms, and better collaboration with the personnel, including NGOs, working in the field, was needed to enable seizure information to reach either the CITES Management Authorities or TRAFFIC as soon as seizures occur.

[30] Daniel Stiles (2009). *The elephant and ivory trade in Thailand* TRAFFIC Southeast Asia, Petaling Jaya, Selangor, Malaysia, p.5.
[31] Nkoke, S.C. Lagrot J.F. Ringuet, S. and Milliken, T. (2017). *Ivory Markets in Central Africa – Market Surveys in Cameroon, Central African Republic, Congo, Democratic Republic of Congo and Gabon: 2007,2009,2014/2015*. TRAFFIC. Yaoundé, Cameroon and Cambridge, UK.

.

# Part VI

## Surveying the Ivory Markets

Now we come to the surveys. These play a vital role because they provide key data about what is actually happening on the ground. How much illegal trade there is and how it is being carried out. The results are presented as reports to the CITES Secretariat, the Standing Committee and/or the CoP. The data provides facts and evidence not opinion, and it plays an important role in the decision-making process.

# 11

# The Confusing Nature of Ivory Markets

The market in elephant ivory is extremely complicated. Yet we must try to understand it if we are to make further progress, for there is a loophole that enables illegal ivory to be laundered more easily. The fact is that not all sales in elephant ivory are illegal. CITES only regulates international trade in elephant ivory.

When, in 1987, all populations of African elephants were moved up from CITES Appendix II into CITES Appendix I where all Asian elephants were already listed, this, in effect, created a total ban on international sales of elephant ivory, with certain exemptions. And of course, it is the exemptions that have caused and still continue to cause problems. So what are they? At their most basic:

- The first exemption makes it legal to sell ivory that was imported before 1989.
- The second exemption makes it legal to sell ivory deriving from the one-off sales
- The third exemption makes it legal to sell antiques and antiquities.

In the course of our story, we shall see just how these exemptions provide opportunities for laundering illegal ivory into the system, and once it is in there, this ivory can be very difficult to detect.

In fact, banning international trade in elephant ivory did nothing to solve the problem of how to regulate domestic trade in lawfully held ivory, and there was a lot of this ivory. It came from various sources:

© Springer Nature Switzerland AG 2019
B. Martin, *Survival or Extinction?*, https://doi.org/10.1007/978-3-030-13293-4_11

- Government-held stocks of ivory from elephants who had died naturally or been lawfully killed
- Raw ivory, legally held by private importers, manufacturers or large state-owned enterprises
- Pre-CITES ivory, much of which was in the form of valuable antiques, even antiquities

What could be done? The Parties to CITES eventually agreed that '*all raw ivory importers and exporters, and all enterprises that cut or carve ivory should be registered or licensed, with proposals for recording, and inspection procedures to monitor domestic ivory trade flows*'.[1] Ten years later, this was followed by Resolution Conf. 10.10 trade in elephant specimens, which required the adoption of comprehensive internal legislative, regulatory and enforcement measures.

As the first one-off sale approached, the situation became urgent. It was imperative that any sale of raw ivory outside the banned international sales should be strictly controlled; otherwise the situation could be exploited by those wishing to launder illegal raw ivory into the system. However, until recently, many domestic ivory markets remained largely unregulated. About 2016, things slowly began to change. There could no longer be any doubts about the devastating effect these markets were having. As early as 2004, a TRAFFIC briefing document analysing ETIS clearly showed that '*illicit trade in ivory is most directly correlated to the presence of large-scale, poorly regulated domestic ivory markets around the world*'.

---

[1] *Domestic Ivory Markets: Where they are and how they work*, A TRAFFIC Briefing Document, September 2004.

# 12

# Researchers and Their Methodology

It is essential to see how markets and/or retail outlets of ivory and rhino horn operate and what effect this has had and continues to have on the illegal trade in these commodities. And we can do this by examining some of the investigations that have been carried out into the markets.

Although TRAFIC, the wildlife trade monitoring network, carries out most of these investigations, similar investigations are carried out by, at the behest of, or funded by well-known organizations, NGOs and charities, such as the EIA, Born Free and IFAW (International Fund for Animal Welfare), Save the Elephants, Elephant Family and the Aspinall Foundation.

Many of the researchers undertaking surveys return at intervals to update their findings, wherever possible to the same places and repeating the same methods of research to ensure that comparisons are as accurate as possible.

Where time is limited, a spot-check 1-day survey is a useful research technique, such as the one carried out in the markets in Maputo, the capital of Mozambique, during a survey in 2005. The researchers pretending to be *'casual visitors with no particular intent to purchase ivory'*. A similar one was carried out in Luanda, Angola.

Another type of survey, which produces rather more detailed results, involves the investigators spending longer in the country. This happened in Senegal in June 2002, when the researchers spent 10 days visiting the retail outlets in three cities, counting and recording the number of ivory items openly on display as being for sale. Any items hidden from view were excluded, even if they had been spotted in, for example, a slightly open drawer.

The nationality of the researchers can also be important. Sometimes it can help them to extract information that would normally be withheld. A good

© Springer Nature Switzerland AG 2019
B. Martin, *Survival or Extinction?*, https://doi.org/10.1007/978-3-030-13293-4_12

example of this happened during a survey in China, when the Chinese researchers, local experts who could identify ivory on sight, were able to chat with the sellers while they pretended to be looking for presents. They even managed to take some photographs. A different kind of evidence.

Wherever possible, the same methodology is used in repeat visits as this makes it easier to work out what is going on. For example, illegal sales are going up? Illegal sales have started to decrease? All of which is essential data, evidence needed to work out the next steps to be taken, the next Resolutions to be drafted in this constant battle to protect elephants and other threatened wildlife.

Surveys in Nigeria provide a good example of just how difficult it is to assess what is happening in the field when it has not been possible to use identical methods of research. Three separate surveys had taken place in 1989, 1994 and 1999, and concern had been growing about possible illegal trading in elephant ivory. So another survey, an in-depth one, was carried out in June 2002. There should have been plenty of evidence available to determine whether and/or by how much the situation had deteriorated. However the researchers came to this somewhat Sphinx-like conclusion: *at the very least Nigeria's ivory trade is not decreasing*. It was the only one they could come to, despite the fact that the results themselves actually indicated an upward trend in the availability of ivory. And it was because the research methods used in the 2002 survey had not been identical to those used in the earlier research.

The resulting reports are published by the NGOs that sponsor/commission/carry out the research.

# 13

## Markets in Africa

In each case, before looking at the results of the survey, we will look at relevant legislation, to see just how well each country is protecting its elephants and rhinos.

We start in Africa where most producer countries are found, in southern Africa, Angola and Mozambique. Their ivory markets have caused concern for many years and unfortunately still continue to do so. And as we shall see later, neither country comes out well in the Great Elephant Census. As both of them were former Portuguese colonies, the legislation protecting their wildlife has much in common.

## 13.1 Angola and Mozambique

**Relevant Legislation: Angola** It has had a most convoluted relationship with CITES, because although its government formally approved the country's membership in 2002, it didn't become a full member until 2013, some 15 years after it had ratified the CBD. So between 2002 and 2013, Angola's legal position was unusual. Instead of implementing CITES, it used the treaty's own provisions to prosecute any terms or clauses of CITES that had been contravened.

Angola has had a troubled past. At war with its colonial masters from 1961 to 1975 when independence was declared, it then immediately plunged into a civil war which continued to rage until 2002 when peace was finally restored.

© Springer Nature Switzerland AG 2019
B. Martin, *Survival or Extinction?*, https://doi.org/10.1007/978-3-030-13293-4_13

At the time of the survey, much of the legislation protecting wildlife consisted of the legal framework that was set up in its colonial era. The Decree No. 40.040 of 1955, Ruling on the Protection of Land, Flora and Fauna applied to the protection of animals, plants and soil in all Portuguese colonies. Containing two Annexes, Annex 1 listed animals of significance that could not be hunted unless a special license had been granted. It included both savannah and forest elephants and rhinos. Annex 2 listed animals that could be hunted.

The Hunting Regulation, Decree 2:873 of 1957, regulated wildlife hunting in line with the earlier Decree. It confirmed both the classification of elephants into forest and savannah forms in Annex 1 and the requirement of a special licence to hunt them. And it also set out a number of different classes of hunting licences, including subsistence (local) hunting, agricultural (crop protection) hunting and sport hunting for Angolan residents and visitors, categories that were to be found in most of the African range States.[1] Annex 3 made it possible for special (supplementary) hunting licences to be issued for hunting elephants, a maximum of two forest and three savannah elephants, permits for forest elephants being the most expensive. However, only big tuskers could be shot, those whose ivory weighed more than 5 kg, though it is unclear whether this applied to each tusk or to the combined weight.

According to Article 86, anyone killing an elephant or rhino (both listed on Annex 1) had 30 days to inform the relevant authorities what they had done. Articles 90–92 set out a number of provisions relating to commercial activities in elephant ivory and rhino horn (and other hunting-related products), requiring information on names of hunters, permits and so on.

Although updated on a number of occasions, the Decree, which also contains provisions relating to possession and trade of wildlife products, essentially remains the same. Both pieces of legislation imposed criminal liability for wildlife offences.

In 1999, another Decree relating to the protection of wildlife was issued.[2] Both elephants and rhinos continued to be listed on Annex 1, and anyone caught illegally killing the listed animals would be fined. The fines, which were difficult to calculate, were high, and if they remained unpaid after 3 months, the courts could take further punitive action. Penalties for illegally killing elephants were very high.[3]

---

[1] Milliken, T. Pole, A. and Huongo, A. (2006) *No Peace for Elephants: Unregulated Domestic Ivory Markets in Angola and Mozambique*. TRAFFIC International, Cambridge, UK.

[2] The Decreto executive conjunto n. 0 36/99 of January 1999, a Combined Executive Decree as both the Ministry of Finance and the Ministry of Agriculture and Rural Development had input into its administration.

[3] See Milliken, Pole and Huongo (n.1), p.9.

Newer legislation shows Angola's intention *to ensure that wildlife manage-ment contributes to social and economic development, particularly local commu-nities through the promotion of tourism and the participation of communities in the benefits derived from protected areas management*.[4] In other words, any new regime of wildlife management will incorporate sustainable use.

**Mozambique:** Like Angola, it also experienced a long period of strife after gaining its independence and also inherited wildlife legislation from its colo-nial days. However, it has been a Party to CITES since 1981 and its legislation is very different.

In colonial times, Mozambique's principal wildlife law regulated hunting, and this regulation continued after independence. Elephants could be hunted, but only under licence and not in the close season. It was forbidden to kill females and young animals, and the use of certain types of weapon was banned. Possession and trade in elephant trophies required certificates of ownership[5] and was strictly regulated, with permits needed for any commer-cial transactions of import or export. Furthermore *all found trophies needed to be handed over to the nearest government administrative unit within specified periods of time*.[6] Unfortunately, although the legislation did contain enforce-ment measures and penalties, it also contained a number of loopholes, its administration was non-existent and its enforcement left a lot to be desired.[7]

Then came the civil war, so it was not until 1997 that the National Forestry and Wildlife Strategy was formally adopted. Its long-term objective was *to protect, conserve, utilize and develop forest and wildlife resources for the social, ecological and economic benefit of the present and future generations of the Mozambique people*.[8] Here again we can see some of the objectives of the CBD including the all-important sustainable use.

The Forestry and Wildlife Law[9] came 2 years later, followed, in 2002, by the Regulation of Forestry and Wildlife Decree No.12/2002.[10] All this legisla-tion should have provided a sound basis for the protection of its wildlife. So did it?

Under the Forestry and Wildlife Law, national parks, in theory, give total protection to wild animals and list activities such as hunting (including illegal

---

[4] Maria Teresa Cirelli, Elisa Morgera *Wildlife Law and the Legal Empowerment of the Poor in Sub-Saharan Africa*, FAO Legal Papers Online 77, May 2009.

[5] de Klemm and Lausche 1986, in Milliken, Pole and Huongo (n.1), p.26.

[6] Ibid, Issufo 2005.

[7] Milliken (n.1) comments that this *essentially negated the practical impact and efficacy of these laws*.

[8] Milliken, Pole and Huongo (n.1).

[9] Law 10/99 of 7 July 1999.

[10] Of 6 June 2002.

hunting or poaching) that are strictly forbidden unless there are scientific reasons or management needs for them to be carried out. Forestry or agricultural exploitation, mining or livestock farming are also forbidden because they open up the forest and create access for potential poachers (Article 11). National reserves offer similar protection, but only to certain species of plants and animals that are rare, endemic, becoming extinct or in decline or to fragile ecosystems (Article 12).

However three forms of hunting are permitted:

- Hunting under a simple licence
- Game hunting
- Commercial hunting[11]

The Regulation of Forestry and Wildlife Decree No. 12/2002 details not only how hunting must be carried out but also punishments, mainly fines, for people found guilty of committing offences. Besides poaching, other offences include:

- Illegal possession, which in our case would include ivory in its many forms as well as rhino horn
- Illegal processing of animal material, such as grinding rhino horn into powdered form
- Illegal export and import
- Illegal supply and sale
- Illegal consumption

Further legislation, Anti-Corruption Law (Law No. 6/2004, of 17 June) imposes harsh penalties. Prison sentences of anything between 2 and 8 years for corruption, tax evasion, fraud and money laundering.

In theory, this legislation should have been a deterrent, but was it?

Because of growing concern about Angola's involvement in the illegal ivory trade, in June 2005, researchers from TRAFFIC visited both Luanda, Angola's capital city, and Maputo, Mozambique's capital, because although elephant numbers seemed to be growing, the Mozambique authorities did not appear to be implementing the requirements for internal trade in ivory.[12] They had failed to submit an important report to the CITES Secretariat as part of the CITES *Action plan for the control of trade in African elephant ivory*. It concerned their unregulated ivory markets and was due to be considered at the 53rd meeting of the Standing Committee.

---

[11] Art 12(3) and other articles.
[12] As noted in CITES Resolution Conf. 10.10 (Rev. CoP12).

Tom Milliken, Alistair Pole and Abias Huongo were the researchers, and their findings were documented in the report *No Peace for Elephants: Unregulated Domestic Ivory Markets in Angola and Mozambique*. This is what they found.

**The Results of the Survey: Angola** Although people from many countries, including Angola itself, and China, were buying ivory, the main customers were Americans and Europeans (including Spanish, Italian, Portuguese and English). And the ivory artefacts on sale had obviously been chosen for their appeal to these particular customers. Some traders were bulk-buying chopsticks and name seals, probably for resale either locally or in other countries.

There was no doubt that at least some of the vendors knew only too well that it was illegal to sell ivory to tourists. Apparently, small pieces of ivory could be got out of the country without difficulty, while larger ones *could be sent in a diplomatic bag*. One tourist shop had an estimated 7.8 kg ivory goods on display, while a curio shop in the duty-free section of the Luanda International Airport, its customers on their way out of the country, was selling all sorts of ivory carvings, whose total weight was estimated at 5.6 kg. There was an astonishing lack of interest on the part of the authorities.

**Mozambique:** The artefacts were similar, namely, human and animal ivory figurines, jewellery and trinkets or ivory used as inlay, all products designed to attract international buyers. The surveyors estimated that about 3254 ivory items were being offered for sale, and, like Angola, some were in shops located in the departure lounge of the airport terminal building. They noted some 588 pieces.

Researchers also try to find out whether it is possible to buy raw ivory. As this ivory can only be illegal, its presence poses a number of questions. For example, Are elephants being poached locally? If not, where is the ivory coming from? Which elephant populations are at risk?

**Raw Ivory Tusks:** In the Mercado do Artesanato, Angola, a market just south of Luanda, a number of stall holders indicated that, given sufficient time, they would be able to obtain large raw tusks. One owner even indicated that he had two sets of tusks in his house, the smaller one weighing 25 kg per tusk and the other 40 kg per tusk. The owner of another market, the Sonol, indicated that he too could get raw ivory within a few days, mainly from the north of the country. He quoted prices for both small and large tusks, and they were cheaper than those quoted by the owner of a curio stall in yet another open market. A number of shops were also visited. Although most of them did not deal in raw ivory, two owners indicated that they could get some, but it was difficult to obtain. One of them said that because of this he no longer did much carving, but the white shavings seen on his work bench rather suggested he might have been somewhat economical with the truth, as their presence could only mean that someone had recently been carving ivory there.

> Some Conclusions: *While southern Africa as a whole, strives for global acceptance of sustainable use policies predicated upon a record of wildlife management and conservation achievement, Mozambique and Angola stand as worrying anomalies within SADC (Southern African Development Community).*[13]

These grim words show just how concerned the team were with what they found. They described *'an abundance of elephant ivory products'* openly on offer for sale apparently without any of the regulatory safeguards required under CITES. In Angola's case, it was clear that much of the ivory was coming, ready carved, from neighbouring countries particularly the DRC and Congo (Brazzaville). These Central African countries are where the endangered forest elephants live and are poached.

Roughly 400 elephants must have died to account for the amount of ivory products seen on sale, though this figure was *a bare minimum*, reflecting only *a static point in time in only two locations in these two vast countries*. And in both Angola and Mozambique, the ivory trade seemed to have increased, which indicated that business was good.

Although the researchers could only draw conclusions from the places they had visited, there were a number of other suspect locations which had not been surveyed, and there were almost certainly other, undisclosed stocks of both raw ivory and worked ivory products. The scale of the ivory trade in the two countries concerned them because:

- Their domestic markets were unregulated.
- They were currently subverting the CITES international trade controls.

## 13.2   Central Africa

This area is important because it is where so many of the endangered forest elephants live and where so many have been slaughtered to fuel the illegal wildlife trade. In 2017, a TRAFFIC report was published which collated and reviewed a

---

[13] Milliken, Pole and Huongo (n.1).

series of surveys into domestic markets in key countries in Central Africa.[14] The baseline was set by the 1999 research carried out by Martin and Stiles.[15] Surveys were then carried out in 2007 and 2009 and from October 2014 to June 2015 in Cameroon, the Central African Republic, DRC, Gabon and the Republic of the Congo, though different countries were surveyed at different times.

Although the results of the early surveys were shocking, though perhaps not unexpected, the encouraging thing was that as the years progressed, most of the markets declined.

Between 2007 and 2015, a total of 2006 kg of ivory in the form of 12,964 items was observed to be for sale in 244 retail outlets, with 68% of them being sold in open markets in Kinshasa, DRC, where the market was still thriving.

Furthermore, *'More than 53,700 kg of ivory from Central Africa was seized globally between 2007 and 2015. Equivalent to approximately 5712 elephants'* with *'4x as much ivory … getting out of Central Africa before being seized than is being seized in-country'*. However, in 2007, 971 kg ivory was on sale in 93 outlets, but in 2014/2015, only 401 kg ivory was on sale in 32 outlets.

And although the number of items for sale remained roughly the same, the recent artefacts were much smaller. Raw ivory was becoming scarce. Wood carving had taken over as *'the main craft activity in retail curio markets'*. Specific orders for ivory were undertaken, but only if the material was available to carry it out. Raw ivory was becoming scarce.[16]

Over the years, the main buyers changed from Europeans, both expatriates and tourists, to Chinese, with some West African traders as well.

Unfortunately, because of the instability and civil unrest in some of the countries, the involvement of organized crime gangs, as well as improved law enforcement, the trade was moving underground, *'with a focus on the export of raw ivory to foreign markets, especially China'*. The ivory continued to be smuggled out by traditional routes, roads and rivers, with the Congo River providing a most convenient method of transport, especially for ivory coming from northern and eastern Congo and DRC.

The current legislation in all the countries except Cameroon prohibits domestic ivory trade, and both Congo and DRC have recently updated their wildlife laws. But everyone, from the enforcers to most of the others involved, tends to interpret it loosely and ambiguously, and, in some countries, people are confused about who is responsible for controlling and monitoring the ivory trade.

---

[14] Nkoke, S. C. Lagrot J. F. Ringuet, S. and Milliken, T. (2017). *Ivory Markets in Central Africa – Market Surveys in Cameroon, Central African Republic, Congo, Democratic Republic of the Congo and Gabon: 2007, 2009. 2014/2015*. TRAFFIC. Yaounde, Cameroon and Cambridge, UK.

[15] Martin & Stiles 2000.

[16] Nkoke, Lagrot, Ringuet and Milliken (n.14), p.xiv.

There is enforcement, but there is also corruption in high places, *'insufficient human and financial resources, mismanagement and weak political will'.*[17]

Ten recommendations were proposed. None of them were new or additional, none of them unexpected, for example, Recommendation VIII: *Permanently close down the Kinshasa ivory market by the DRC.* In fact, most of the proposals were in accord with the countries' own legislation, regulations and commitments at all levels, including global treaties. And, with goodwill and collaboration, all are achievable, which is essential to effective elephant conservation.

The first three were concerned with scaling up law enforcement performance, particularly in DRC, as this is already paying dividends in Cameroon, CAR, Congo and Gabon. And particularly enforcement at DRC's exit points which link with other countries, with *'enhanced enforcement checkpoints along the identified trafficking routes such as the Souanke-Congo and the Douala/ Yaoundé route; the CAR-Cameroon and Gabon-Cameroon routes, the Congo-Gabon and DRC-Congo routes, and the DRC-CAR routes'.* This would also involve fighting corruption, as well as capacity building and enhanced collaboration at both national and regional level, with continuing support from partner organizations such as TRAFFIC and WWF.

A key component is harmonization of wildlife legislation of both national governments and regional organizations such as COMIFAC (Central African Forest Commission) and ECCAS (Economic Commission of Central African States). Traffickers are only too aware of different levels of protection and different punishments and are adept at exploiting available loopholes. So the third recommendation dealt with inter-agency and regional collaboration. Agencies such as the police, customs, gendarmes, judiciary and others, if they are involved, must be part of the solution. Information should be shared at the regional level as well, and, where possible, there should be joint field actions (such as patrols in cross-border protected areas) to curb the movement of raw ivory. Governments *should take full advantage of* programmes such as AFRICA-TWIX (the information exchange platform).

ECCAS already has an Urgent Anti-Poaching Action Plan (PAULAB), *'which covers the savannah zones of northern Cameroon, southern Chad and northern CAR'*, one of its objectives being the signing of accords of collaboration, while the African Union-led African Strategy on Combating Illegal Exploitation and Illegal Trade in Wild Fauna and Flora in Africa, Decision 15/2, proposes:

---

[17] Ibid.

- *'The establishment of an inter-regional co-operation mechanism and partnerships to fight illegal trade in wildlife and share expertise and knowledge to further reduce demand for illegal trade in wildlife'*
- *'The establishment of national, regional and sub-regional networks of regional wildlife enforcement to promote increased sharing of best practices and experiences, communication and co-operation links'*

Recommendation IV is concerned with more effective reporting and use of ETIS. Recommendation V with NIAPS, which *'should be seen as an integral part of the conservation strategies of the different countries to enable proper functioning through finance from the national budgets and the implication of State agencies'*. Indeed, *'commitment and implementation'* is *'imperative to address the illegal ivory trade'*. All the countries, except CAR, had already been requested to develop and implement a NIAP, and it was recommended that CAR and Chad, both of whom had previously attended a relevant TRAFFIC workshop and had said they were interested in the NIAP process, *'should consider creating their own action plans'* regardless of whether they were formally invited to be part of the process, or not. Recommendations IX and X stressed the importance of developing programmes of regular surveys, adapting the methods where required if the trade had gone underground and monitoring the amount of progress the countries were making toward their commitments. COMIFAC was suggested as an independent monitoring body which could also look at plans such as PAULAB and priority decisions such as Decision 15/2.

## 13.2.1   TRAFFIC's Minimum Standards for Effective Ivory Stockpile Management

Stockpile management had always presented a problem to range States, so TRAFFIC's minimum standards were very welcome. They included:

- Information on the source of ivory
- How to measure and mark each piece of ivory in the stockpile
- Registration of the information
- Centralization of the ivory in a national government stockpile
- A range of security issues to consider
- Procedures for audits and periodic verification exercises
- Legal provisions and other matters[18]

---

[18] Ringuet & Lagrot, 2013, in Nkoke, Lagrot, Ringuet and Milliken (n.14), p.80.

To safeguard the stockpiles over time, where possible, there should be quarterly, certainly regular inventories, with independent audits every year.

To protect seized ivory or ivory collected in the field, there should be a clear procedure in place to facilitate its timely transport to the stockpile.

To prevent ivory leakage, which continues to fuel ivory trade, field staff need to be instructed in the system of marking seized ivory under CITES specifications and national procedures, which should be stuck to.

Recommendation VI deals with effective stockpile management. The governments of all the countries surveyed should comply with TRAFFIC's minimum standards.

Recommendation VII suggests a *paradigm shift by international bodies, national governments, development partners and other wildlife conservation stakeholders'* because:

- *'There is a dire need for more political will (by the national governments) to ensure the development and implementation of good wildlife policies'.*
- Organized crime syndicates, not just local poachers, are involved in a lot of the illegal wildlife trade, so state agencies, NGOs and IGOs (intergovernmental organizations) need to think more broadly.
- Elephant populations throughout Central Africa are steeply declining and *real concerted efforts are needed to address* (this) *problem.*
- Because of the number of rebel groups involved in international ivory trade, this had now become a security issue as well.
- Chinese nationals, *especially those living long-term in Africa*, need to be made aware that international trade in ivory is banned by international law (CITES). Fortunately, this is already being done under the auspices of FOCAC (the Forum on China-Africa Cooperation).

FOCAC's Johannesburg's Action Plan 2016–2018 *needs to be properly implemented.* Cooperation is the key word, from providing capacity building for rangers to training opportunities on environmental and ecological conservation and various projects, as well as jointly fighting the illegal wildlife trade, especially poaching endangered species in Africa *with an emphasis on elephants and rhinos.*

## 13.3   Egypt

The savannah elephant became extinct during the time of the Ancient Egyptians, but that ancient civilization fully appreciated the beauty of ivory, and its craftsmen were carving objects out of the material some 8000 years ago. The ivory they used came from both sub-Saharan Africa and the horn of Africa. In one tomb in the Valley of Kings, there are wall paintings showing men from foreign countries carrying elephant tusks to the Pharaoh. There are also texts dating back to circa 1500 BC, from the reign of the remarkable Hatshepsut, a woman Pharaoh whose many achievements included negotiating treaties with neighbours. They describe an important trading expedition to the almost mythical Land of Punt, to bring back all manner of ancient and exotic goods. Including *pure ivory*. As we have seen before, love and appreciation of ivory goes deep into the past.

Although Egypt was one of the earliest countries to sign up to CITES, becoming, in 1978, the 41st Party to the treaty, this far-sighted action was followed up by complete inaction. Only after considerable pressure had been exerted by both the CITES Secretariat and the CoP did Egypt's Ministry of Agriculture and Land Reclamation finally issue the Ministerial Decree No. 1150 of 1999 that provided the legal framework for the treaty.

Egypt has been surveyed on a number of occasions because *'there is little doubt that, with the complete closure or even reduction in scale of the Egyptian ivory market, elephant conservation will benefit throughout Africa'*.[19] In other words, it is one of the countries that holds the key to the survival of elephants in the future.

Although there was a significant drop in the illegal trade between 1999 and 2003, mainly because of improved law enforcement and some shopkeepers realizing that they were involved in illegal activity, a comprehensive assessment in 1998 had concluded that Egypt was one of the largest domestic ivory markets in Africa. A country to keep an eye on. The 2005 survey was carried out because of evidence that trade was picking up in Khartoum, the capital of neighbouring Sudan, the country which had always supplied Egypt with most of its illegal ivory.

The cities of Cairo, Luxor and Aswan, previously surveyed in 1998, were all revisited in 2005, together with two new tourist hotspots, Sharm El Sheikh and Hurghada. The field work was carried out in just under a month by Martin and Milliken, two of the most experienced researchers in the field, who visited both retail outlets and carving workshops. This is what they found.

**Ivory Carving:** Shrouded in secrecy and only happening in Cairo, many retailers described it as a *dying art* as so many of the expert carvers, who had produced some exceptionally fine ivory sculptures, were either dead, retired or elderly. Young people were not being trained to take their place, so the products were mainly small figurines or cheap jewellery.

---

[19] Martin, E. and Milliken, T. 2005 *No Oasis: the Egyptian Ivory Trade in 2005*. TRAFFIC International, Cambridge, UK, p.1.

**Retail Sales and Consumers:** Cairo had the most retail outlets, including the ancient market of Khan al-Khalili, and correspondingly the most sales. The market attracts thousands of foreign visitors every day. Ivory items on sale included figures of humans, Gods and animals, as well as pieces of jewellery, particularly pendants. This being Egypt, scarab beetles were popular. Some ivory was antique. Some newer items had obviously been made in other countries especially China, Japan and France.

Alexandria had virtually no ivory for sale, because even though it is the second largest city in Egypt, very few tourists were visiting it. Sales were obviously aimed at international tourists, particularly Europeans, and the researchers observed that *'many Egyptian sales personnel are very savy operators on the international tourist scene and would often quote prices for ivory products in US dollars (USD) or euros once they ascertained the nationality of prospective buyers'.*

In fact, the trade appeared to be purely international as *'reportedly, Egyptians do not buy ivory objects'*, most of the buyers coming from Europe, particularly Italy and Spain. with a few from France. While others were from Latin America and the Gulf States, almost none came from China, which was perhaps why the artefacts on sale would not have appealed to this group of consumers.

**The Indian Connection:** There was also a very small amount of trade in Asian ivory. Although there are ivory stockpiles in India, some of the privately owned ivory seemed to be making its way onto the Egyptian market, which was unexpected. But it was observed in three surveys, and one trader, who also owned an ivory workshop, produced an elephant tusk weighing 28 kg which, he claimed, was from India. The tusks in the 2005 survey probably came in by air, through the Gulf States.

Because Asian elephants are so endangered, any action that might encourage illicit trading sales must be taken very seriously. Fear of poaching is never far away.

**Conclusions on the Situation up to and Including 2005:** The results of both the 1998 and 2005 surveys were almost identical. Enforcement was a major problem.[20]

Although plenty of ivory was for sale, with some shops openly displaying *'hundreds of ivory products'*,[21] there was a 43% reduction on the combined number of ivory objects for sale between 1998 and 2005. Furthermore, the

---

[20] Ibid, p.15.
[21] Ibid, p.21.

government was trying hard to implement its CITES legislation, a fact noted and commended by the researchers.

That was the situation in 2005, so what happened next?

**After 2005:** Two important things. The Arab Spring came and went and Chinese buyers came and stayed. In fact the Chinese became the principal buyers of ivory in Egypt.

So in March and April 2011, Esmond Martin, commissioned by Tom Milliken for TRAFFIC, and this time accompanied by Lucy Vigne, another experienced researcher, returned to Egypt to see what changes there had been, if any, in Cairo and Luxor.

The results were disappointing. Although the number of ivory objects for sale had decreased by a further 10% on the 2005 figure, this had mainly been in Luxor, where the number of tourists had dropped. Furthermore, nearly a third of the objects on sale had been produced recently, that is, within the last 5 years. Cairo remained the carving centre.

All the sales were illegal as each ivory item required a special permit before it could be sold and no permits had been issued.

**The Consumers:** These had changed radically. Now the Egyptians themselves were buying ivory, so were Gulf Arabs, and they wanted walking sticks and rosary beads. However, most of the buyers were Chinese, and they were big spenders, one seller telling the researchers that *'Chinese buyers would sometimes spend USD50,000 on ivory during one bargaining session'.*[22] Getting their purchases out of the country apparently did not present a problem either, with some sellers prepared to, even offering *'to write a receipt indicating that a piece was an antique or made of camel bone'.*

There seemed to be no attempt at enforcement, even though the Egyptian Management Authority for CITES was holding training courses for personnel involved in implementing Egypt's legislation at the same time the survey was being carried out.

**The Conclusion:** That although Egypt's illegal ivory markets had grown very much smaller between 1998 and 2005, this trend had now reversed, and they were once again thriving. Once again Egypt had become a leading trader in illegal ivory. So why is it of such importance to our story? Because most of the

---

[22] *Illegal ivory trade blooms during Egypt's Arab Spring*, TRAFFIC – Wildlife Trade News – Illegal ivory trade blooms during… See http://www.traffic.org/home/2012/1/30/illegal-ivory-trade-blooms-d... Accessed 28/10/2014.

ivory on sale came from forest elephants, from elephants from the Central African Republic and the DRC, as well as some from Kenya and the Cote D'Ivoire. From elephants threatened with extinction.

## 13.4   West Africa

The story here is little better, investigations carried out in 2002 reaching the grim conclusion that:

West Africa's ivory markets continue to pose serious challenges to the implementation of CITES and to the effective conservation of elephants throughout the African Continent.[23]

Côte d'Ivoire, Senegal and Nigeria were all surveyed in 2002, because Abidjan (Côte d'Ivoire), Dakar (Senegal) and Lagos (Nigeria) had been identified in an earlier survey as *the most significant ivory carving centres in West Africa*.[24] Nigeria continues to have problems with its ivory markets. Most of the ivory comes from either the rare forest elephants or the even rarer West African elephants.

### 13.4.1   Côte d'Ivoire

Although domestic trade in elephant ivory was officially banned in 1997 by the Decree No. 97-130, sales only declined slightly, as the researchers discovered when they carried out the 2002 survey. And the industry was still very active.

The survey was carried out in and around the capital Abidjan and Grand Bassam.

**The Results:** They found 11 workshops, employing about 88 carvers, although most of these were part-time workers, and 68 retail outlets, mainly selling worked ivory.

Most of the trade was international, hence illegal. The customers were mainly tourists and expatriates, particularly Americans and French, though there were also Chinese and Japanese buyers. It was alleged that some Asian diplomats were buying ivory products for resale abroad. Senegalese traders were among the local African businessmen who bought ivory.

---

[23] Courouble, M., Hurst, F. and Milliken, T. 2003. *More Ivory than Elephants: domestic ivory markets in three West African countries*. TRAFFIC International, United Kingdom.

[24] Martin and Stiles 2000, ibid. p. i.

And the source of the ivory? Once again, it was forest elephants. They were being slaughtered for these unlawful sales. Rather than local sources, most of the ivory came from Central Africa, particularly DRC, Cameroon, Gabon and CAR. Some was also being filtered in from privately owned stocks within the country itself.

Another fact worth noting is that when this particular survey was conducted, there was political unrest in the country which was affecting the trade in two ways:

- Some tourists were reluctant to visit the country.
- It was difficult to replenish stocks of raw ivory.

## 13.4.2   Senegal

Three cities were surveyed in this country, Dakar, the capital, and two others, Saly and Mbour.

**The Results:** There were 6 ivory workshops employing 31 carvers and 54 retail outlets. In Dakar, more shops were selling ivory, up from 30 (in the 2000 survey) to 44, and although less ivory was on display for sale, the pieces being sold were larger which is never a good sign. It is often related to poor law enforcement. Despite this, the researchers concluded that the situation was probably relatively stable.

Unlike most other countries, there were no factories and/or workshops as such. The trade in ivory was carried out by families, as family businesses. Small household workshops carved the ivory which was then sold, often by members of the same family who might even sell it in neighbouring countries, particularly Côte d'Ivoire.

There are two points to note here:

- Senegal had not got enough worked ivory of its own to supply demand, so in all probability needed to import more than it was producing.
- *'There is evidence to suggest direct linkages to the production and trade in ivory with Côte D'Ivoire'.*[25]

Perhaps the one follows from the other.

---

[25] Ibid, p.iii.

The customers were different again, for although both European tourists and Chinese bought ivory artefacts, so did Koreans and the local Marabouts. Some traders claimed Chinese diplomats were frequent customers. Another difference was that a few of the European traders were actually buying ivory for resale.

**What Was the Source of This Ivory?** Although most of the raw ivory was illegally imported from Central Africa, particularly Cameroon, DRC, Gabon and CAR, four other countries were mentioned. Kenya and Sudan make sense, but China is puzzling. The fourth country was Côte d'Ivoire.

While worked ivory, like some of the raw ivory, came from Cameroon, DRC and Gabon, as well as from Guinea, more than half of it, reportedly, came from Côte d'Ivoire.

**The Senegal/Côte d'Ivoire Connection:** More than half of all the worked ivory for sale in Senegal was reported as coming from Côte d'Ivoire, and this was partly corroborated by the 2002 survey in Côte d'Ivoire itself, which reported that Senegalese traders were some of their buyers. A possible reason for this could be that Côte d'Ivoire is closer to the sources of raw ivory than Senegal.

The situation regarding raw ivory was somewhat different. Although ivory was coming in from Côte d'Ivoire, here it was probably merely acting as a transit country.

**Conclusions:** The legislation was weak and out of date, passed in 1986 before the ban on international sales (in 1989) came into existence. The major players included some Marabouts who are important and powerful religious leaders. Not only ivory but many different kinds of wildlife were being traded. The traders in the Soumbedioune market had organized themselves into a guild *significant economic and political clout*. And there were some serious allegations of corruption suggesting that customs, businessmen, religious leaders and government officials were all implicated in illegally trading ivory. In other words, '*The trade in ivory is protected by powerful lobbies in Senegal*'.

Neither Côte d'Ivoire nor Senegal was compliant with CITES Resolution Conf. 10.10 (Rev.), and although the former country did occasionally submit information to ETIS, the latter never had. The researchers' commented '*whether this reflects poor administrative engagement or a total lack of law enforcement with respect to ivory remains to be seen*'.

## 13.4.3   Nigeria

The situation in Nigeria, which is something of a hotspot for the illegal trade in elephant ivory because of its large-scale unregulated ivory markets, is complicated by the fact that Nigeria is a Federal Republic, made up of some 36 states. This has an effect on both legislation and enforcement. CITES has been implemented in Nigeria by the Endangered Species (Control of International Trade) Decree No. 11 (1985). This is federal legislation, so it applies to the whole country. However, some of individual states have also enacted their own, compatible legislation. Where they haven't, Federal Decree No. 11 should be applied.

Trade in ivory is controlled, but unfortunately the African elephant (*Loxodonta africana*) is only listed on Schedule II of the Decree, which means that it (and/or its products) can still be hunted, traded, imported and exported provided the requisite licence has been obtained. The Decree was under review. Furthermore, although no licences were reported to have been issued since the trade ban in 1989, the researchers actually found trading in ivory was widespread.

Like many of the countries involved in this illegal trade, Nigeria has a tradition of carving and selling this beautiful substance. Three earlier surveys had been carried out because of growing unease, the first in 1989, at just about the time the total ban on international sales was coming into force, followed by two others, in 1994[26] and 1999,[27] so there was plenty of evidence available to determine whether and/or by how much the situation had deteriorated.

The researchers conducted surveys in Lagos, Abuja, Kaduna and Kano between June 15 and 26 in 2002.[28]

**The Results:**  In the short time they were there, 2 workshops were identified in the Lagos area, with possibly others in Onitsha, Benin City and Port Harcourt, and some 42 retail outlets, Lagos continuing to be the main centre for ivory sales.

Once again, Americans, Chinese, Europeans (this time Italians) and some Koreans were the buyers. The ubiquitous diplomatic staff were also mentioned. It was alleged on no fewer than 11 occasions that they could be used to export large items of ivory such as whole tusks. Lebanese buyers were men-

---

[26] Dublin et al. 1995; see Esmond Martin and Lucy Vigne *LAGOS, NIGERIA: One of the Largest Retail Centres for Illegal Ivory Surveyed to Date*, TRAFFIC Bulletin Vol. 25 No. 1 (2013), p.35.

[27] Martin and Stiles, 2000, ibid.

[28] See Courouble, Hurst and Milliken (n.23).

tioned as well. Allegedly, sometimes they and Chinese buyers would engage in 'large-scale "container" exports of ivory presumably for re-sale purposes'.

Where did this ivory come from? The names are again familiar. Most of the raw ivory entering this West African country was coming from Central African sources, DRC, CAR and Gabon, but also Cameroon and Congo (Brazzaville).

**The Conclusions:** Despite the fact that the results themselves actually indicated an upward trend in the availability of ivory objects for sale, the researchers reached the cautious conclusion that 'at the very least Nigeria's ivory trade is not decreasing'. The reason for caution was because the survey methodology was different.

**Nigeria's Non-compliance with CITES and Its Result:** Like both Côte d'Ivoire and Senegal, Nigeria was found to be non-compliant with the conditions set out in Resolution Conf. 10.10 (Rev.) and its participation in ETIS poor. In fact, this 2002 survey was carried out at a time when Nigeria was required to demonstrate its compliance with the Resolution (under CITES Decision 12.39).[29] The researchers described their results as very worrying and concluded that 'unless rapid action to amend the situation is taken, Nigeria could see sanctions imposed on its wildlife trade at the next meeting of the CITES Standing Committee in March 2004'.[30]

In fact, CITES did end up having to impose a trade ban on Nigeria. In July 2005, because of its continuing failures at law enforcement, CITES banned Nigeria from selling any CITES-listed species.[31] This was followed up by two missions from the CITES Secretariat. The first one, in January 2010, was led by the Secretary-General himself. Under the Secretariat's guidance, Nigerian officials finally began to inspect the wildlife markets and make ivory seizures.

**The 2011 Inspections and the 2012 Survey:** The Secretariat returned to Nigeria in February/March 2011, where it carried out inspections of markets and other relevant locations in Kano and Lagos. The results were encouraging. There was insufficient worked ivory for sale to give rise to concern.[32] In May that same year, the Nigerian government strengthened its legislation. And in August, following the results of its earlier inspection, the CITES Standing Committee rescinded the trade ban it had imposed on Nigeria.

---

[29] A procedure (an intersessional process) carried out under CITES Decision 12.39.
[30] See Courouble, Hurst and Milliken (n.23), p.v.
[31] Milliken et al., 2009; see Martin and VIgne (n.26), p.35.
[32] CITES, 2011a,b, ibid.

The 2012 survey was carried out in Lagos (only), that September, in other words, about a year after the new legislation had come in and the trade ban had been lifted. To keep the methodology consistent with the 2002 and other surveys, only objects openly on display were recorded, even if it was obvious other items had been stored or hidden.

As before, the researchers concentrated their efforts on the main souvenir market (the Lekki), other markets, hotels, arts and craft shops and the international airport, places where they would expect to find retail outlets selling ivory. They recorded what kind of items were for sale, which were new and which were old (i.e. pre-CITES ban), the origins of the tusks, where they had been carved and the prices being asked for their sale. And of course they counted the items.

They also visited two workshops in different parts of Lagos. In one of the locations, there were two other inactive workshops close by and, in the other, a workshop that was closed. In other words, out of five workshops, they visited the two that were operating. The carvers came from different countries.

**The Workshops:** In the first one, the carvers, from Burkina Faso, were working on cow bone because *'it was illegal to use ivory'*.[33] The researchers saw no trace of ivory, unlike in the second workshop. Here again the craftsmen claimed they no longer worked ivory, hadn't done so for the last 5 years because its sale and possession were banned. Although they too were carving cow bone and there was no sign of ivory in the workshop, the researchers noticed that one of the craftsmen was wearing *a newly made ivory beaded bracelet that he was seen to remove, presumably so as not to be photographed wearing it*. Of the 11 craftsmen there, all of whom were self-employed, only 3 came from Nigeria, and the remainder were from Guinea.

Although the closed workshop displayed a number of figures carved from cow bone, there was also some ivory. Two busts, in fact. These were larger than the figures, and were sitting on a shelf, or rather they were, until a few hours later when they had somewhat mysteriously disappeared. The shop remained locked.

**The Retail Outlets:** No ivory should have been for sale, of course. But it was, although not in the shops at the international airport. In the Lekki market alone, there were 33 retail souvenir outlets, which between them had 14,200 ivory artefacts on display. The items varied very considerably in size, from jewellery to polished tusks. Even more worrying was the fact that *'over 99% of the*

---

[33] Ibid.

*items appeared to be recently made'.* The number of objects for sale at each outlet ranged from 2 to 1250, the average being 430. There were also *'thousands of smaller items such as jewellery, name seals and chopsticks'* being stored *'under the counter',* so not openly on sale. This meant they could not be included in the count. The researchers observed that *'during the survey, it was evident that more ivory was being sold than any other souvenirs in the market'.*[34]

The Arts and Crafts Centre, situated in the grounds of a large hotel, was the other place still selling ivory, although the number of objects for sale there, 149, was tiny in comparison with those in the market. Over half of them were items of jewellery, with some animal and human figures, the obvious target buyers being foreign tourists and visitors. And so it proved to be European, American, Japanese and Lebanese buyers, in that order. Only the Chinese were seen buying ivory at the Lekki market, where they preferred the smaller objects, ones *'that they can more easily transport to China in their personal luggage'.*[35]

**Where Did This Ivory Come From?** Nigeria is still a range State, mostly for savannah elephants which live in the Yankari Game Reserve. And their numbers have been dropping dramatically. Although it is not easy to count elephants, particularly those hidden by the forest canopy, they were down from an estimated 600 in 1993[36] to 348 in 2006,[37] with possibly another 105 elsewhere in the country.[38] Excluding the hidden ones, only 82 were counted in the Game Reserve in 2011, and poaching is a significant threat.[39]

Although some of the ivory on sale in Nigeria came from elephants poached in the Game Reserve, according to the vendors, most came from Cameroon, the Central African Republic, Congo Brazzaville, DRC and Gabon, from the beleaguered forest elephants. Some was smuggled in from Kenya.

**Sellers' Views on the Ivory Trade:** These are not always recorded in reports and they make for interesting reading. Not surprisingly, these sellers were deeply suspicious because the researchers were not buying, but asking questions and they wanted to know why. Were they for a report? None of them admitted the trade was illegal, and they were all prepared to

---

[34] Ibid, p.38.
[35] Ibid.
[36] Said et al., 1995, ibid., p.36.
[37] Omondi et al. 2006, ibid.
[38] Blanc et al. 2007, ibid.
[39] Bergl et al. 2011, ibid.

write whatever the customers wanted on their receipts *'to make it easy to take ivory items abroad'.*

Vendors knew it was illegal to sell ivory, so no wonder they were uncooperative. Some *'closed their shops … or covered their ivory items over with cloth or hid them to avoid their being seen or counted'*, making it very difficult to collect accurate data. It took four visits to the Lekki market to collect the details needed to complete the survey.

**The Conclusions:** These were grim. Despite the special measures we saw earlier, the illegal trade was still flourishing. In Lagos, retail sales of ivory had increased by almost a third since 2002, and even the total weight of ivory was slightly higher. Furthermore, almost all the items had been carved recently, the ivory was new. They found that this *'availability of both raw and worked ivory'* was *'contributing to the serious impact that such trade is having on the elephant populations of Central and West Africa'*, the demand coming from Chinese buyers.

They reached the depressing conclusion that *'the Nigerian Government has done very little over the years to reduce'* their illegal ivory trade and recommended:

- *'Parties to CITES should consider reinstating the CITES trade ban on Nigeria until the government takes appropriate action to address the country's blatant ivory trade'.*
- Implementing the 2011 legislation (which bans the ivory trade).
- Better law enforcement and ivory seizures.
- Regular inspections of the retail outlets with all ivory found confiscated.
- Together with the Chinese Embassy, a campaign to raise awareness.
  With periodic surveys to see what progress had been made.

# 14

# Markets in Southeast Asia

Our story now moves to the farther side of the planet, for although some of the illegal ivory is sold in the domestic markets of many African countries, most of it is smuggled, often by circuitous routes, to southeast Asia where we must follow.

Once again we will only look at some key players, Thailand, Myanmar and China.

## 14.1 Thailand

Thailand, with its three-headed elephant god, Erawan, has had a long relationship with elephants and ivory. First used for warfare and transport in the seventeenth century, then for logging in the late nineteenth century, at one time an estimated 100,000 domesticated elephants worked mainly in the logging industry, and although the numbers had plummeted by the end of the twentieth century, there were still 3074 in 2005.[1] Now logging has been banned, so the remaining domesticated elephants have either been absorbed into the tourist industry or are unemployed.[2]

Thailand is both a producer and consumer country because it also is a range State, having its own population of elephants, Asian elephants. However, unlike African countries, Thailand has both wild and domesticated elephants, and this has caused problems, one being the complexity of its legislation. This

---

[1] Dublin et al. 2006; see Daniel Stiles (2009). *The elephant and ivory trade in Thailand* TRAFFIC Southeast Asia, Petaling Jaya, Selangor, Malaysia, p. 1.

[2] Ibid.

© Springer Nature Switzerland AG 2019
B. Martin, *Survival or Extinction?*, https://doi.org/10.1007/978-3-030-13293-4_14

was a direct result of the fact that although trade in wild elephants and their products was illegal, it was not illegal if the elephants were domesticated. And the distinction gives rise to another, rather more sinister problem, that of ivory laundering, for who can tell whether that beautiful ivory carving comes from wild or domesticated ivory or from smuggled African or domestic Thai elephant.

**Thailand's Legislation:** Essentially, two statutes protect wild elephants, the Wild Elephant Protection Act of 1921 and the rather more recent Wild Animal Preservation and Protection Act of 1992 (WARPA), with some protection provided in the National Park Act (1961). Together, these made it illegal to capture or kill wild elephants unless the government gave official permission. WARPA (1992) also banned the internal trade in Thai wild elephants and their products, as well as ivory from international sources. So wild elephants could be fully protected.

Domesticated elephants, however, are classified as draft animals, like cows and water buffalo, so come within the Draught Animal Act of 1939. Their protection is limited. The sale and possession of ivory from these elephants were quite lawful, as was trade in the animals themselves. In fact, most of the legal raw ivory in Thailand *comes from domesticated elephants that have had their tusks pruned*.[3]

Acts passed in 1926, 1956 and 1979 related to import and export of goods, and these were used to implement CITES when Thailand became a Party in 1983. Under one of them, the Commercial Registration Act of B.E. 2499 (1956), any place processing ivory for crafts or handicrafts or engaged in selling ivory, either wholesale or retail, had to be registered with the government. Since 2007, this legislation has been used to register all wildlife entrepreneurs (Anon, 2007a).[4]

The legislation is both complicated and confusing, not an ideal situation. For example, although someone could be prosecuted under the Wild Animal Reservation and Protection Act of 1992 for being in possession of an illegally imported piece of ivory, there is a condition imposed, namely, that the accused must be shown to have been personally responsible for the illegal import, export or re-export. And this can be difficult to prove, especially where the specimen has been in the country for some time or where it has changed hands.[5] Furthermore, the fact that someone could lawfully possess

---

[3] Former Director of the CITES Division quoted in the *Bangkok Post*, Anon. 2001; see Stiles (n.1), p. 8.
[4] Anon. 2007a, ibid.
[5] Anon. 2006a, ibid., p. 9.

and sell ivory from domestic elephants provided any potential ivory launderer with what must be a most tempting loophole.

**CITES and Thailand:** As can be imagined, CITES and Thailand have not always had an easy relationship. As we already know, once a country has agreed to be bound by a treaty, it is supposed to implement it. In the case of CITES and elephant ivory, this means that the country must not only determine the origins of its existing stockpiles of ivory but must also prevent any exports of worked ivory (international trade) or imports of new ivory (again, international trade), unless they are authorized. Thailand has been slow to do this and has suffered some consequences, one of which was a trade ban.

**Thailand's Ivory Trade:** Like many of the countries selling illegal ivory, lawful trading in the commodity goes back centuries. Thailand is no different. It has been exporting ivory to China since the thirteenth century and in large quantities to Japan since the seventeenth century.

However, it was only after the Second World War and the development of its own internal ivory manufacturing industry that Thailand began to import raw ivory, perhaps because *the local elephant population was no longer large enough to satisfy the demand.* We know where it came from because the Royal Thai Customs kept records. Taxes had to be paid. It came mainly from Africa, particularly Sudan and Kenya, but also from Hong Kong and Laos PDR.

But it was also smuggled in from neighbouring Asian countries. During an interview in one of the many surveys investigating the ivory trade in Thailand, an ivory carver *'recalled that in the early 1960s, tusks were smuggled in from Myanmar and Lao PDR, one thousand pairs at a time of all sizes, up to two metres long',* and this continued until very recently, even though both the quantities and tusk sizes were much smaller.

Myanmar and Laos PDR only became Parties to CITES in 2004, in other words, just over 20 years later than Thailand, and until that had happened, the rules relating to trading with non-CITES Parties applied to them. Smuggling was not necessarily just a way to avoid paying tax.

Added to this rather unsatisfactory situation is the lawful raw ivory, the ivory sourced from domestic elephants. Although there is still a large population of domestic elephants, until recently very much larger, not all their ivory is available. Many elephants are used in the tourist industry, and the tourists prefer large bulls with impressive tusks. When these animals die, their tusks are often put in dedicated spirit houses as the owners believe it is unlucky to sell them. Every few years after becoming 15 years old, the remaining tuskers have their tusks pruned, these days as short as it is possible to trim them back, because the owners are worried that their animals might otherwise be stolen. This *tusk tip* ivory is not registered with the government, but some of it is used as a legal supply of ivory for carvers and sellers.[6]

Now we begin to understand just how complicated it is to distinguish between legal and illegal ivory in Thailand, so we can turn to the surveys, and there have been many of them, to see what they tell us about possible illegal trade. We shall only look at the most recent ones.

## 14.1.1   Surveys in Thailand 2008–2009

**The Carvers:** The history of ivory carving in Thailand is unusual, and unlike most other countries selling illegal ivory, commercial carving is comparatively recent in origin. Originally, the Royal carvers were the only people permitted to carve ivory, but this tradition ended in 1910. In the 1930s when carving resumed, the carvers either taught at the Fine Arts School in Bangkok or were Chinese carvers working in Chinatown.

The researchers visited both workshops and retail outlets. This is what they found.

**Ivory Workshops (The Results):** They started at Phayuha Khiri, once a centre of ivory carving workshops.

**Phayuha Khiri:** Craftsmen began to carve ivory here in the late 1930s. There were estimates in 1989,[7] in early 2001[8] of between 50 and 100 carvers and about 100 in 2002.[9] Then came ivory seizures by the Thai authorities. By 2006, the carvers had become unhelpful, but what little information could be gleaned from them implied that they no longer carved ivory, only wood. This was possibly due to a scarcity of raw ivory, as the researchers received information that ivory from Africa was either not currently available or was almost used up. Despite this however, there still were some ivory carvings for sale in the nearby shops, as well as some items carved from hippo and pig teeth, and mammoth ivory, all or any of which might, or might not, have been produced locally.

---

[6] Doak, N. (2014). *Polishing off the Ivory Trade: Surveys of Thailand's Ivory Market*. TRAFFIC International, Cambridge, UK, p. 12.

[7] Luxmoore, see Stiles (n.1), p. 24.

[8] Martin and Stiles 2002, ibid.

[9] Ivory carvers in Phayuha Khiri, pers. comm. to Daniel Stiles, 2003, ibid.

Phayuha Khiri's importance was diminishing; it only had one workshop, producing trinkets for the tourist market, but other centres, some nearby, were expanding their ivory industry.

**Uthai Thani, Chai Nat and Bangkok:** There were at least eight workshops in Uthai Thani, three in Bangkok and one in Chai Nat. Here a different and expanding ivory industry was discovered, one that specialized in knives and swords with ivory handles, silver and ivory belt buckles as well as jewellery, and it too seemed to be aimed at international buyers. The researchers estimated that in February 2008, there were probably between 50 and 60 craftsmen carving ivory, most of them in the Phayuha Khiri/Uthai Thani area.[10]

**Retail Outlets:** In the earlier surveys of 2001 and 2006/2007, when the researchers visited Bangkok, Phayuha Khiri, Chiang Mai, Uthai Thani and Mae Sai, the results appeared to show that although the total number of outlets selling ivory had increased very slightly, this could have been due to differences in sampling. Disregarding this however, there appeared to be a startling drop in the number of ivory artefacts on offer, down from 88,179 in 2001 to 23,258 in 2006/2007.

In the 2008 survey, three possibly new outlets were found in Phayuha Khiri, and there had been a significant increase in the number of retail outlets in Bangkok. Once again though, the number of ivory items displayed for sale had decreased dramatically, a drop of almost 77%, despite the fact that the amount of worked ivory had actually increased in three of the centres. The same situation was found in the markets.

Ivory substitutes seemed to be increasing, including worked mammoth ivory from China, although there was no substantive data on the amount of that, bone or indeed resin pieces painted to resemble ivory that were being sold. And it was becoming common for shopkeepers to have *mixed ivories* on display, thus adding to the difficulties of wildlife enforcement officials.[11]

By 2008 there was a significant change in the nationalities of the buyers, from Europeans, Americans, ethnic Chinese from Malaysia, Singapore, Taiwan, Hong Kong and mainland China and Japanese to the Thais themselves.[12] The ethnic Thais preferred amulets and good luck charms, and the Chinese Thais, Chinese-style figurines, chopsticks and mounted polished tusks.

---

[10] Ibid, p. 26.
[11] Ibid, p. 50.
[12] Luxmoore, 1989; Martin and Stiles 2002, ibid.

**Where Did the Ivory Come From?** The report concluded that most of it was African in origin and had probably been imported after 1990, because informants claimed that there was no raw ivory left from pre-1989 African stocks. In other words, it was probably illegal, not only under CITES but also under Thailand's own legislation.

Although it was hard to work out exactly which ivory artefacts were legal, the estimation was 25% of them. These would have been made from either Thai elephant tusk tips, in other words, domesticated not wild elephants, or ivory that had been imported before the 1989 ban. Some could have been genuine antiques. It was considerably more difficult to determine the origin of the ivory used in the belt buckles, knives and ivory and silver jewellery. While some of it would have come from Thai elephants, some would also have been smuggled in from neighbouring countries. In other words, although all of it would have come from Asian elephant tusks, not all of it would have been legal.

**Conclusions:** The results were thought-provoking. For example, although there was a big increase in the number of shops in Bangkok selling ivory, which would indicate that *controls over the internal ivory market have been inadequate*, there was also a dramatic drop in the number of ivory items for sale. Perhaps there was not enough raw ivory, but it was very difficult to work out exactly what was going on without further detailed research. Raw ivory was increasing in price, and ivory artefacts were selling very well, so Stiles concluded that this would *suggest that lower demand was not a significant factor*.

These findings made it imperative that Thailand should continue to be monitored, so yet another series of surveys of known ivory retail outlets was undertaken between January and March 2013, the month when CoP16 was held in Bangkok, and again, from October 2013 to May 2014.

## 14.1.2   Surveys in Thailand 2013–2014

Using the same methodology, monthly surveys of 4–5 days, of retail outlets (including newly identified ones) selling ivory, were carried out in Bangkok. The surveyors, native Thai speakers and trained to identify ivory by sight, visually estimated both the quantity of ivory and the numbers of each type of artefact for sale. The shops' locations were noted, as were any that were closed. Any additional locations and shops were identified, recorded and then re-visited in the repeat surveys. New and additional retail outlets were identified in every month of the survey, and although there is a legal requirement to register and monitor any retail outlet selling or processing ivory, the traders did not appear to be complying with it.

The surveyors had a hard time of it. They were forced to act *without the vendors' knowledge or prior warning to ensure items were not removed from view and that surveyors were not chased away from the retail outlets.* This hostile attitude on the part of the vendors made it difficult to obtain photographs which would have made it easier to assess the amount of ivory on offer in the shops, and although some sellers were informally interviewed about prices, they were reluctant to provide other details such as the source of the ivory and who bought their products. Unfortunately, most of them refused to give any details at all.

This meant that some data had to be omitted from the final comparison, but despite these difficulties and the fact that CoP16 was taking place during the first set of surveys, the results were stark. They *'revealed a disturbing increase both in the number of retail outlets offering ivory as well as the quantity of ivory available'.*[13] What was even more disturbing was the fact that this *growing market* could in no way be satisfied by the legal ivory coming from the domestic elephants. The numbers didn't add up. Not only that, many of the artefacts must have made out of larger tusks, tusks from African elephants, tusks smuggled into Thailand. And this grim finding was upheld by the seizure data, which confirmed *'attempts to move large quantities of African Elephant ivory to Thailand from Africa'.*[14]

Thailand was failing elephants badly. It had been described as *'one of the most problematic countries in the illegal trade'.*[15] Like Nigeria, it had one of the world's largest unregulated ivory markets, and it was already under an obligation to comply with its own CITES Ivory Trade Action Plan, because it was in breach of its international obligations. The survey results did nothing to disprove this. So the sales would continue to be closely monitored until the situation improved.

## 14.2  Myanmar

Like so many countries in our story, Myanmar has a troubled past, and its elephant population has suffered both at the hands of poachers, who kill them, and smugglers, who traffic the living animals out of Myanmar and into

---

[13] See Doak (n.6), p. 1.

[14] Ibid.

[15] Ibid.

Thailand to be used in the tourist trade. Despite this however, Myanmar has the potential to become a haven for the beautiful and hard-pressed wild Asian elephants.

A Party to CITES since 1997, illegal trading in ivory (and other elephant products) had continued unabated, and its elephant numbers had continued to decline, so it became essential to see what was happening. In 2006, TRAFFIC investigated.

**How Many Elephants Are There in Myanmar?** This might seem a curious question to ask, but because Myanmar was, to all intents and purposes, a closed country for a number of years, we simply did not know. There had never been an accurate nationwide census. However, in the year 2000, the number was estimated to be between 4000 and 5000 animals, which would mean that, outside India, it had the largest remaining population of wild Asian elephants in Southeast Asia, with potential to be *'a major stronghold for Asian Elephants'*.[16]

**How Are Its Elephants Protected?** Although elephants have been protected there for many years, the measures remain complex and confusing. Early legislation included the Elephant Preservation Act 1879, which was amended in 1883 and replaced in 1935 by the Wildlife Protection Act, which was amended in 1956.[17] Now the relevant legislation is the Protection of Wildlife and Wild Plants and Conservation of Natural Areas Law (State Law and Order Restoration Council Law No.583/94.1994), which lists the Asian elephant as a *Totally Protected Species*. This makes it an offence to kill, hunt or wound, possess, sell, transport or transfer a wild elephant or any part of it, without permission. Punishment can be as much as 7 years in prison, or a maximum fine of 50,000 MMK (USD1490). Live elephants are protected from illegal trade by the Elephants Registration Act 1951.

Until quite recently, it was thought that, like Thailand, it was legal to buy tusk tips from domestic elephants or tusks from domestic elephants that have died naturally and transport and sell carved ivory.[18] But the Nature and Wildlife Conservation Division, Forest Department, Ministry of Forestry denied this and stated that any ivory sale was illegal.[19] The situation was unclear. Meanwhile, the Forest Department and the state-owned Myanmar

---

[16] Leimgruber et al. 2003; see Chris R. Shepherd and Vincent Nijman (2008): Elephant and Ivory Trade in Myanmar. TRAFFIC Southeast Asia, Petaling Jaya, Selangor, Malaysia, p. 2.

[17] Aung 1997, ibid., p. 4.

[18] Martin and Stiles 2002; Naylor 2005, ibid.

[19] Tin Tun, Deputy Director, Nature and Wildlife Conservation Division *in litt.* to TRAFFIC 2007, ibid.

Timber Enterprise had stated that before Myanmar signed up to CITES, the company *used to export tusks from dead elephants*. And it would seem that the Ministry still gave some permission to Myanmar Timber Enterprise, which allowed them to auction *extra ivory*, in local markets, the *extra ivory* presumably meaning the *sawn-off tips and tusks from elephants that have died of natural causes*.[20]

The situation was different again for privately owned elephants whose deaths had to be reported to the Forest Department so that their registration could be cancelled. If they had tusks, the owners could sell them to local traders, despite the fact that there were no clear rules laid down to govern the process.[21]

## 14.2.1   The 2006 Survey

Following a number of limited surveys, this one, carried out by Shepherd and Nijman together with a local consultant who preferred to remain anonymous, took place between January and March 2006. It is perhaps indicative of the dangerous nature of their undertaking that in Mong La, they had to employ the *obligatory State tourist guide* to help them find those places selling wildlife products, while in Three Pagodas Pass, they were assisted by *an individual working for the Ministry of Agriculture and Irrigation. 'Neither of these government officials was aware of the true nature of this survey'.*[22]

The retail outlets were located in both the interior and on the borders of the country and the large dealers in the cities of Mandalay and Yangon.

**Yangon (Formerly Rangoon):** Both the largest city in Myanmar and an important centre for wildlife trade, ivory was plentiful, and the shops were selling it quite openly, although the dealers were only too well aware this was illegal. In fact, because of this, they *'stated that tusks were kept in their homes and only brought into the shop if someone ordered them'.*[23] One dealer would only get them carved once a customer had placed a specific order.

Prices were in both US dollars and Burmese currency, so it was obvious that some of the sales were to foreign nationals, tourists probably, as well as to local people.

---

[20] U Than Myint, WCS Myanmar Program, *in litt.* to TRAFFIC 2007, ibid., p. 5.

[21] Anon., WCS Myanmar Program, *in litt.* to TRAFFIC, 2007, ibid., p. 5.

[22] Ibid., p. 11. The illegal trade in ivory was only part of this research.

[23] See Shepherd and Nijman (n.16).

The researchers took the opportunity to visit the head office of the Myanmar Timber Enterprise, which, as we know, was allowed to sell *extra ivory*, and although the staff confirmed that there were tusks stored there *'presumably all from dead timber-working elephants'*, they did not say how many tusks there were.[24]

**Mandalay:** This is Myanmar's second largest city, and there were workshops as well as retail outlets.

The researchers visited two workshops. One owner, who was also the carver, was the fourth generation of her family to carve tusks, and although there were six large tusks waiting to be carved, she carved to order. Most of the orders came from Japanese wholesale buyers, who only wanted chopsticks and name seals. She was afraid Myanmar was losing its art of carving ivory.

There were 17 retail outlets, and 1821 ivory artefacts were openly on sale. The buyers came mainly from China, Japan and Thailand, and much of the business was wholesale. However, two large-scale dealers with in-house carving operations were also selling to European buyers, from France and Italy. Market traders were selling to tourists, both local and foreign and also to business men, quoting prices in both Burmese currency and US dollars (USD).

Ivory was the only wildlife product for sale in the shops selling ivory.

Myanmar, however, borders ivory trading countries such as China and Thailand, and this potentially opened up all sorts of opportunities for illegal traders. So in order to discover what was happening in the border regions, four border towns were visited:

- Tachilek: in Myanmar, *'on the Myanmar side of the Thai-Myanmar border'*
- Mae Sai: in Thailand, but *'situated immediately on the Thai-Myanmar border'*
- Ruili: in China, *'immediately on the Myanmar border'*
- Jiegao: in China, *'near the Myanmar border'*[25]

**Tachilek (Myanmar):** The researchers described this town as *notorious* and *well-known for the availability of a wide variety of illegal wildlife products*, which were sold in its large wildlife market. The specimens came from India and China as well as Myanmar itself, and although most were trophies or ornaments, there was also some trade in medicinal products. Disturbingly, it was in this place, out of all the locations they surveyed, that the researchers found

---

[24] Ibid.
[25] Ibid.

the most elephant products for sale. They surmised it could have been because Tachilek was located so close to a city, even though that city, Mae Sai, was in Thailand.

Most of the ivory came from Myanmar, with perhaps some from India, and, as in the cities, most was sold wholesale to buyers from Thailand, China and Japan for re-sale in those countries. Some was sold to tourists.

Because most of these sales would have been illegal, let us try to work out just how blatant this illegality was:

- We know that CITES bans most international sales of ivory.
- That Myanmar, Thailand, China and Japan are all Parties to CITES.
- The researchers observed that *virtually all customers visiting this market are foreigners.*
- There were large numbers of Thai nationals.
- All the prices were quoted in both Thai currency and US dollars.
- Of even more concern were the facts that *'all dealers stated that smuggling ivory in any quantity from Tachilek into Thailand was simple and that taking it by air out of Bangkok was also not a problem'* as *'many of their customers did this'.*

The answer has to be 'quite blatant' or perhaps 'completely indifferent'. So how did this compare with Mae Sai, the Thai city at the other side of the border?

**Mae Sai (Thailand):** Apparently ivory, but only ivory, was for sale there, and because of the loophole regarding domestic elephants (the one we looked at earlier), it was being sold quite openly, both to Thais and foreign tourists. In the eight shops selling it, all prices were in Thai currency. The researchers counted 446 items for sale. Customers requiring other wildlife products were referred to Tachilek.

The last two towns in the survey were in China.

**Ruili (China):** The first of these, Ruili, almost spans the border with Myanmar. The researchers found 35 shops selling ivory and saw 268 ivory artefacts for sale. Many of the other items were made of bone which, the sellers claimed, was Myanmar elephant bone, a claim that could not be verified. All the customers were Chinese and all the prices were in Chinese currency.

As we shall see later, the Chinese government has strong legislation in place to make sure that only lawful ivory is being sold, but in Ruili, apart from one of the larger dealers, little attempt had been made to comply with the law.

The dealers believed that the authorities would only take action against them *'if they sold any other protected wildlife species'* and that ivory itself was *'safe to sell'*. And although the larger dealer did provide each customer with the required laminated card on which was a printed photo of the item together with its registration with the *Ruili Jade and Jewellery Association, Yunnan Province*, it failed to say anything about the legality or origin of the ivory.[26]

**Jiegao (China):** In the final town to be visited, the results were not unlike those of Ruili. However, there were more shops selling ivory, 44 out of the 45 surveyed, and the total number of pieces for sale, 283 items, was slightly higher. Most buyers were Chinese and all the prices were in Chinese currency. When asked where their ivory came from, the sellers that answered said:

- Myanmar: 37 shops
- Thailand: 3 shops
- India: 1 shop

One dealer claimed to own an ivory carving business in Myanmar although he would not disclose its location, and two of the bigger dealers provided laminated cards bearing the legend *'registered with the Ruili Jade and Jewellery Association, Yunnan Province'*. However, as in Ruili, essential details remained undisclosed. The researchers wryly commented that *'the function was purely to assure buyers that the ivory was genuine'*.[27]

**Conclusions:** (There were a number of these, but we are only concerned with those relating to ivory.)

Despite the fact that Myanmar is a Party to CITES, it was continuing, quite openly, to sell ivory to foreign buyers and also to smuggle it out of the country. Although some of the ivory came from elephants illegally killed in Myanmar itself, some raw ivory might have been smuggled in from India, another neighbouring country. The researchers also concluded that *'illegal cross-border trade is basically carried out with minimal risk of detection'*.[28]

It was quite clear that the dealers understood that international trading was illegal unless the ivory originated from the legally auctioned lawful tusk tips. However, they appeared to be well aware of the ambiguities in the law and quite prepared to make good use of any potential loopholes and shortcomings

---

[26] Ibid.
[27] Ibid.
[28] Ibid.

these provided to mask any illegal trade in ivory. Furthermore, not only did they not attempt to hide any pieces of ivory they were selling, they were also quite ready to *disclose smuggling techniques and other illegal activities with potential buyers*, all of which indicated a severe problem with enforcement.

Myanmar's wild elephants were becoming rare. Even the dealers recognized their numbers were decreasing and that ivory was harder to come by, which led to the conclusion that *illegal hunting and trade in Asian elephants and their products are still widespread in Myanmar and are thought to be a significant cause for the decline in Myanmar's wild Asian elephant populations*.

# 14.3   China

A country that plays such an important part in our story, China, is steeped in a culture that recognizes both the beauty and the value of ivory. It has produced ivory artefacts of great beauty and complexity for thousands of years but is only now realizing where ivory really comes from. And, despite its best efforts, it is still heavily involved in its illegal trading.

A rather grim statement was made in 2009: *'China remains the most important contemporary player in the illicit trade in ivory'.*[29] According to ETIS, *from 1990 to 2008, seizures of tusks and worked ivory en route for China and within China were the most in the world by weight and third by number of pieces'.*[30] But China deliberately strengthened its legislation and introduced a new system so that it could be considered as a potential buyer in the second of the one-off sales. It was successful. In 2008, it bought 62 tonnes of lawful African elephant ivory, one of the two countries (Japan being the other) CITES considered suitable buyers.

**Legislation and the Leaflet Produced by the State Forestry Administration:** Under the then current legislation, ivory manufacturers and retailers all had to be designated by the government, whose State Forestry Administration published lists of the names and places of the licensed facilities. Ivory could only be sold in these specific shops, all of them registered, and there had to be a framed certificate on display stating that the shop was permitted to trade in elephant ivory.[31] In May 2004, another condition was added: every item of elephant ivory was

---

[29] Milliken et al. 2009; see Esmond Martin and Lucy Vigne (2011) *The Ivory Dynasty: A report on the soaring demand for elephant and mammoth ivory in southern China* published by Elephant Family, the Aspinall Foundation, and Columbus Zoo and Aquarium 2011.

[30] Ibid, p. 6.

[31] See State Forestry Administration notice regarding ivory processing factories (2004) #85.

given its own identity. In 2008, after the sale, the old identification card was replaced by a new one stating that *'this ivory carving is made of raw ivory imported from Africa in 2008 and has been registered. It can be traded and used within PR China only if accompanied with the certificate and the special mark. Export is prohibited'.*[32] Thus, not only was each item clearly identified as ivory originating from the legally imported 62 tonnes, but it was also made absolutely clear that this ivory could not be exported.

As an additional safeguard, each ivory product's identity card had to be marked with a unique alphabetical code by the factory where it was carved and the wholesaler who sold it. Each factory and wholesaler could supply any number of retail outlets provided they were licensed to sell the legal ivory. Direct sales of ivory products from carving factories are forbidden.

Taken together, this meant that:

- Every ivory object had to have its own identity card, which was set out beside the item displayed for sale.
- If the object weighed over 50 g, there were further requirements. The identity card for each of these artefacts had to carry a photograph of the object on it, together with its size, weight and other descriptors that could identify it, its serial number and the name of the factory where it was produced.
- In the case of smaller items like pieces of jewellery or name seals, where space was limited, the identity card could be kept in a drawer.
- The retailer had to notify the government every time an item was sold and provide details of the buyer, although this did not apply to most of the smaller items (weighing less than 50 g) as these were normally mass-produced and only required a numbered card.[33]
- Only licensed shops could sell the ivory.
- Carving factories could not sell their products directly to the public.

In theory then, under this legislation, it should only have been products from the legal stockpile that were sold.

**The Leaflet:** Produced by the State Forestry Administration and targeted at any potential buyers of elephant ivory, it was supposed to be freely available

---

[32] Grace G. Gabriel, Ning Hua, Juan Wang, Making a Killing: A 2011 Survey of Ivory Markets in China IFAW (The International Fund for Animal Welfare), p. 9.

[33] Wan Ziming pers. comm. February and March 2011; in 2010 Wan Ziming worked in the CITES office in Beijing; see Martin and Vigne (n.29), p. 6.

in all retail premises selling ivory and was another attempt to deter illegal sales. Possible buyers were warned to take great care when purchasing ivory products.

Packed full of information, the leaflet was distributed to all shops selling ivory. It was of course, only written in Chinese, because unless certain conditions were met, it was illegal to sell the ivory to other nationals.

It contained the following:

* Customers were requested *'to buy ivory only in designated ivory shops and to ask for the ID card'*.
* The government's *'standardisation of domestic ivory management measures'* was set out.
* It stated that an ivory item could not be separated from its *'ivory collection card'* or ID card, guaranteeing its legal status.
* It recommended there should be a photograph on the card if an item was more than 50 g and was a unique artistic piece with high value, but this was not mandatory.

Each ID card also required descriptors, which included:

* Hand-painted floral patterns and shading lines
* Laser anti-counterfeit labels for security
* Drip disappeared technology
* Security lines
* Relief shading
* Double 's' anti-lift incisions
* Colourless fluorescent security ink
* Microfilm text

These had to be identified, and the ID card matched up with the item. Foreign tourists and expatriates were informed that:

* *without permission to import elephant ivory and its products [it] is illegal. Similarly [for] PRC citizens without permission to carry ivory and its products from abroad [it] is illegal [through] immigration also.*
* *Citizens from abroad to carry ivory and ivory products [through] immigration, need to get permission from relevant departments of the exporting country and issue export certificates, while government departments need to get the consent of the appropriate department for an import licence [for] customs before release.*

Perhaps the most unexpected information on the leaflet was that it explained why some ivory sales were legal that *'only Botswana, Namibia, Zimbabwe and South Africa have their elephants on Appendix II of CITES, which allows controlled and limited trade in tusks under certain circumstances'*. And it proudly proclaimed the fact that *'the international community'* had praised *'the management, maintenance and development of this ivory tag information system, commissioned by the Chinese State Forestry Administration'*.

It was a helpful and informative leaflet. What it failed to do though was carry a warning, a picture of a dead elephant with its face hacked off by poachers removing its tusks and the legend 'This is what happens if you buy illegal ivory'. A stark message but one that would leave both buyers and sellers in no doubt about how illegal ivory is obtained.

## 14.3.1   The First 2011 Survey

China was obviously trying to prevent illegal sales of elephant ivory. But did it work? Our story now turns to two surveys carried out in 2011. They are illuminating because they highlight some unexpected attitudes. Things were starting to improve.

The idea behind the second one-off sale was that such a large quantity of ivory would flood the market, thus eliminating the need for any more illegal trading, but, in 2011, demand for elephant and mammoth ivory was described as *soaring*.[34]

### 14.3.1.1   The First Survey

The object of the research was to examine:

- The impact of the 62 tonnes of ivory on the existing market in China
- Any evidence of illegal trading
- The sales of mammoth ivory, which were continuing to increase and were adding confusion to the sales of elephant ivory
- The effectiveness of law enforcement

Two cities were visited, Guangzhou, important in the retail ivory trade and near Hong Kong a possible source of illegal ivory, and Fuzhou. Both located in southern China, these cities were main ivory manufacturing centres with a

---

[34] Ibid.

high reputation for their carving skills, had rapidly increasing populations and had previously been surveyed by Martin, in 2004.[35] Therefore, in addition to visiting new locations, the researchers were able to re-visit premises that had previously been surveyed, to see whether any changes had been made and if so were they beneficial.

We must bear in mind the fact that the elephant tusks being carved came from a number of sources. Some had been illegally smuggled into the country, and some of them were from the legally bought 62 tonnes, while others came from old privately owned stocks or government supplies.

**Guangzhou:** Of the three ivory factories (workshops) visited, one of which was government-owned. This factory had twice as many carvers when compared with the earlier, 2004 survey (up from about 20 to 40/45). Most of the artefacts being made were figures and figurines, though there were also some *magic balls, a Guangzhou speciality*.[36] Some inexperienced craftsmen were being trained by master carvers. The manager told the researchers that the factory had four shops in the city. These sold ivory products, and 60% (by weight) of the items they sold every year came from elephant ivory.

One of the privately owned factories only came into existence after 1990, in other words, after the total ban on international sales of elephant ivory had been imposed. The government had recently refused this factory permission to carve elephant ivory.

The researchers were not welcome at the third factory, but they got in just the same. The manager had not wanted *'his ivory carving "secrets" exposed'*, and indeed, this did seem to be the case, for once he realized they were not Chinese, he let the researchers take some photographs and look around. There were no master craftsmen, and the carvers, who were all young, were mainly carving bridges and bangles, all of which were sold in their main retail outlets.

The main ivory retailers were based in markets, shopping centres and even luxury hotels situated in central Guangzhou. The researchers found 80 outlets selling elephant ivory. They were mainly selling small items of jewellery such as pendants, beads, bangles/bracelets as well as charms and some figurines, nearly all of them carved in China. Most of the sales-people *tolerated* the researchers' questions and allowed them to take photographs. This was even the case where the items were carved out of ivory from Thai elephants, and because this ivory was rare, the prices were expensive. The seller was quite

---

[35] Ibid.
[36] Ibid.

happy to use a torch to illuminate the pinkish glow which identified this as authentic Thai ivory. He was also unconcerned that the ivory was illegal.

The researchers counted 6437 items on display, only a few of which were made of old ivory. About 3947 items did not have ID cards, so were illegal. The researchers were sometimes directed to a verification office when they could ask for an ID card. The office would prove the item was authentic. But verification offices were really there to verify gemstones.

Even more worrying was the fact that although ivory could only be sold in specific registered shops, there were still many unregistered outlets that had not been officially inspected, and some sellers had no qualms about telling the researchers that their ivory was illegal. None of it had been confiscated.

The researchers found that Guangzhou had increased both its production and retail sales of carvings of elephant ivory; fortunately however, the *vendors told us that their ivory sales were presently slow*.

**Fuzhou:** Famous for its carving, especially carvings made from its local stone, both wood and ivory carvers travel to Fuzhou to seek work.

The researchers visited three factories and found things very different.

The first factory employed 40 carvers, 28 of whom were working mostly on elephant ivory. Recently, 10 other carvers had left to carve wooden items as these were more profitable. In fact, the owner had been in dispute with the government as a result of the CITES ban on ivory in 1990. The government had removed some 600 kg of elephants' tusks, paying only a small amount of compensation in return. The owner had then disputed this sum and been lucky not to end up in prison. After that incident, he had had to go through a long and complicated procedure before he could obtain new supplies of ivory and even those were strictly controlled. Some of the actions he had to take were that the factory had to submit its designs to the government and photograph each of its items for the compulsory ID cards. The carvings were sold to shops and private customers in China.

The researchers counted 20 carvers at the next factory, although some were on holiday. This owner had actually been with Chinese officials buying tusks in southern Africa when the 2008 sale had taken place. However, even he had only been granted limited supplies of ivory by the government.

The third factory had stopped carving elephant ivory in 1990 and had diversified into other materials instead, particularly mammoth ivory and bone. This owner pointed out that the one-child policy was causing factory closures because parents wanted their only children to go to university rather than become artisans.

There were 39 outlets selling ivory in Fuzhou, although none of these specialized in it, and 282 items for sale. This time the majority were old or pre-1990 items. The most common objects were name seals, figurines and belt ornaments. Business was far from brisk, and many of the vendors were not going to bother to replace their dwindling ivory stocks.

However, business was booming in carvings made from mammoth ivory, hippo teeth and bones.

**Mammoth, Other Ivory, Bones and Synthetic Substitutes:** The first thing to remember is that it is quite lawful to export, import and sell mammoth ivory in any form. There are no restrictions. Nor are there restrictions on other ivories or bones. In fact, after the ban on international sales of elephant ivory, Chinese craftsmen began to carve more bone and mammoth ivory, which was why the leaflet (produced by the State Forestry Administration) set out the differences between elephant and mammoth ivory and explained how to distinguish the two.

Siberia exports mammoth ivory, and China imports it, often via Hong Kong where there is no import tax. Indeed, some factory owners even go to Russia to buy mammoth tusks. All the ivory is graded and it fetches good prices.

As items made of mammoth and hippo ivory and bone were included in this survey, we will briefly return to Guangzhou and Fuzhou to examine this trade.

In Guangzhou, the ratio by weight per year of the ivory items produced by the first factory was:

- 60% of elephant ivory to 40% of mammoth ivory.

In other words, almost half and half. Most of the items produced by the second factory were carved from cow bone and consisted of elaborate landscape scenes, magic balls and some composite pieces similar to those of elephant ivory. In 2007, this factory bought a supply of mammoth ivory and in 2010, hippo teeth. No mammoth ivory was observed in the third factory although the manager said that they did use it.

However the situation was very different in the retail sector. There were 30 outlets selling mammoth ivory, and 13 of these only sold mammoth ivory. There were 6541 items for sale. In other words, there were more items of mammoth rather than elephant ivory for sale, 104 items more. Most of these had been carved in China since 2004, and because of the nature of this ivory, most carvings were unique.

Some shops had put up signs explaining that although trading and use of elephant ivory were only allowed within China itself, items made from mammoth ivory could be taken from the country. The shops that specialized in mammoth ivory had posters in their windows describing mammoths. Unfortunately, there were also sellers who pretended their elephant ivory was mammoth ivory *'which does not require an ID card and which foreigners can export'.*[37]

The researchers did not see anyone buying items made from mammoth ivory. However, although the whiter elephant ivory was still the most popular form, they were clear that *'with increased publicity, mammoth ivory has become an accepted substitute for elephant ivory'.* In fact, the expansion in items for sale of mammoth ivory predicted by some vendors in the earlier research in 2004 probably exceeded even their expectations, up by more than 100%.[38] Was this good news? We shall see.

Mammoth tusks were also being carved in the three factories in Fuzhou. In the first factory, only about 10% of the sales were exports, and these were mainly to Europe and the USA. In the second factory 40% were exported, again to the USA and Europe, this time to France in particular, and in the third factory which only carved mammoth ivory and bones, most of the ivory was exported. The ivory carvings, of human and animal figures, were particularly fine. Some of the tusks were merely polished.

Of the 39 shops selling ivory, only 100 items were of mammoth ivory and 90 of those were in the third factory's shop. Again, no sales were observed. However, as Martin and Vigne, the researchers, pointed out, most of the items made in Fuzhou from mammoth ivory had always been sold in other places in China, or had been exported.

Hippo teeth were another substitute for ivory. Although hippos, like elephants, are listed in CITES, they are listed in Appendix II, which means that some trade, strictly controlled, is permitted. These items were not popular as they were fairly small, could be very expensive and had a tendency to crack.

In Guangzhou, the researchers found several shops selling *quantities of inexpensive fake ivory beads, bracelets, necklaces, bangles, figurines and other common items* made from a mystery material. The sellers called it *resin, plastic, elephant bone, camel bone, re-constituted ivory powder and sometimes ivory.* However it was not plastic as it did not melt, nor did it smell like bone. Furthermore, it had lines running through it, not unlike ivory, and its temperature, texture and weight also resembled ivory, which enabled some sellers to confuse it with

---

[37] Ibid, p. 10.
[38] Up from 3064 items in 2004 to 6541 in 2011.

ivory *accidentally and on purpose*. There were even some *ivory* necklaces for sale, which consisted of genuine ivory pendants surrounded by fake beads.[39]

**Problems with Sales in Retail Outlets:**  In addition to the problems caused by selling elephant ivory substitutes, others were connected with the wrongful use of ID cards. Some items were on display without display cards. *'Several vendors openly said their ivory was new and illegal and occasionally pretended new items were old. This suggests official inspections and confiscations have not taken place in most shops'.*[40]

Even in registered shops, sometimes mammoth and ivory items were mixed. For example, in one shop, the seller claimed she did not know which items were made of elephant ivory and which of mammoth. Another shop was selling ivory statues of the Buddha, and although there were two cabinets, one for each type of ivory, the Buddha statues had not been separated into the appropriate cabinet. They were all together and all looked identical. Some had ID cards beside them, while others were in a drawer. As the researchers noted, it would have been all too simple to reuse any ID cards that were not taken by a buyer or for a buyer mistakenly to think that his carving was of mammoth ivory when it was really elephant. Another occasion when ID *fraud* might occur was where the item bought was smaller than 50 g as the ID cards of such items were usually kept in a drawer, thus making it all too easy for the purchaser to leave without one. And indeed, the researchers saw this happen. In other words, at the time of the survey, it was comparatively straightforward to launder-in and sell illegal items of elephant ivory if the seller was so-minded.

All these incidents occurred in shops registered to sell ivory, but most of the outlets visited by the researchers in both Guangzhou and Fuzhou were unregistered and were still *selling items, both old and new, all illegally (with no ID cards).* Despite this, the researchers concluded that tighter controls and better law enforcement would reduce illegal trading.[41]

## 14.3.2   The Second 2011 Survey

This was conducted by IFAW, the International Fund for Animal Welfare, which has also been investigating sales of elephant ivory in China over a number of years.

---

[39] See Martin and Vigne (n.29).
[40] Ibid, p. 14.
[41] Ibid, p. 16.

In 2007, the year before the second one-off sale, IFAW commissioned a public opinion poll to get some indication of who, in China, was actually buying ivory. They found that most of it was sold to well-educated adults of between 26 and 45 years old who had a steady income, in other words, those Chinese consumers who had the most buying power. Had anything changed by 2011?

Of the three researchers, Grace Gabriel, Ning Hua and Juan Wang, two were Chinese. They were local experts, had experience in monitoring the ivory trade and could identify elephant ivory by sight. They surveyed ivory markets in five coastal cities in September and October 2011, and their findings, written in Chinese, were submitted to China's wildlife management and enforcement authorities in November 2011. (They also monitored some online sales, but we will return to this later.)

They visited 158 ivory trade facilities in Beijing, Shanghai, Guangzhou, Fuzhou and Putian, including two ivory carving factories. Only 57 of the retail premises were licensed, which meant that the rest were illegal. The researchers pretended they were looking for presents for companies, and because they spoke in Chinese, they were able to glean a considerable amount of information from chatting to the traders. They even managed to take some photographs.[42] We will restrict ourselves to Beijing, apparently *a hotspot for illegal ivory trade*.[43]

**Beijing:** Of the 98 shops discovered, only 32 were licensed, so 66 were operating illegally.

Out of the 32 licensed shops, 22 were also cheating the system, for example, by misusing the ID cards or using their licence as a cover to trade in smuggled ivory. Fourteen shops actively discouraged buyers from taking their ID card(s), which could then be used again and again with any number of items. The team even found some retailers selling *identification cards to unlicensed traders to help launder illegal ivory*. In other words, the ID card itself had become a commodity.

Illegal outlets were often found in the same location as legal ones, as in the suburban antique mall where only 1 of the 22 shops selling ivory was licensed. Even more worrying was the fact that, since IFAW's previous visit only 2 years earlier, several new markets had been established both in and around the city.

Most of the items for sale were small, *bracelets, necklaces, chopsticks and brush holders*, and there was no information on the nationality of the customers.

---

[42] See Gabriel, Hua, Wang (n.32), p. 7.
[43] Ibid.

The researchers asked the traders about government inspections to check whether they were complying with the legislation and about enforcement.

Their conclusions were bleak. Like Martin and Vigne, they saw the flaws in the registration process, the widespread misuse of the ID cards making *a mockery of the registration system*. They identified the problem as one of rising demand for ivory and not enough legally imported ivory to satisfy it. The result was a mushrooming of illegal traders and illegal activities by lawful traders, with *'large numbers of licensed facilities … white-washing ivory illegally smuggled into China'*, although they did not consider that China was wholly to blame *for the escalation of elephant poaching and ivory trafficking worldwide because the trade chain spans continents and oceans from Africa to Asia'.*

They recommended:

* Frequent and more effective inspections of the markets, suggesting they *'should be carried out in a more clandestine manner by undercover law enforcement officers'*.
* That the evidence and intelligence thus garnered *'could support large-scale, multi-agency operations to crack down on illegal ivory trade'*
* That enforcement should be intelligence-led
* Punishments severe, such that 'wildlife trafficking becomes a high-risk crime like drugs and arms trafficking'[44]

Controversial proposals perhaps, but ones that would probably get results, results that were urgently needed.

Therefore, despite the injection of lawful ivory from the second cone-off sale and the Chinese Government's very considerable efforts to tighten up the legislation, problems continued, particularly with regard to property inspections and enforcement. Not enough inspections and poor law enforcement, but these were problems that could be solved.

---

[44] See Gabriel, Hua, Wang (n.32).

# 15

## Markets in the USA

The USA has always been a big consumer of ivory and is one of the largest ivory retail markets in the world. Once again this is partly down to history.

Culture plays a pivotal role in our story, for even in the USA, the tradition of carving ivory goes back over 2000 years. This time though, there is a difference, for the ivory carved by the indigenous peoples, and later by the Europeans, came from marine animals, not from elephants. This specialized carving, which is produced by etching, is called scrimshaw, and examples of it continue to command high prices when they are offered for sale.

## 15.1 The History of Ivory in the USA

Although scrimshaw is of ancient origin, it became very popular in the late eighteenth and nineteenth centuries, when American sailors needed to pass the time during long voyages collecting sperm whale oil. They carved whatever caught their fancy, exotic scenes of tropical Paradise, fishes, seabirds, their homes and sweethearts onto walrus tusks, sperm whale teeth or whale bones.

Most scrimshaw came from New England and Hawaii, and it was not until the Endangered Species Act was passed in 1973, making it illegal to use either walrus ivory or the teeth of sperm whales, that the industry started carving elephant ivory instead.

However, during the late eighteenth century, a parallel carving industry using elephant ivory began to grow in New England, in the Connecticut River Valley. Eventually, two large family firms emerged, their factories

© Springer Nature Switzerland AG 2019
B. Martin, *Survival or Extinction?*, https://doi.org/10.1007/978-3-030-13293-4_15

*'the biggest ivory factories in the USA and perhaps in the world'.*[1] This ivory was shipped from Zanzibar, via Europe, to either Massachusetts or New York, where it passed through customs, carried by smaller boats to Essex and Deep River, transferred into horse-drawn wagons and taken to the factories where it was carved into artefacts. Initially combs, then a whole range of small, useful items as diverse as billiard balls, cuff links, umbrella handles and, the most famous product of all, ivory piano keys.

The first set, made in 1839, started a sort of gold-rush demand for these pianos, which reached its peak in 1910 when some 350,000 instruments were manufactured. It is hard to exaggerate the popularity of the instrument. The extraordinary fact is that in each year from 1900 to 1910, *'one out of every 260 Americans'* bought a piano.[2] Although demand fell during the Great Depression of the 1930s, it revived from 1946 to 1951 and then continued declining until manufacture of both keys and ivory piano actions ceased by the 1960s.

Unlike the carvers in southeast Asia, the American manufacturers preferred working with the soft ivory from Savannah elephants. Between 1891 and 1903, the USA imported 1,767,024 kg of tusks, 1,165,844 kg of which was bought by the two family firms.[3] This cost a lot of money so substitutes were considered. Although mammoth ivory was unsuitable, being too hard and brittle, celluloid keys, which first appeared in the late 1870s, gradually replaced ivory as they became more and more acceptable.

Ivory use was dropping. A survey published in 1979 estimated that the USA only consumed 15 tonnes of raw ivory a year during the 1970s. A small quantity. So why, some 40 years later, did America get the unenviable reputation of being the second largest consumer of ivory in the world? We have a puzzle here.

The survey showed the scrimshanders, some 1000 of them, were still mainly based in New England, Hawaii and Alaska, and they got their ivory from 12 raw ivory importers found mainly in New England, California and Seattle. The products had changed, and only Alaska was producing scrimshaw. Scrimshaw with a difference, however! The Seattle importers bought in small tusks for the carvers, tusks weighing between 1 and 3 kg, which were first carved to resemble walrus tusks, then etched with scrimshaw designs. That accounted for most of the raw ivory.

---

[1] Esmond Martin and Daniel Stiles *Ivory Markets in the USA* Published by Care for the Wild International and Save the Elephants, 2008, p.13.
[2] Conniff 1987, ibid.
[3] Ibid.

This ivory was imported from Hong Kong, the UK or Belgium, but one company in New York imported some 230 kg each year directly from Johannesburg, to carve into *'figurines, gavels, bracelets, rings, earrings, chess pieces, dominoes and draughtboard counters'*, in other words, the sort of artefacts we saw in the other surveys. A commonality of artefacts.

It also found that, in 1978:

- There were 317 businesses and individuals importing ivory.
- It is worth USD7,382,624.
- 16 firms bought 90% of it.
- Most of these were based in New York, Honolulu, Miami and Seattle.
- 82% of the carved pieces came from Hong Kong.[4]

A TRAFFIC survey in the mid-1980s found that the annual consumption of raw ivory had again declined, down to 7 tonnes a year, and the craftsmen, many of whom worked part-time only, were experimenting with substitute ivory including mammoth tusks, walrus ivory, mastodon tusks and hippo and warthog teeth.[5] Their 80 tonnes stock of ivory was priced at less than USD1000 per kg.

While imports of raw ivory were declining, imports of worked ivory were booming. These artefacts were no amateur productions from local talent. They were not even from Africa. They were the best Europe and Asia could offer. Figures for 1986 showed:

- More than 200 companies involved, the owners of the five largest being of Chinese origin.
- They imported 8,497,135 items.
- 65% of which came from Hong Kong.
- The items were worth USD26,000,000 wholesale.
- The five 'Chinese' companies imported jewellery, figurines and netsukes.[6]

Was this massive rise in imported worked ivory the answer to the puzzle? No, because in 1989, the world suddenly woke up, became aware of the horrendous killing of enormous numbers of elephants and imposed the ban on international sales of ivory. The USA imposed a moratorium on all new ivory imports, with two exceptions:

---

[4] John Hallagan 1979, ibid.
[5] Thomsen 1989, ibid.
[6] Ibid.

- Sport-hunted trophy tusks
- Antiques over 100 years old

And the market collapsed.

Only to rise, and rise again, this time triggered by the rise in auction sites on the Internet. Before we consider this however, we will look at some relevant legislation, most of it also applying to rhino horn.

## 15.2   Legislation

Once again we need to bear in mind the fact that like South Africa and Nigeria, the USA is a federation of states, so is governed by both state and federal legislation.

We start with the Lacey Act. Passed in 1900, it is very thorough. It regulates trade in wildlife, both interstate and from outside America by, in effect, making most of such trade unlawful or strictly controlled. Illegally imported ivory is illegal, so possessing and/or selling/attempting to sell it will be illegal.

It was followed by the Endangered Species Act 1973, which defines both *endangered species* and *threatened species*. The Act also implements the import/export regulations required by CITES when the USA became a Party to the treaty in 1975. In 1976, the Asian elephant was listed as *endangered* under the legislation, and the African elephant as '*threatened*' 2 years later. This protected them from trade by persons 'subject to US jurisdiction'.[7]

The African Elephant Conservation Act passed in 1988, the year before the American moratorium on all imports of new ivory, reflects the concern about the high levels of poaching in Africa. Under this legislation, it became illegal to:

- Import raw ivory from any country other than an ivory-producing country that is a Party to CITES.
- Export any raw African elephant ivory from the USA.
- Import raw or worked ivory that was exported in violation of an ivory-producing country's laws.
- Import worked ivory other than personal effects, unless the exporting country has certified that the ivory was from legal sources.

But there were two exemptions applying to antique elephant ivory and African trophy tusks, which provided loopholes for anyone wanting to exploit the illegal trade.

---

[7] Hoover and Tarr 1997, ibid.

Unusually, in the USA, the functions of the CITES Management Authority and Scientific Authority are carried out by a single body, the US Fish and Wildlife Service (USFWS) which is responsible to the Secretary of the Interior. Two bodies, the USFWS and the US Department of Agriculture, Animal and Plant Health Inspection Service, share responsibility for enforcing all the US laws related to CITES and wildlife conservation.

CITES is implemented by the *use after import* regulations.[8] Sales are illegal unless the seller can show either that the ivory was imported before 1990 (the year of the ban on international sales of elephant ivory) or else is pre-Convention ivory, for which a special pre-Convention certificate is required. This shows that the ivory was acquired legally and removed from the wild before 26 February 1976 (when the African elephant was first listed on CITES). The date for pre-Convention Asian elephant ivory is earlier, 1 July 1975. The special CITES certificate is required for certain noncommercial purposes, such as museums, education and musical instruments.

The regulations, which apply to both interstate and intrastate transactions, contain another important exemption, that of elephant sport-hunted trophies.

However, their domestic ivory markets were not regulated until February 2014.

As a federation, each state can have its own ivory legislation. California plays a key role in America's ivory trade, and its law is contained in its Penal Code Section 653o. Passed in 1976 it has had unforeseen consequences. Although the section makes it unlawful to import, possess with intent to sell or sell any elephant part, other wording creates a loophole, making the law unenforceable.

Another problem arises from the fact that although the USFWS enforces all the US laws related to CITES, the Californian Department of Fish and Wildlife refuses to enforce 653o because it is contained in the Penal Code and it is the job of the police, the sheriff deputies and other peace officers to enforce this. Furthermore, there is absolutely no reference made to elephants or elephant products in the California Fish and Game Code or in state wildlife regulations enforced by the Department of Fish and Wildlife.[9] California's law has now changed.

US legislation is complex and hard to administer and to enforce.

---

[8] 50 CFR 23.55. These set out the rules governing domestic commerce in African elephant ivory.
[9] Stiles 2015.

## 15.3   The 2006/2007 Survey and Its Results

The first survey carried out in the USA to collect and report on quantitative data on the nature and scale of the internal US ivory market took place between March and December 2006 and March and May 2007. In other words, after the first one-off ivory sale in 1999, but before the second sale in 2008. Some 16 towns and cities were surveyed, including San Francisco, an important centre of the ivory trade.[10]

The Canadian city of Vancouver was also included because a year earlier, a Chinese Canadian had been caught smuggling some 30,000 items of worked ivory into it. Home to a large community of Chinese immigrants, the city has links to Hong Kong. The case, which turned out to be the largest involving elephant ivory in recent Canadian history,[11] posed the question: was this ivory *destined to supply a large Vancouver ivory market with links to the USA*?[12]

The methodology they used was similar to the surveys we have already looked at.

**The Results:** The ivory market had shrunk since 1989. There were fewer outlets and workshops, and many former businesses had closed. However, with regard to raw ivory, the craftsmen still preferred their tusks either to be small or cut into blocks whose sizes ranged from 0.2 to 1.0 kg. This meant that 3–5 kg tusks were more expensive per kilo than 10 kg tusks, while the cut pieces were sold for a very high price.

The numbers of craftsmen had plummeted from an estimated 1400 in 1989 to a minimum 120, including both part- and full-time workers. They normally worked at home, traditional work, making new knife, gun and walking stick handles, scrimshaw pieces, jewellery and parts for cue sticks, and they continued to use legal raw ivory (most of it old, some cannibalised from broken or damaged ivory items) for restoration work.

At the time of the survey, the USA was estimated to consume less than 1 tonne of raw ivory a year, and the UNEP-WCMC CITES Trade Database showed that between 1990 and 2007, some 3530 tusks and about 2400 raw ivory pieces were legally imported into the country, although the investigators

---

[10] See Martin and Stiles (n.1).
[11] Canada Newswire 2007a and b, ibid.
[12] Ibid., p.7.

*'found evidence suggesting that some of this material was illegally sold into the commercial market'.*[13]

Out of the 16 places visited, 3 cities were selling most of the ivory. These were New York (an early *ivory city*) with 11,376 items and two Californian cities, San Francisco with 2777 items and Los Angeles with 2605 items, where nearly a third of the items they saw for sale might have been produced after 1989, thus making them illegal to import. The researchers however, qualified this bombshell by pointing out the difficulty in dating ivory artefacts, so it really was an estimate.

The USFWS recorded over 40,000 worked ivory items (excluding personal effects) legally entering the USA between 1995 and 2007. Presumably therefore, these were all antiques, as only ivory antiques would have been legal. Unfortunately, many of the mainly Chinese netsukes, items of jewellery and figurines that the investigators examined were not antique at all, but had been made recently. Some of the African items also looked dubious.

**Enforcement:** Although none of the CITES Resolution Conf. 10.10 (Rev. CoP14) recommendations had been implemented, according to ETIS, the country had a good record of enforcing CITES regulations in respect of international wildlife trade. It had also reported the largest number of seizures of illegal ivory in the world. Despite this, shops and Internet sites were rarely inspected by US federal and state agencies to look for illegal raw or worked ivory, and the majority of craftsmen and people selling ivory thought the authorities ought to establish a regulated, legal international trade in ivory.

The investigators concluded that:

* Some ivory was successfully smuggled into the country.
* There were no effective internal ivory transport and retail market controls.

However, because the sales were the result of illegal worked ivory being imported for resale, rather than the country's own industry, they would only have had a small detrimental effect on elephant populations.

### Canada

These results were encouraging. The researchers found only 234 ivory items for sale in Vancouver's 45 outlets, and despite the sizeable Chinese community, there was no Chinese ivory market.

---

[13] Ibid., p.6.

## 15.4　The 2014 Survey of Elephant Ivory Trafficking in California

This month-long survey took place in the spring of 2014, just before some of the federal American legislation was strengthened, and was sponsored by the Natural Resources Defense Council of the USA who were worried that the country might be acting as a driver in the recent upsurge in elephant poaching. It concentrated specifically on California. How much ivory was actually being traded there and how much of it was possibly illegal?

The Natural Resources Defense Council required the following data:

- Number of outlets seen selling elephant ivory, both legal and illegal
- Number of elephant ivory items seen for sale in each outlet, both legal and illegal
- An estimate of the number of items seen that were likely illegal under California law (i.e. post 1977)
- Representative prices of ivory items
- Methods that outlets were using to source ivory, the origin of the ivory and techniques they might use to camouflage illegal ivory'[14]

**Fig 15.1** *'Elephant, hippo, mammoth ivory and carved bone mixed together in a display'.* Dan Stiles poses the question *'Which pieces are illegal?'* Copyright Daniel Stiles

---

[14] Daniel Stiles *Elephant Ivory Trafficking in California USA* 2014, p.8.

**Surveying the Retail Outlets:** The first task was to find them. Starting with the 2006/2007 survey list, the researcher, Dr. Daniel Stiles, used a variety of methods including Internet sources and websites to identify antique dealers and auctions that sold ivory. Over 1200 possible outlets were identified in Los Angeles (the largest urban area in the USA), while in San Francisco there were about 450.

Fortunately, many could be eliminated at an early stage. They had closed down. They no longer sold ivory. The rest could be tackled by a telephone call. The answers were mixed however, and frequently the call had to be followed up by a visit. If the vendor only sold a few pieces of ivory, it was possible to take down prices and descriptions during the conversation. Vague, uncertain or answers indicating there were several ivory pieces for sale all had to be followed by a visit. Then there were those markets that took place every so often, flea markets, antique markets and craft markets. All of them open-air markets, there were hundreds of people selling huge numbers of items, some of which were made of ivory or had ivory inlay somewhere in them. And some of that ivory was real, and some was fake, and some was legal and some illegal. A Herculean task! But information a government body urgently required if it was to take further action to try to protect elephants from poachers.

To get the information, the researcher pretended to be *a potential buyer*, which could then be extended into *writing an article* if the seller was inclined to be helpful. This meant he could ask about the age and possibly the origin, generally elucidate far more detail about the piece/s and even ask more detailed questions in a friendly chat.

Time constraints compelled the researcher to confine himself to business outlets only. There were just too many listings of ivory for sale on auction websites across the USA, some 4000, and even though not all of this was elephant ivory, it was a sensible and practical decision to exclude individual sellers. To complicate matters even further, *'some of the items displayed on websites were not physically present in California, or even in the United States, but they were for sale in California online and notices stated that they could be shipped anywhere in the world (including California)'*.[15] These obviously would have had to have been included.

**The Results**
They were disappointing.

---

[15] Ibid.

**Los Angeles:** There were 77 sellers in 32 stores and 10 multivendor malls or markets. Although 777 ivory items were for sale, many of them were also for sale on websites, so people could buy them without visiting the outlets. Some sellers had gone over to eBay and only used that to sell their ivory artefacts.

Antiques are deemed to be 100 years old and none of the items had provenance to show that age. In fact, any sort of provenance was sadly lacking. Most were the kind of items we have seen earlier, items such as buttons, penknife handles, letter openers and 'old-style jewellery', although the few figurines did look as though they had some age.

The difference between state laws and federal laws was factored into the findings:

- Under Californian law: between 77% and 90% of the ivory was probably illegal.
- Under federal law: between 47% and 60% may have been illegal.

One caveat was the possibility that some of the items could have been made using old ivory, ivory already in the USA.

### Fake African Ivory Antiques

Many of the alleged antiques were obviously fraudulent, recent fakes. Although one seller had 96 pieces of African ivory, 41 of which, he claimed, had been made by a particular ethnic group, this assertion was rapidly debunked by the expert called to verify the claim. In fact, he concluded that as few as 5, possibly 6, of the 96 were genuine. This could even have come as a shock to the seller.

In fact the researcher, himself very experienced and knowledgeable, was later able to identify with some degree of certainty the very market in DRC the pieces had come from. Bikeko Market, which he had visited during the autumn of 2014 when he had discovered a potentially disastrous new development since his previous visit in 1999.

When he had talked to the sellers, he had discovered that as the Ministry of Culture provided them with permits to sell antiques, they were inclined to produce/fake these desirable objects themselves. Not only could they get the requisite permits to sell/export them, the artefacts were proving very popular in their own right. Indeed, they were attracting dealers from America and Europe who visited the market and bought many pieces at a time. Attractive business to sellers in a poor and unstable country. Some items were shipped to the USA.

Unfortunately the profit margin for the fake antiques was proving to be much higher than for the modern pieces. Another incentive to produce more.

And the elephants being slaughtered to produce this ivory? Probably the endangered forest elephants.

**San Francisco:** Despite its much smaller population, the researcher still found 473 ivory items for sale. He visited 30 outlets, 28 of which were selling ivory. There were 30 sellers.

He was quite unable to determine the number of elephant ivory items in two of the stores he visited, because of the complete jumble of different ivory pieces. Elephant ivory, mammoth ivory, hippo ivory and bone pieces, all mixed up together. It had been the same when he had visited the shops during his 2006 investigation. On that visit, the estimated total of elephant ivory artefacts had been calculated at over 800 items, so, working on the same principle, he considered it *'likely that they still possess large quantities numbering in the hundreds'.*[16]

Like Los Angeles, ivory pieces could be found that had probably been made of pre-ban ivory, the sort of ivory used by the scrimshanders. One outlet contained 42 such pieces, and they were probably correctly labelled. Once again though, there was a complete lack of documentation, no provenance for any of the alleged 'antiques'.

These results were:

- Under Californian law: 80% of the ivory seen was probably illegal.
- Under federal law: 52% could have been illegal.

Once again there were discrepancies. For example, two vendors claimed that every piece of ivory they had on display had either been imported before the ban in 1989 or was not elephant ivory. There were *'several hundred ivory items'* displayed. The researcher observed the curious fact that if this claim was accurate, it would mean *'that they had all been in inventory for at least 25 years without selling'* and it was inconceivable none of them had been sold.

Once again there were plenty of what were obviously fake 'antiques', many so similar that they had obviously been produced by the same 'craftsman'. And they were of recent origin. In fact, of the 473 pieces seen, only 5 or 6 appeared truly old.

In his **discussion**, the researcher observed that:

- The number of ivory pieces that appeared to have been recently manufactured had increased by a quarter, in other words, up 25% from his previous 2006 visit, and now accounting for about 50% of all elephant ivory pieces for sale.

---

[16] See Stiles (n.14).

- Most of the ivory was coming from China.
- Demand from Chinese consumers was particularly high, but they wanted Asian pieces.
- Many of the so-called *'antique ivory'* pieces were not only fakes but had been made recently and probably smuggled into the country.
- There were fewer outlets and fewer items for sale, which was partly due to a growing awareness on the part of the sellers that there could be legal complications.

This is understandable, as we already know just how complex and technical the legislation is and that there are differences between state and federal laws which make it difficult to determine which ivory actually is legal.

**Enforcement:** In this survey, the researcher learnt that *'not a single one of the 107 ivory outlets in California reported ever having any ivory items seized by law enforcement'*,[17] which, in a country like the USA, was really rather shocking. Nor did the outlets report any ivory seizures. While New York and some other states had experienced large-scale ivory 'busts', this again had not happened in California. In fact, it seemed that the only enforcement taking place was at ports of entry, by customs agents, possibly because of the 'dispute' over who should enforce the California Penal Code Section 6530.

However, sellers in a few, though only a few of the outlets surveyed, told the researcher that they had recently received visits from agents from the USFWS, who had *'warned them that they would only be able to sell antique ivory in the future'* and they would have to prove these pieces were antique.[18]

To return to our puzzle. While these surveys showed there was a market for ivory in the USA, it did not appear to be large unless people were mainly buying from auctions and the Internet. However, before we examine that possibility, we must return to the law.

**Developments in the Law:** In 2013, while President Barack Obama was visiting Tanzania, he announced a National Strategy for Combating Wildlife Trafficking which was followed by Director's Order No. 210. Amongst other things, this strengthened US trade controls for elephant ivory and rhino horn. (The Director is the Head of the Fish and Wildlife Service.)

Early in 2014, the USFWS proposed new federal regulations which would apply to auction houses and antique dealers, putting the burden of proof onto

---

[17] Ibid, p.16.
[18] Ibid.

the seller, who would have to produce the required documents to show that the ivory artefact they were offering for sale was indeed antique. Sales of non-antique ivory were to be severely restricted.

## 15.5   Live Auctions

The International Fund for Animal Welfare, IFAW, and TRAFFIC have made a major contribution to understanding the role live and Internet auctions are playing in the illegal sales of both elephant ivory and rhino horn.

In 2012, IFAW used America's Freedom of Information Act to get some numbers, some data that the USFWS had put onto LEMIS, the Law Enforcement Management Information System. The numbers were large. America had legally imported a total of 13,221 ivory artefacts and 430 kg of additional tusks and ivory pieces between 2009 and 2012, most of which had been imported by auction houses and art galleries. There was more. In the same period, 6753 ivory objects had been legally exported.

And in that same period, there were seizures as well, some 1000 ivory products seized upon entry and 250 on export. INTERPOL estimated this to be roughly 10% of the illegal ivory and that 90% was getting through. Therefore *the estimated illegal ivory that is smuggled into and out of the United States may be more than 3.000 specimens per year'.*[19] In fact, it was partly because of this data that the new regulations had been proposed.

Because most of the importers were antique galleries and auction houses, IFAW decided to investigate. They wanted *'to better understand the market forces driving the elephant poaching crisis'* and *'to fill an information gap'.*[20]

**A Brief Interlude:**   First, we must find out what happens at an auction. We will look at two situations.

You, **S**, want to sell a tiny ivory elephant that you bought 10 years ago from a shop that sells all sorts of things. It was not expensive and there were no documents. You are tired of it and want to sell it. A friend says it might be worth more money than you paid for it as it looks really old, so take it to an auction, which you do.

---

[19] *Bidding Against Survival: The Elephant Poaching Crisis and the Role of Auctions in the U.S. Ivory Market* IFAW 2014, p.8.
[20] Ibid., p.1.

Someone at the auction house looks at your elephant and tells you that it is old. It could fetch a lot of money. It is included as a 'lot', which is identified by a number. The auction produces an online catalogue for the sale. Some buyers just look at the catalogue, but others like to visit the live preview where they can see and examine any items they might be interested in buying.

Then comes the auction itself. The auctioneer takes bids for each lot from the floor, from the telephone, perhaps some from the Internet. The price is right. The bidding stops. The elephant is sold.

**Or**

You, **S,** want to sell the tiny ivory elephant your aunt left you in her will. She told you it was valuable and it comes with various documents. You love elephants and want to sell your elephant to raise money to send to the elephant orphanage.

You take it to the auction house where your documents are examined very carefully. It is valuable. A valuable antique. The documents show its 'provenance', its previous history. The auctioneers sell it for you, and the new owner acquires both the elephant and its provenance.

## 15.5.1   IFAW's Survey and What They Found

Taking place during February, March and April 2014, it looked at live auctions and their previews, with the investigators personally attending both. This enabled them to *sound out* the views of the auction houses on the lots and look at those buyers who were physically present in the room, their age, ethnicity and so on. Even without the telephone and Internet buyers, it would be a useful indication.

The investigators visited 14 auctions pretending to be potential customers interested in buying or selling ivory artefacts. Some, but not all, registered as bidders. They studied catalogues where these were available, and this enabled them to pick out items that might be of particular interest. The data collection was standardized.

During the surveyed period, there were 833 ivory lots on offer including statues, netsukes, accessories, paintings, carved tusks and jewellery. The values placed on the items ranged widely, from $10 to $25,000, although in practice they sold for much less. The highest bid, for a Russian samovar, only reached $18,000, while a pair of two large tusks sold for $13,000 and some smaller items went for under $100.

**Sellers and Auctioneers:**  Like sellers in many of the countries we have looked at, sellers in the auction houses in the USA seemed to have a very 'gung-ho' attitude. They did not appear to be worried about the legality of the ivory they were selling and tended to assume all of it was legal.

At only one of the auctions and previews visited were there documents accompanying the ivory lots to show their provenance. There were no CITES import permits to indicate that the ivory had legally entered the country, nothing. The auctioneers never questioned the sellers about whether their ivory was legal and just went along with whatever the seller told them. In some cases apparently, they were even prepared to go by the stamp on the base of a statue indicating a particular dynasty. To many people this would seem naïve, but where the market is unregulated so that, in effect, anything goes, unless you were an auctioneer with a particular interest in ivory, perhaps your attitude might be 'why bother'.

The general consensus was that it was up to the buyer to make sure they had the information. Caveat emptor—buyer beware! The customers, not themselves who should bear the legal responsibility for following endangered species legislation. Not a customer-friendly attitude. After all, the auctioneers are regularly selling ivory, so they should be very familiar with the workings of the legislation and what should be done to conform with it, not the poor punter who might only buy one piece of ivory in his life, thus heavily dependent on what should be informed advice, expertise, experience and wisdom of the auctioneer.

One *'prestigious auction house investigated indicated that it trusted what its sellers say about the provenance and legality of their ivory, rather than requiring paperwork or any proof of provenance or age, and the house relies on staff expertise to estimate ivory age… (it) offers "any necessary paperwork", yet it does not require any proof of origin when it receives an ivory item for sale'.*[21]

Another problem was how to identify the many fake modern/old ivory pieces. Paperwork/provenance is essential and the auctioneers should have recognized this fact. Based on the regulations in place at that time, they simply did not believe documentation was necessary for sales inside the country, provided they did not ship internationally. Interestingly, they no longer considered ivory to be such a safe investment.

There was also considerable ignorance about the need for the new regulations, with many employees simply not believing they would come into force. A new administration might drop them. One gallery owner apparently believed that *'elephant populations are breeding a lot in Africa right now, so maybe*

---

[21] Ibid, p.20.

*it [selling ivory] will be legal again sometime soon*.[22] It is obvious that even in the USA, education still has a vital role to play.

**The Customers:** The investigators noted an Asian connection. By observation and questioning, it seemed that *'a significant proportion'* were *'males of Asian descent'*. The owners of two of the galleries they visited were Chinese, and *'several auction websites posted their catalogs and other promotional materials in Chinese'*. Chinese, rather than Japanese, ivory artefacts were preferred.[23]

There seemed to be two types of customers, the collectors and the resellers (those who buy to sell on). Again, the collectors fell into two broad categories, those who collected older, pre-ban or antique artefacts, which would normally be highly priced, and those who liked collecting small pieces of ivory like netsukes that would give them pleasure while not costing large sums of money. Some customers bought for resale, perhaps to put in their shop or perhaps to sell on the Internet. All this is fairly typical of customers in many Western countries, although they would probably not be buying ivory. And it was becoming obvious that in America the general public were becoming uneasy about this.

**Sotheby's and New York Auctions:** Sotheby's is an old and respected auction house. It is one of the world's largest auctioneers of fine arts, antiques and collectibles. Therefore, from time to time, it sells important pieces of ivory, some of which are expected to fetch large sums of money.

LEMIS data from 2009 to 2012 showed that Sotheby's imported 456 items of legal ivory and one of their export shipments was seized. We don't know why, only that it contained 14 ivory specimens.

The change in mood of potential buyers who no longer seemed so interested in buying ivory was beginning to affect the auction houses. The investigators found ivory lots were down in number, and one New York auctioneer who specialized in selling antique ivory alleged he had withdrawn ivory artefacts worth an estimated USD500,000 because *'he believed the loss of their resale value would discourage bidders'*, although there were still 20 ivory lots advertised on their website for May 2014.[24]

---

[22] Ibid, p.22.
[23] Ibid.
[24] Ibid, p.26.

## Conclusion

- The legislation had to be simplified.
- Anyone who had to use it must be taught how to do so.
- Documentation and education were needed.

Although all the proposed legislation permitted the sale of antique ivory provided there was proper documentation, lack of it would make it much more difficult to sell any new/old pieces.

It seemed that the investigators in America were getting the same sort of answer to questions about what could or could not be sold, what documentation was required and so on, as the researchers in China.

# 15.6    Citizens Battle to Improve Their Legislation

There was no doubting the impact of the 2014 survey. Some states decided that the time had come for them to take matters into their own hands. They must do something. By April 2015, 25 states were either considering or had already considered legislation that would so severely restrict intrastate sales of both elephant ivory and rhino horn, as to almost amount to a complete ban on sales.

New York and New Jersey acted with great speed, perhaps because the New York City-New Jersey metropolitan area, as we have seen, comprises one of the three areas responsible for selling most of the illegal ivory. An estimated 80% of New Yorkers supported the proposed ban. Rather surprisingly, Sotheby's opposed the new legislation. A ban would have *a negative economic and cultural impact*. They wanted a legal market for ivory antiques, with built-in appropriate safeguards.

The state of Hawaii, another of the areas wanting new legislation, was unsuccessful. Together with Iowa, Maryland, Oklahoma, Virginia and Washington, their bills were withdrawn, because fierce opposition severely dumbed-down the provisions. They decided they would try again later.

The District of Columbia and seven other states including Colorado, Delaware, Georgia, Kentucky, Missouri, Pennsylvania and Texas were considering whether or not to take action. And in ten others, Arkansas, Connecticut, Florida, Illinois, Massachusetts, Nevada, Oregon, Rhode Island, Vermont and California, bills were before the legislature. Vermont was particularly interesting. There the fight was being led by a *'homegrown elephant activist and mother of two'*, a passionate advocate for elephants, who, as well as giving evidence to the state's House Committee on Fish, Wildlife and Water Resources, was making sure that school children in her neighbourhood were also made aware of the loopholes in the legislation. California, shocked by the findings of the survey, also reacted swiftly, producing the Atkins Bill.

In July 2016, new regulations amounting to an almost complete ban on commercial trade in African elephant ivory came into effect. They made enforcement easier by imposing more restrictions on sales and exports across state lines, and trophy hunters were limited to just two ivory trophy imports a year.[25]

People power was at work, with good effect!

---

[25] Jani Actman *U.S. Adopts Near-Total Ivory Ban* National Geographic, 3 June 2016. See https://news.nationalgeographic.com/2016/us-ivory-ban-regulations/ Accessed 15/11/2018.

# 16

## Markets in the UK

Ivory is also highly desirable to Europeans and the UK, which prides itself on its love of elephants and is only too aware of the horrors of poaching and is another country with one of the largest ivory markets in the world.

A 2014 study of ivory trading in the EU showed a clear rise in legal trade, particularly since 2007, in:

- Commercial re-exports of pre-Convention ivory products (old artefacts)
- Re-exports of worked ivory
- Re-exports of raw ivory[1]

This trade is strictly controlled by regulations which are legally binding on all the Member States, which, at the moment, includes the UK. This is how it works.

Commercial trade in ivory is permitted under prescribed conditions for:

- Antique ivory: *worked* specimens acquired before 3 March 1947
- Pre-Convention ivory: *raw* or *worked* items, acquired before the date on which CITES or the EU Wildlife Trade Regulations started to apply in the country in which they were acquired

Unfortunately, both the CITES Trade Database and ETIS indicated problems within the UK, which had always been a major trader in ivory. Indeed, up to 500 tonnes of it were imported annually between 1910 and 1914, much

---

[1] Lau, W., Crook, V., Musing L., Guan, J., and Xu, L. (2016) *A rapid survey of UK ivory markets.* TRAFFIC, Cambridge, UK.

of it being made into *'decorative items and household and personal objects'.*[2] And legal commercial exports of pre-Convention ivory products from the EU increased between 2003 and 2012, the re-exported ivory including both carvings and raw tusks (either whole, pieces or scraps) *'with re-exports of tusks showing a particularly pronounced upward trend after 2007'.*[3] The main destination of the ivory was China and Hong Kong SAR.

So in 2016, TRAFFIC carried out a rapid assessment of the current state of the ivory market in the UK using an earlier survey by Martin and Stiles in 2004 as a comparative baseline, re-visiting selected shops and markets, but also including new data researched from online offers for sale on UK-based antique and auction websites. The survey included antique shops in 2 areas of London as well as 13 markets, and the researchers were Chinese, European and North American nationals, all experienced in detecting ivory.

Although the target products were mainly antiques, *all ivory items being offered for sale in antique markets/shops were recorded*, as were the specific types of products, such as Cantonese magic balls and netsukes. After the results had been analysed, they were grouped into four categories:

* Jewellery
* Figures
* Household goods
* Personal items

Nearly half the figures were medium-sized (10–20 cm); the jewellery items included small brooches, bangles/bracelets and pendants, while handles for cutlery and magnifying glasses accounted for most of the household goods. There were a few personal items for sale, mostly walking sticks with ivory handles and medium-sized fans.

Of the items for sale:

* 56% were figures.
* 27% were household goods.
* 9% were jewellery.
* 8% were personal items.

### Survey Results

Although these were encouraging, the surveying team urged they be treated with caution because it was not possible to make exact comparisons.

---

[2] Ibid.
[3] Mundy 2014, ibid.

**The Markets:** Only 200 stalls were offering ivory items for sale, compared with about 640 in 2004. Of those, 6 offered more than 90% of the items, the same as in 2004. The number of items on offer for sale had also declined, down to 3200 from about 6000 in 2004. But the data suggested that the traders, even though there were fewer of them, were offering more ivory, each trader, on average, having a larger stock. The cheapest prices were very low, as little as GBP10, but depended on the item being sold.

**The Shops:** In one of the areas, Kensington Church Street and the Kensington Antiques Centre, both the number of shops and the number of items for sale appeared to have declined. Visits were 'by appointment only' in Mayfair, which is a very exclusive part of London. In this rapid survey, there was no time to make appointments, but there were 4 shops offering antique ivory for sale, about 40 items in total, and very highly priced. An ivory cabinet was priced at GBP30,000. Harrods no longer sold antiques, so it had no ivory items for sale.

**London Chinatown:** No ivory items were seen displayed for sale in either the 2004 or the 2016 survey.

**The Buyers** They had also changed. In 2004, they were mainly American and European tourists, whereas in 2016 they were mainly from East Asian countries/territories (including mainland China, Japan and Hong Kong SAR), with some Europeans, from Germany, Austria and Portugal. What came as a surprise though was that the Asian buyers were not necessarily interested in buying ivory. They were looking for skilful *craftsmanship, style of carving and the era and/or provenance of antiquity*, so *there was little difference in price between products made with ivory or other materials.*

**The Middlemen:** Two *Chinese-speaking scouts* were seen *'recording information on antiques, including ivory items, for sale at Portobello Market and sending this on to prospective buyers'*.[4] This was an interesting observation.

**The Sellers:** They generally knew something about the objects they were selling, those at the high-end being very knowledgeable. Apparently most of what they knew came from reading the *Antiques Trade Gazette* which they were obviously found very helpful.

Like the buyers, or perhaps because of the buyers, the traders were *'generally more concerned with craftsmanship, style of carving and the era and/or provenance of antiquity as indicators of quality and value rather than the material the item*

---

[4] Ibid.

*was composed of*.[5] Similar objects of ivory or plastic therefore showed very little price difference. And perhaps it follows that the dealers were not at all clear on how to identify ivory either. *Colour, brightness, how it felt*, their *expertise* and their *experience*, only one mentioned the importance in the grain, and no one mentioned the tell-tale crossed pattern/Schreger lines.

They were questioned on, among other things:

- *Their knowledge/perception/awareness of the status of the UK ivory market*
- *The buyer demographic*
- *Regulations concerning cross-border trade in ivory*[6]

Their knowledge of applicable legislation was mixed. As perhaps would be expected, the high-end dealers knew a great deal about those areas of national and international legislation applying to ivory sales. They knew their customers required an invoice and verification of age to take ivory out of the EU. While some offered to acquire export permits for their customers, many told the customers they were themselves responsible for that.

The other dealers generally had some understanding of the requirements, and all of them knew there was a specific cut-off date for lawful trade in antique ivory. The problem was only some of them knew when an ivory item is considered antique by law, namely, ivory acquired and modified before 1947. This ivory can be traded within the EU without CITES documentation. Dates ranging from the 1920s to the 1960s were suggested. Some dealers knew it was illegal to trade in unworked tusks within the EU.

While many dealers had some knowledge of ivory legislation in the USA, generally there was concern about selling to the researchers from China and America. They were unsure of the law, particularly regarding China, with several traders thinking it risky to export ivory there. Some thought it illegal; others insisted the customer take responsibility for checking what could or could not be done. One even refused to sell ivory to Chinese nationals because of *regulatory changes*. A few though were not so troubled. Although they knew there had been changes, they still suggested that *'purchasing small ivory items was not an issue and that the buyer would be able to transport these items in their personal luggage without detection'*.[7] One Chinese citizen stated that *'she had successfully transported many ivory items by post'*.[8] Some dealers suggested:

---

[5] Ibid.
[6] Ibid.
[7] Ibid.
[8] Ibid.

- *Declaring the items as personal/household items that the buyer is travelling with*
- *Using women to take items as they were less likely to be checked by the authorities*
- *That a receipt was adequate proof of age*
- *That the trader could state on the receipt the item was made of animal bone*[9]

## 16.1   The Online Antique Ivory Market in the UK

Online surveys were also carried out. Two types of website were surveyed: antique auction consolidation websites and online antique market places. Information came from two main sources: the Saleroom and Barnebys. The Saleroom provided more specific details of the auction houses that were involved, whereas the results from Barnebys included more guide price information (96 lots had guide prices), so it could be used for price analyses.

The auction lots came from over 70 different UK-based auction houses, 65 using the Saleroom which offered 578 lots containing ivory during the period of the survey. Barnebys had 10 extra auction houses, which were not using the Saleroom, and had 117 lots on offer, making a grand total of 695 lots. It was possible to identify the main auction houses involved, and although 53 auction houses offered between 1 and 9 items each, it turned out that 22 auction houses were responsible for offering nearly 75% of the ivory items.[10]

Several traders used many online market places, so that although 6 online market places were sampled, once the duplicate offers had been subtracted, the total number of items offered for sale was considerably fewer, 1710 rather than 2008. Three platforms, Antique Atlas, Online Galleries and British Antique Dealers Association, were responsible for 85% of the items, with Selling Antiques responsible for another 10%.

**Products and Prices:**  All the items on offer for sale came from a number of countries and fell within the UK's definition of antique ivory. Japanese figures and netsukes were quite common, but there were also *'French and English household items and miniatures (paintings), German, French and Spanish jewellery; and various other decorative items such as panels and vases from China, Japan, Africa and India'*. Dates ranged from the seventeenth century, with most coming from the early twentieth century.[11]

---

[9] Ibid.
[10] Ibid.
[11] Ibid.

Let us look at some of Barnebys guide prices for items offered for sale. (Remember, these were the antique auction consolidation websites.)

- 42% were under £500.
- 37% were between £1000 and £5000.

The cheaper items included *'chess sets, gambling chips, cutlery, fans and a wide range of figures'*.[12]

While most items were similar to those sold in the physical markets, a few were unique. These included a cricket cage which Christie's was auctioning, the estimated selling price being between £15,000 and £25,000. Another of their items had as its estimated selling price £20,000–£40,000. This was the highest guide price the researchers saw, and the item was an ebony table cabinet, circa 1650, containing a mix of materials, including ivory, tortoiseshell and cedar wood.

Prices on the online antique market places were much lower, as would be expected. Antique Atlas offered the lowest price range. Its most expensive item was a pair of Japanese carved vases, on offer for £7500. Most items were small, things such as needle cases, letter openers and brooches. In fact, 60% of their ivory items were priced at under £100, while most fell within a price range of £100 and £500. This included, for example, Chinese fans, portrait miniatures and small- to medium-sized figures.

On both types of website, most ivory items were offered at under £500.

Both online market places and the various auction consolidation sites offered different information on the age and legality of the ivory for sale, so it is interesting to compare a few examples of declarations concerning age, provenance and legality:

**Selling Antiques:** Dealers must include a formal declaration giving the date of manufacture. The date is shown on the product detail page.

**Online Galleries and BADA:** Their websites include a date of manufacture after the title, but it is not clear whether dealers must provide this.

**BADA:** Includes a paragraph making clear that a CITES certificate must be obtained if the item is to be exported outside the European Union. The paragraph is found at the bottom of an ivory product advertisement webpage. *'We can obtain these on your behalf…'*, and they give details, including price. The paragraph concludes *'Please ensure you contact us prior to purchase for a quote'*.

---

[12] Ibid.

**Christie's Auction House:** Has a special notice about international trade on its website. However, it is not as clear as BADA's paragraph:

*Several countries prohibit the importation of property containing materials from endangered species, including but not limited to coral, ivory and tortoiseshell; prospective buyers should familiarise themselves with relevant customs regulations prior to bidding if they intend to import this lot into another country.*

Sworders and Other Auction Houses (Have Similar Wording in Their Own Terms of Reference): *buyers intending to export goods should ascertain (a) whether an export licence is required and (b) whether there is any specific prohibition on importing goods of that character because, eg, they may contain prohibited materials such as ivory.*[13]

Apart from BADA, the onus was therefore on the buyers to sort out their own export requirements. Sellers were not even required to produce evidence that their items fell within the worked specimen derogation, and the survey found very few dealers able *to provide proof of age or documentation to prove legal acquisition.* As might have been expected, the high-end dealers and auction houses were far more knowledgeable about the legal requirements, which are complex and technical.

The evidence coming from this survey was disturbing. Seizures had increased in recent years, over 150 postal packages containing ivory carvings in 2015 alone, all of them destined for China. Worse still was the discrepancy in the CITES trade data for elephant ivory.

# 16.2 More Shocks and Some Urgent Action

Then in August 2017, the EIA published yet another shocking report. It showed that the UK was not only *'the world's largest exporter of legal ivory'* but also *'the largest exporter of legal ivory to Hong Kong and China',* with exports to them increasing *'dramatically'.*[14] It was time for the UK to take urgent action. Michael Gove, Secretary of State for the Environment, did just that. Caring deeply about the situation, he proposed an outright ban on ivory sales, and a government consultation was launched in November. About 60,000 people responded.

In January 2018, *The Times* published the results of an undercover investigation into the auction industry. These too were disturbing. Several auctioneers

---

[13] Ibid.

[14] *UK is the largest supplier to the world's ivory markets* EIA, 10 August 2017. See https://eia-international. org/uk-largest-supplier-worlds-ivory-markets Accessed 17/08/2017.

on the-saleroom.com, *the UK's largest site for auction house listings*, refused to provide an age for the more than 1000 pieces of ivory they were selling that week, not just to British but also foreign buyers, although it is not a legal requirement to provide such provenance. Not only that, it was also providing a platform for blatantly dubious transactions, such as the raw ivory tusk which, in 2017, the seller was willing to export to Hong Kong SAR and, in January 2018, several raw ivory tusks which an American auctioneer from Texas, USA, was selling (it is legal to sell ivory in Texas).[15]

*The Times* also passed on details of one dealer to the National Wildlife Crime Unit. He was prepared to dispense with the CITES permit needed to send ivory to Japan, offering instead to send it using tracked Royal Mail, something he had done before. He wrote to the buyer *'I will bypass the form (the CITES permit) as it takes too long to complete/authorise'* and *'I will complete a customs form and there should not be any issues (as with previous deliveries to Japan)'*. When confronted by the reporter, he admitted he had made mistakes, claimed ignorance of the law and sent documents to support proof of age.[16]

That same week, a man was arrested on suspicion of offering prohibited species for sale. Police had raided two properties, where they found and seized more than 40 items, including elephant tusks and rhino heads and horns. A rare species identification expert was brought in, and after being interviewed, the man was released, while enquiries continued.[17]

---

[15] Georgie Keate *Dealer offered to bypass ivory law* The Times, 20 January 2018.

[16] Ibid.

[17] *Convention on International Trade in Endangered Species (CITES) enquiries in Lancashire* Legal Eagle (The RSPB's investigations newsletter), Spring 2018, No 84, p7.

# Part VII

## Illegal Trading in Rhino Horn

Results from all the surveys showed illegal trading in elephant ivory becoming more and more threatening to the future of elephants, so our story must return to the rhinos, to see what has been happening to them.

# 17

# The History of Rhinoceros Listing on CITES

As early as 1977, in other words, only 4 years after CITES came into existence, all species and populations of rhinos were listed in CITES Appendix I, and since then controls over the exploitation of these animals have been gradually strengthened. Until 1994, when South Africa's population of white rhinos was downlisted to Appendix II. This meant there could be international trade in the live animals, provided they went to appropriate and acceptable destinations, and in hunting trophies, an exemption that has caused, and continues to cause, so many problems for the animals.[1,2]

A series of Resolutions have been passed at the CoPs to provide protection for the animals.

**Resolution Conf. 3. 11:** In 1981, a moratorium was requested on the sale of all government and parastatal stocks of rhino products under the control of both CITES Parties and non-Parties. At CoP3, it was included in Resolution Conf. 3. 11, so was the recommendation that all non-Parties prevent the commercial import and export of rhino products across their international borders, even though non-Parties could not be bound by CITES.

---

[1] Annotation to the downlisting.

[2] CONVENTION ON INTERNATIONAL TRADE IN ENDANGERED SPECIES OF WILD FLORA AND FAUNA< 61st meeting of the Standing Committee, Geneva (Switzerland), 15–19 August 2011, *Interpretation and implementation of the Convention, Species trade and conservation, Rhinoceroses*, CONSERVATION OF AND TRADE IN AFRICAN AND ASIAN RHINOCEROSES, SC61 Doc. 45.2 – this document was submitted by Hungary on behalf of the European Union and its Member States, p.2.

© Springer Nature Switzerland AG 2019
B. Martin, *Survival or Extinction?*, https://doi.org/10.1007/978-3-030-13293-4_17

**Resolution Conf. 6. 10:** Adopted in 1987, this urged all Parties not only to prohibit international and domestic trade in rhino products, but also improve the awareness of their law enforcement agencies, increase penalties for trading in rhino products and take effective action against both poachers and middlemen. And because the 1981 moratorium had failed to stop the trade, Resolution Conf. 6. 10 again urged all Parties to destroy all government and parastatal stocks of rhino horn and recommended action be taken against any countries still allowing trade in rhino horn, by applying pressure to them.

**Resolution Conf. 9. 14:** Passed at CoP9, it directed the Standing Committee to pursue actions aimed at reducing illegal trade in rhino horn, and down-listed the South African population of White rhinos to Appendix II (with an annotation). Now Resolution 9.14 (Rev. CoP15), it has been amended at a number of CoPs and continues to be the current Resolution regarding the *'Conservation of and trade in African and Asian rhinoceroses'.*[3]

**Resolution Conf. 9. 14 (Rev. CoP15)—Conservation of and Trade in African and Asian Rhinoceroses:** A key piece of legislation, this Resolution is also concerned with conservation of the species.

This was because of the developing situation *'that some rhinoceros populations have continued to decline drastically and that four of the five species are threatened with extinction'.* And although some African and Asian range states were successfully managing and protecting their rhinos, *'often under difficult circumstances',* and were taking measures to control and reduce the use of rhino horn by consumer countries, particularly those where the cultural use went back centuries, it concluded that *'the above measures have not arrested the decline of all rhinoceros populations'.* This was *'a global law enforcement problem'* extending beyond the rhino range states and the traditional consumer countries, and unfortunately *'emphasis solely on law enforcement has failed to remove the threat to rhinoceroses'.*

There is an appeal to range states and implicated states for cooperation and funding, and all governments and intergovernmental organizations, international aid agencies and NGOs are *'called upon'* to provide funds. These are not just to be used for conservation and measures to prevent the rhinos being poached and their horns trafficked but also to enable the IUCN and TRAFFIC to effectively report their findings to the Secretariat before a CoP.

---

[3] It was amended at the 11th, 13th, 14th and 15th meetings of the Conference of the Parties and further amended by the Secretariat in compliance with Decision 14.19 and with the Decisions adopted at the 61st meeting of the Standing Committee.

In other words, without more money, it would be very difficult for both TRAFFIC and the IUCN to carry out their monitoring/reporting activities, and without those, it would be very difficult to determine whether or not the range States were in fact actively trying to conserve rhinos, as well as taking measures to stop both poaching and the illegal trade in their horns. Lack of funds continues to be a major problem.

The Resolution also *calls for constructive engagement among all Parties to the Convention* and for *synergy between the Convention and the IUCN/SSC Rhino Specialist Groups to achieve the aims of this Resolution*. An appeal for everyone involved to work together.

**The CITES Rhinoceros Indicators Process:** This is a 'MIKE equivalent' for rhinos.

In 1999, the Standing Committee discussed a report produced by the SSN (Species Survival Network), the result of a workshop whose participants included people from TRAFFIC, the IUCN African and Asian Rhino Specialist Groups and the CITES Secretariat. Four systems were assessed, and the most thorough system recommended, but as it was the most expensive, the Standing Committee, did not approve the expenditure. Not enough money.

At CoP13, in 2004, three Decisions, 13.23–13.25, were adopted. It was hoped these would lead to more effective reporting by the IUCN/SSC African and Asian Rhino Specialist Groups.

**Decision 13.23:** This was concerned with data collection for the next CoP. Every 2 years, the IUCN/SSC African and Asian Rhino Specialist Groups were to compile and produce a summary of the information they had received on the status, trends and developments in rhino conservation in both African and Asian range States. The report from each country should contain the following information in time for CoP14:

- National plans;
- Rhino committees;
- Rhino numbers;
- Translocations;
- Mortalities;
- Horn stocks;
- Horn seizures; and
- Criminal cases.

**Decision 13.24:** This was another call for funds to enable the IUCN/SSC African and Asian Rhino Specialist Groups to carry out the work required under Decision 13.23. Unfortunately, only two sources delivered. Thankfully, the funding they provided was generous.

**Decision 13.25:** This was also concerned with data collection, data that was essential to the effective conservation of such incredibly rare species. The data should be arranged in eight sections:

- Section 1: Management strategies;
- Section 2: Status of populations;
- Section 3: Legislation;
- Section 4: CITES' Decisions;
- Section 5: Illegal killing;
- Section 6: Stockpiles;
- Section 7: Trade routes; and
- Section 8: Recommendations.

It was to be compiled by the IUCN/SSC, with the final report, which was to be translated into three working languages, submitted to the CITES

**Fig. 17.1**  Black rhinos in Kenya. Copyright Chester Zoo

Secretariat prior to CoP14 in 2007. But funding was still a problem so a simpler reporting system was adopted. The data was essential, and this system would have to do.[4]

The ultimate irony was that although rhino horn is worth more than gold, and there was no shortage of people finding the money to buy it, the Parties to CITES and other possible donors were either unable or unwilling to fund the operations essential to their survival. Without this sort of information, how do we know how to help rhinos. Without enough money, how can we help them?

---

[4] SC54 Doc.27 Annex. *Conservation of and trade in African and Asian Rhinoceroses*. A report from the IUCN Species Survival Commission (SSC) Africa and Asia Rhino Specialist Groups and TRAFFIC, to the CITES Secretariat.

# 18

# The Illegal Trade in Rhino Horn

Although there is an almost total ban on trade in rhino horn, people can still buy beautiful artefacts carved out of it, provided they are antique, while hunting trophies have caused major disruption to the protection of these animals. And we have already seen how difficult it is to investigate what is happening on the ground. We do, however, have some pointers.

We are going to consider two countries, South Africa, home to most of the world's white and black rhinos and which, despite its best efforts, still provides most of the rhino horn that is illegally traded, and Viet Nam, which, for a time, became the most important consumer country as China managed to reduce though not eradicate demand. Now demand in Viet Nam is showing signs of decline, but unfortunately, in recent years there has been an upsurge in demand in China, whose consumption may well be outstripping that of Viet Nam.

But first some legislation.

## 18.1 South Africa's Legislation

This is where we begin to see the important role sustainable use plays in our story.

Although South Africa was the 15th country to become a Party to CITES, signing in 1975, it was very slow to implement the treaty provisions into its national legislation. The result was that each of the nine provinces had its own laws, and the whole system was fragmented.

© Springer Nature Switzerland AG 2019
B. Martin, *Survival or Extinction?*, https://doi.org/10.1007/978-3-030-13293-4_18

Twenty years later, the National Environment Management Act 107 of 1998 (NEMA) was passed. This was the first Act to establish the principal of sustainability. Nature conservation activities were prescribed, and it was clear that the use of renewable resources must not exceed the level *beyond which their integrity becomes jeopardised*. Applying this to our story, both elephant ivory and rhino horn could be used to generate income, but that use had to be carefully controlled.

Two other pieces of national legislation are important in the regulation of wildlife management, the National Environment Management: Biodiversity Act 2004 (NEMBA), and the Threatened or Protected Species Amendment Regulations 2008. They apply to both black and white rhinos, which are listed in NEMBA as *an Endangered Species* and *a Protected Species*, respectively.

### 18.1.1    Viet Nam's Legislation

Viet Nam became a Party to CITES in 1994. However, its provisions were only implemented in 2006 in a very thorough piece of legislation, the Decree 82/2006/ND-CP on management of export, import, re-export, introduction from the sea, transit, breeding, rearing and artificial propagation of endangered species of precious and rare wild fauna and flora. This law applies to all nonindigenous species of rhinos, and any allowances for exceptional trade require CITES permits. These permits, which are used to import white rhino trophies from South Africa (they fall within CITES Appendix II), have now been strengthened because of the involvement of some Vietnamese in exploiting the system.

Punishments for breaking the law are set out in another decree.[1] Or the revised Penal Code if a serious or criminal offence has been committed. The severity of a crime is determined by the value of the goods seized plus a number of other factors such as:

- The offence is organized.
- The offender has abused his position.
- The offender has abused the power resulting from his position.
- The hunting has occurred in a prohibited area or in prohibited seasons.
- It has caused severe or exceptionally severe consequences.

The result can be a maximum fine of 500 million Vietnamese dollars (USD29,000) and up to 7 years in prison.

---

[1] 99/2009/ND-CP on Sanctioning of Administrative Violations in the Domain of Forest Management, Forest Protection and Forest Product Management.

## 18.2   The 'Surveys'

Because rhino horn cannot be sold per se, the closest approximation we have to our elephant ivory surveys is the kind of undercover investigation the EIA carried out and that we looked at earlier in our story (Case Study 3). When they visited pharmacists in southeast Asia, they found that the horn was more valuable per ounce than cocaine, and every bit, even the tiniest pieces, was being bought up.

They followed the trail of the smugglers to Taipei in Taiwan, working undercover and pretending to be potential buyers. One dealer they spoke to had a stock of incredibly rare horn from Asian rhinos. He claimed the government didn't have its own policy on rhino horn but merely reacted to outside pressures. Furthermore, pharmacists were warned before police carried out raids on their shops, and even when rhino horn was found on any premises, the owners weren't prosecuted.

When the investigators visited the market in China, although no rhino horn was openly on sale, they were told yet again that it was available, although the traders were very suspicious of potential buyers. And here as elsewhere, any available horn was immediately bought up by the traders because of its great value and potential to increase in value.

By the mid-2000s, the action had moved to Viet Nam where rhino horn suddenly became an important part of the trade in both traditional uses for medication and newly found uses such as *'cures for cancer'*.

A survey carried out in 2004 of the main markets and shops in and around Hanoi that were selling animal-based traditional medicines found three traders who could not only sell rhino horn, provided it was ordered in advance, but could also guarantee its quality. Furthermore, at least half of the shops in the traditional medicine centre in Hanoi, and about 70 shops with several similar businesses in adjacent streets, most of them both wholesale and retail, were correctly registered and held business licenses with the Department of Trade for wholesale and retail traders or held practicing licenses issued by the Department of Health for traditional medicine practitioners, sometimes both.[2] Unfortunately though, this was no guarantee of either quality control or of restricting/banning the sale of endangered species.

Ho Chi Minh City, which is the major distribution centre for traditional medicine products in the south of the country with more than 500 businesses

---

[2] Nguyen and Nguyen 2008, see Milliken, T. and Shaw, J. (2012) *The South Africa – Viet Nam Rhino Horn Trade Nexus: A deadly combination of institutional lapses, corrupt wildlife industry professionals and Asian crime syndicates*. TRAFFIC, Johannesburg, South Africa, p.125.

based there, was also surveyed in 2004. It was found that *'medicinal products from wild animals were conspicuously and routinely available for sale, including…rhinoceros horn…provided the price negotiated was high enough'*. This was, despite the fact that it was illegal to sell all the species on offer. In fact, larger cities all have traders in traditional medicines. *'Rhino horns are traded secretly in Viet Nam, even if it is a rather 'open secret' in many local markets'*.[3]

More recent research by TRAFFIC discovered that there were at least two major wholesalers of rhino horns who were reported to be supplying them to shops selling traditional medicines, hospitals and clinics throughout the north of the country.[4]

Nor were traditional medicine dealers the only sellers. Indeed, local environmental groups *'are increasingly of the opinion that rhino horn marketing is taking on a whole new dimension'*.[5] Much of the horn was fake. Some outlets seemed bizarre, with trade in rhino horn bearing absolutely no resemblance to the actual purpose of the shop. A curious example was the badminton racket shop discovered by TRAFFIC market researchers in May 2009. Its main business was re-stringing badminton rackets, but it also carried a very obvious sign advertising *'rhino horn bowls'* for sale. You bought your rhino horn and then ground it to a powder in these special bowls!

Unlike the ivory outlets, which were only there to sell ivory, legal or possibly illegal, often these shops were contact points where potential consumers could be put in touch with suppliers. Or they could use rhino horn *touts*, who targeted the very sick and the terminally ill, often with cancer, and who were normally to be found around certain hospitals, particularly in Hanoi and Ho Chi Minh City. Some were even on the staff,[6] although that did not necessarily mean that the patient's primary doctors were either involved with such contacts or even agreed with them.[7]

---

[3] Ibid.
[4] Ibid, p.126.
[5] Ibid, p.128.
[6] TRAFFIC market research, May 2009; Smith, 2012b in Milliken and Shaw(n.337).
[7] Milliken, T. and Shaw, J. (2012) *The South Africa – Viet Nam Rhino Trade Nexus: A deadly combination of institutional lapses, corrupt wildlife industry professionals and Asian crime syndicates.* TRAFFIC, Johannesburg, South Africa, p.128.

# Part VIII

## Other Problems Revealed

Now it is time to consider other problems. We start in Europe, with the Rathkeale Rovers who weren't worried about laws.

# 19

# In Which We Meet Our First Organized Crime Gang and the Law Is Changed

## 19.1  A Confiscation and Rhino Horn Thefts

**The Confiscation:** This was the first link in a chain that was to lead, eventually, to the arrest of some members of what turned out to be a notorious organized crime gang. Though the event passed unnoticed at the time.

In January 2010, at Shannon Airport in Ireland, 8 rhino horns were confiscated from the luggage of two passengers, Jeremiah and Michael O'Brien, who were flying in from Portugal. Although the brothers were Irish, they lived mainly on travellers' sites in France and Germany, from where they carried out their trade as travelling antique dealers. They were not arrested, and the horns were sent to Dublin Zoo.[1]

**The Thefts:** The first robbery took place in December 2010, when a rhino horn (1) was stolen from Allwetterzoo in Germany, and although thefts of rhino horns had been occurring in South Africa since 2002, there was nothing to indicate that this was the start of an 'epidemic'. Here is a selection of some of the other thefts that took place.

The second theft occurred 2 months later. This one was nothing if not audacious. It was from the respected auction house Sworders Fine Art Auctioneers, in the UK, and took place the night before a Country House Sale. After forcing open doors into the premises, the thieves took away the mounted head of a black rhino, whose 2 horns were attached by staples. The horns were impressive, some 49 cm long, and the item had an estimate on it

---

[1] Adam Higginbotham *The Irish Clan Behind Europe's Rhino-Horn Theft Epidemic* Bloomberg, 2 January 2014. See http://www.bloomberg.com/bw/articles/2014-01-02/the-irish-clan-b Accessed 26/01/2016.

© Springer Nature Switzerland AG 2019
B. Martin, *Survival or Extinction?*, https://doi.org/10.1007/978-3-030-13293-4_19

(that it would sell for) of £20,000–30,000. There were 7 other black rhino trophies included in the sale, but as these were smaller, all shield-mounted horns, they had been stored overnight in the strong room.[2] That was February 2011.

A month after that, in March, a rhino horn (1) was stolen from the Museum of Natural History, in France. The next theft took place in April, in Portugal, in the Museum of Natural History. On that occasion the robbers took 2 rhino horns. After that, it was the UK's turn again. At the end of May, Haslemere Educational Museum, which houses one of the largest natural history collections in the UK, was broken into, and a stuffed and mounted rhino head with 2 horns was stolen.[3]

In June, the pace speeded up. This time there were six thefts, four of them in Germany, where the villains took the horn (1) of a white rhino from the Natural History Museum, another horn (1) from the museum in Oerrel and the entire upper jaw of a rhino with 2 horns attached together with 4 more horns, thus making a grand total of 6, from a museum in Hamburg. The final theft took place in daylight, at the Gifhim Museum of Hunting, and the robbers got away with 2 horns. There was a good haul from the La Specola Museum in Italy, which lost 3 horns, but the thieves who tried to rob the Liege Natural History Museum in Belgium were not so lucky. They were caught by the police.

July saw thefts in Germany, Belgium, France, but also Sweden, its Museum of Natural History, where, it was reported, the thieves were so laid-back they sawed the horn off the rhino's head while still in the museum. 3 horns were stolen from a castle in the Czech Republic. In Belgium, although the attempt on the Museum of Africa was unsuccessful, the attempt on the Brussels Natural History Museum was blatant. It took place in broad daylight; 2 horns were grabbed and dropped out of a window into a waiting vehicle.[4]

Then Rosie the Rhino lost her horn. She was a greater one-horned rhino, and she had lived in Ipswich Museum in the UK since 1907, where she was a favourite of the children. Even that didn't save her. In the early hours of 28 July, two thieves broke into the museum, tore off her horn (1), which was 45 cm long, and grabbed the skull of a second rhino (number of horns

[2] Antiques Trade Gazette *Thieves target rhino horn at Sworder's sale* 28 February 2011. See http://www. antiquestradegazette.com/news/2011/feb/28/thieves-targe Accessed 26/01/2016.

[3] Esther Addley *Epidemic of UK rhino horn thefts linked to one criminal gang*. See http://www.theguardian. com/environment/2011/aug/08/rhino-horn-the Accessed 26/01/2016.

[4] Milliken, T. and Shaw, J. (2012) *The South Africa – Viet Nam Rhino Trade Nexus: A deadly combination of institutional lapses, corrupt wildlife industry professionals and Asian crime syndicates*. TRAFFIC, Johannesburg, South Africa.

unknown) before escaping in a car. Fortunately Rosie has now been fitted with an artificial horn, which hopefully will not be a temptation to thieves.

The villains were also busy in August, and this time they mainly struck in the UK. Unluckily for them, the 2 horns they seized from the Tring Natural History Museum were fakes. They were brazen enough when they 'visited' Drusillas Zoo, however, attacking in daylight and taking a horn (1) from a locked cabinet. In fact, the attacks grew bolder, that both the robberies carried out in Austria in November took place in the daytime. Matters took a turn for the worse in December. Two people, with the help of an accomplice, burst into the Museum of Hunting and Nature in the Central Marais district of Paris, France, attacked the museum guards with a stun gun and made off with a rhino horn (1). The guards, though not badly injured, did require some hospital treatment.[5] All this took place at lunch time. And a rhino head hunting trophy (number of horns unknown) was stolen from a pub in Austria in April 2012.

In fact, in all, 56 successful and 10 attempted thefts of rhino horn took place throughout Europe between the beginning of 2011 and May 2012,[6] and it had been obvious for some time that these were no isolated incidents but had been carefully coordinated. Take, for example, the theft at Sworders. The alarm went off at about 8.15 pm, and although it only took the police 10 min to get there, the robbers had already fled. Their tyre tracks were seen running across a nearby field. They had struck in the short interval between the salesroom closing and the overnight security staff arriving, managing to lever the mounted head off the wall to which it had been securely bolted, carry it to their vehicle and drive away. The Essex Police had no doubt it was a targeted burglary and were fully aware of the horns' potential value if sold into the traditional medicines market.[7]

It was possible to make large sums of money even from long-dead rhinos. The remaining rhino horns in the sale sold for over £368,000. In July 2010, a year before Rosie lost her horn, the National Security Advisor had reviewed security arrangements at Colchester and Ipswich Museum and found that they conformed to the guidelines laid down by the UK government's Department of Culture, Media and Sport.

Furthermore, a number of warnings had been issued. The same week the theft occurred, *'the Metropolitan Police warned that gangs were visiting muse-*

---

[5] BBC News *Rhino horn thieves use stun gun in Paris museum raid* 7 December 2011. See http://www.bbc. co.uk/news/world-europe-16067019 Accessed 26/01/2016.

[6] Anne Taylor *How rhino horn poaching fuels criminal gangs in UK and Europe* The Ecologist,18 May 2012. See http://www.theecologist.org/investigations/natural_world/1380947/ Accessed 26/01/2016.

[7] See Antiques Trade Gazette (n.2).

*ums on reconnaissance with a view to carrying out thefts of rhino horn'*. Although this did not reach Essex, a notice from Essex Police sent via Essex County Council did, as did another warning from the museums sector. The Museum responded. They decided to leave Rosie where she was as she was such a popular exhibit, but they did their best to make sure that the front of house staff kept a keen eye open for any signs of suspicious behaviour.[8] In any case, it would have been difficult to prevent the theft. The thieves ignored such tempting items as the gold burial masks of Titus Flavius Demetrios or the priceless Hawaiian cape made from feathers of the 'o'o bird. Rosie's horn was the target.[9]

By then Europol, the agency responsible for monitoring organized crime and terrorism across Europe, had identified a particular gang from Ireland, who were trading in the stolen horns. The Rathkeale Rovers.

In August 2011, in the UK, the Museums' Journal contained more warnings about the gang and the dangers it posed. The Metropolitan Police were concerned that *'the gang is targeting premises after conducting research'* and *'hostile reconnaissance'*, their modus operandi ranging from *'carefully planned burglaries'* to *'smash and grab raids'*. They had been known to use force when challenged.

The Metropolitan Police Service's Art and Antiques Unit recommended that *'museums should review their security arrangements, remove rhino horns from display and inform the public that items had been removed'*. But this was easier said than done. The Natural History curator at the Horniman Museum pointed out the difficulties not only of removing heavy-mounted heads but also of finding a place to store them. He suggested a possible though expensive option could be temporarily to remove the horns and replace them with artificial ones. He also made the important point that horns that had been subjected to taxidermy procedures might well be dangerous to consume, since arsenic and other toxic chemicals were used in the preparation of specimens.[10]

The Natural Sciences Collections Association (NatSCA) issued a set of guidelines for museums worried about their rhino horn(s).

---

[8] BBC News *Rosie the rhino's horn stolen from Ipswich Museum* 28 July 2011. See http://www.bbc.co.uk/news/uk-england-suffolk-14326670 Accessed 26/01/2016.

[9] Esther Addley *Epidemic of UK rhino horn thefts linked to one criminal gang* The Guardian, 8 August 2011. See http://www.theguardian.com/environment/2011/aug/08/rhino-horn-the Accessed 26/01/2016.

[10] Rebecca Atkinson *Twenty thefts of rhino horn in six months* Museums Journal, 10 August 2011. See http://www.museumsassociation.org/museums-journal/news/100820 Accessed 26/01/2016.

## 19.1.1  The Rathkeale Rovers

In July 2011, Europol went public with the fact that the horns were being stolen by an organized crime gang of Irish and ethnically Irish people. They had been identified as *'significant players'* and the gang who were *'known to use intimidation and violence to achieve their ends…were more commonly associated with drug trafficking, money laundering and smuggling'*. And they were using international auction houses in the UK, France, China and the USA to sell their stolen horns.[11]

Summer 2010. John Reid, the Irish Police Force's liaison to Europol, was working at Europols' headquarters in the Netherlands, dealing with matters of intelligence. He was getting some curious requests. Although they came from different countries, Scandinavia, France and Belgium, and sometimes even from different agencies within the same country, they all concerned the same names and vehicles. And there were many other elements in common. An investigative journalist, Adam Higginbotham, pieced together the early part of this labyrinthine story.

These were some of the requests. They concerned:

- A man (there were a number of different men) driving a vehicle that had been registered in Britain. He was involved in a petty scam and on questioning claimed either to be Irish or British.
- He frequently carried a UK driving licence.
- He gave a vague, temporary address in England. Both the French and German police thought he originally came from 'Raheele' or 'Rackeel' (they wrote it down phonetically).
- All the men had similar names, often multiple identities, and *frequently every word they said was a lie'*.[12]
- They were, without exception, aggressive when questioned.
- They could not be tracked in police databases.

By November 2010, Reid, together with detectives from Dublin's Criminal Assets Bureau, a multi-agency investigative unit part of whose function is to fight organized crime in Ireland, had amassed a wealth of data on a gang called the Rathkeale Rovers, *'part of a network of clans called the Irish Travellers,*

---

[11] See Taylor (n.6).
[12] See Higginbotham (n.1).

*a nomadic and often secretive ethnic group that maintains its own distinct customs and language'.*[13]

The story now moves back in time to summer 2010 and over to the USA. A taxidermist reported John Sullivan to the USFWS, because it was a criminal offence to transport rhino products from America to Ireland without the requisite documentation. Sullivan wanted rhino trophies to decorate his castle in Ireland on an African theme.

The taxidermist, let us call him T, kept in touch with both Sullivan and the USFWS and started acting as an informant for the latter. Sullivan arranged for his 'brother' to fly to Colorado to buy 4 rhino horns from T, and in September T, accompanied by an undercover Special Agent, G, from the USFWS met two Irishmen, Richard O'Brien and Michael Hegarty, his brother-in-law. G told the two men his cousin had 4 horns and would sell them for USD 8500. O'Brien explained that he would ship them to Britain in the sea containers which he regularly used for shipping over the antique furniture they dealt in. He acknowledged he knew that the rhino horns could not legally leave the USA.

Two months later, Richard O'Brien and Michael Hegarty were arrested when they returned to America with nearly 13,000 euros in cash, the money to buy the horns. They stuck to the story that they were merely buying the horns for Sullivan to use to decorate his castle. However, as we know only too well, they had been caught in a sting operation, and the two men were found guilty. After that they become silent partners in the operations.[14]

In September that same year, yet another Irishman, Michael Slattery Junior (and his accomplices), was conducting a similar business transaction in America. This time in Texas. When he was told by the taxidermist who was offering to sell the stuffed rhino head that it could only legally be sold to a Texas resident, he devised a cunning plan. He recruited a homeless man to buy it instead. The price was USD 18,000. They made their way to New York where they sold 4 rhino horns to a Chinese buyer for USD 50,000. The horns came with provenance, endangered species bills of sale, which, of course, were fraudulent. They were sold for a second and then a third time in New York before leaving the country for China. Slattery, however, had long since left America, having boarded a plane out as soon as the first sale was completed—3 days after the arrests of Hegarty and O'Brien.

Our story now returns to the Hague, to Europol's headquarters. There, a few days after the arrest of O'Brien and Hegarty, Reid had convened a meeting

---

[13] Ibid.

[14] John Simpson *No rhino horn was safe from gang* The Times,1 March 2016.

with Europol intelligence analysts and liaison officers from a dozen of the EU's Member States to explain how the crimes they had been reporting all had links with the Rathkeale Rovers. Operation Oakleaf takes shape. The Rathkeale Rovers had just become the target of a pan-European investigation.

## 19.1.2  Operation Oakleaf

In fact it was not just a pan-European operation, although police forces from 33 European countries were actively participating. The USA, in the form of the USFWS, was also deeply involved. And not a moment too soon.

Suddenly the gang was showing up all over Europe. No castle, stately home, auction house or museum containing a taxidermy collection was safe. Burglaries were rife. We saw some of the results earlier. Two gang members were even caught trying to hire gunmen in South Africa, to poach rhinos. Other gang activities, such as stealing valuable Chinese artefacts, appeared to have been neglected. It's now all about rhino horn. Well, the returns are very high value.

But evidence was a problem. There was not enough. It only amounted to:

- The occasional telephone intercept
- Some video footage
- Occasional details from a vehicle licence plate
- Men, described by witnesses as 'Irish', visiting places later robbed, asking to see rhino exhibits, and taking pictures

A lucky break came in April 2011, when the Museum of Natural History in Coimbra, Portugal, lost 2 eighteenth-century horns. The police were able to isolate telephone traffic that occurred during the robbery. One call, from inside the building, was traced to *'an Irish cell phone registered to the wife of a senior member of one of the Rathkeale clans. The thieves apparently were calling for last-minute directions'.*[15] However, most of the people who had been caught red-handed were found merely to be workmen or others employed by the clans.

Eventually though, the tide began to turn. More and more data was collected, and gradually the pieces of the jigsaw came together. In September 2011, an Australian antique dealer and his son were arrested at Lisbon Airport. 6 rhino horns worth approximately USD 500,000 and 100,000 euros cash

---

[15] See Higginbotham (n.1).

were found in their luggage. They were about to fly to Dublin, and the detectives from the Criminal Assets Bureau knew all about them. About the fact that although based in Shanghai, the father/dealer made weekly flights to Europe to meet members of the Rathkeale Rovers in France, Spain and Portugal. They also suspected he was the original middleman in this trafficking ring, and it was him selling the stolen and smuggled horns to customers in Asia. The dealer was charged with smuggling offences.

In February 2012, the gang carried out a spectacular raid on the Offenburg Museum in Germany. While two people distracted the museum staff, two others clambered onto a display case, pulled a rhino head off the wall and smashed its horns (2) off with hammers. The stolen horns were estimated to be worth 50,000 euros.

This time arrests followed swiftly. At the end of February, three men were apprehended after the car they were travelling in was stopped in Munich. The car had been stolen in the UK, and the men bore more than a passing resemblance to artist impressions of the thieves. Investigations followed, and before the end of March, one of the suspects was arrested. All three were held pending extradition proceedings, and the police continued to pursue their inquiries into the other two men. In fact, two of them ended up in a German prison, sentenced to 2 years and 2.5 years, respectively. The third one turned out to be a youngster, so he was sentenced to 4 weeks in a juvenile detention centre.[16] Then in October, a fourth British person was arrested, this time a woman. The Spanish Police used a European arrest warrant to detain her on the island of Tenerife.

In April, the gang struck again, in its own back yard, in Ireland itself. Although the National Museum of Ireland had thought its rhino collection was stored in a safe place, unfortunately this turned out not to be the case. A year earlier, the four stuffed heads had been sent to the Museum's storage facility near the airport. This was raided by three masked men who forced their way in and then tied up the guard who was on duty. Now the criminals could turn their attention to the prize, the four rhino heads. There was a problem though. Because these heads were old, they had been stuffed with plaster, timber, horsehair and straw, and they were very heavy. One of them had a particularly formidable horn, more than 3 ft. long. Not an easy job, but the thieves got away with all their booty within the hour. (Probably 8 more horns.)

After that, things started to unravel.

---

[16] Fiona Govan *British woman arrested in Spain over rhino horns theft* Telegraph, 23 October 2012. See http://www.telegraph.co.uk/news/worldnews/europe/spain/9628603 Accessed 26/01/2012.

Michael Kealy of Rathkeale was arrested. He had been involved in an audacious robbery of a rhino horn from an auctioneer in the UK. Apprehended in Ireland using a European arrest warrant, he was successfully prosecuted and sentenced to 3 months in prison. Another resident of Rathkeale was picked up in Dublin, and then extradited to Austria where he was later released. There was a connection with the two thefts of rhino horn in Vienna.

In January 2013, there was a series of raids across eight European countries. More than 30 people who had connections with Rathkeale were arrested. Eight months later, the kingpin, Richard Kerry O'Brien, self-styled King of the Travellers, was arrested in England, in the Smithy Fen Traveller camp in Cambridgeshire. This was only one of eight raids carried out simultaneously in the UK and Ireland. It was a dawn raid, and both a battering ram and a crowbar were used to gain entry to O'Brien's caravan. He was taken in for questioning and released on bail. Four rhino horns were found.

Michael Slattery Jr. was arrested the following day, as part of the investigation Operation Crash into the trafficking of rhino horn. He had been about to board a plane to London. In November, he pleaded guilty to a charge of conspiracy to violate the Lacey Act (USA) which forbids trade in illegally obtained wildlife.[17]

In 2014, Daniel 'Turkey' O'Brien and an accomplice were sent to prison. They had beaten up an antique dealer in England while stealing a rhino horn (1) from him.[18]

At the beginning of March 2016, a trial concluded, the third of the three linked trials that had taken place in England over the last few months. In this trial, held in Birmingham Crown Court, Michael Hegarty, Richard O'Brien Jr., John 'Kerry' O' Brien and their uncle Daniel 'Turkey' O' Brien, all self-styled 'generals' of the gang because they helped plan and oversee the offences, were found guilty of conspiracy involving six raids in the UK or attempted burglaries relating to works from the Qing and Ming dynasties.

The previous day, four of Richard Sheridan's accomplices had been found guilty of hiring and overseeing burglars across the UK. Sheridan was one of Daniel 'Turkey' O'Brien's nephews. Patrick Sheridan, another relative, was extradited to America in September 2015, charged with trafficking black rhino horns.[19] Richard Sheridan himself faced a 10-year custodial sentence, but as the Americans wanted to prosecute him for smuggling, he is to be sent there. In London 2014, the police had caught him with a libation cup carved

---

[17] See Higginbotham (n.1).
[18] See Simpson (n.14).
[19] Ibid.

from rhino horn that had been traced to an auction house in America.[20] And he was not the only one, other members of the gang faced further international arrest warrants.[21]

Once the trial ended, more details became available. These included the fact that the gang's other name was the Dead Zoo Gang and that in addition to Europe, which includes both the UK and Ireland, there were links with Australia and the USA, even with South America. And despite a number of setbacks, the gang had been spectacularly successful; the total value of their raids estimated to be some £57 m/EU73 m, an almost unimaginable haul of loot, of which very little has been recovered.

They were helped by an antique dealer, Donald Chi Chong, a millionaire living in London. He too was successfully prosecuted.

It had taken Europol, the combined efforts of the police forces of 33 European countries as well as the USFWS to bring the Rathkeale Rovers to justice. And the law had been changed.

## 19.2    Changes in Legislation

**The UK:** Back in the UK, Nevin Hunter, Head of Compliance at Animal Health (part of the UK CITES Management Authority) and his team had spotted the steady rise in prices of rhino trophies that were being sold in the auction houses. For example, in August 2010, Sworders had sold three black rhino trophies. All of them dated back to the 1880s, and all were bought for considerably more money than their estimates:

- 1: weight 2.4 kg, was estimated to sell for £8250 but sold for £30,000
- 1: weight 6.8 kg, was estimated to sell for £15,675 but sold for £57,000
- 1: weight 8.15 kg, was estimated to sell for £16,775 but sold for £61,000[22]

Sales like these would be very attractive to auction houses, because, in addition to their commission, both seller's and buyer's premiums would be considerable.

Another auction house, Tennants, in North Yorkshire, which has an excellent reputation for selling natural history specimens, was also doing well. A month earlier, they had sold 8 mounted rhino horns for £455,000. This

---

[20] John Simpson *Rhino gang leader faces extradition,* The Times, 3 March 2016.
[21] Sean O'Driscoll and John Simpson *Raiders push rhino close to extinction,* The Times, 2 March 2016.
[22] Hunter 2011.

included one from 1930 prepared by Rowland Ward, which sold for a record £106,000. The Chinese were currently spending very large sums of money buying back their own antiques, and Tennant's specialist thought this was part of the same trend.[23] Curiously, research published in March 2010 showed that although buyers/consumers preferred horn from recently killed rhinos, this was not reflected in the selling price. According to intelligence, recent examples showed that antique horn was selling for well above the black market price being obtained for 'fresh' horn in Africa.[24]

By this time, everyone was waiting for a law change that would severely restrict such sales. However the specialist was gloomy. He thought it would drive sales underground.[25] Fortunately for the rhinos, Nevin and his team had not only identified the trends in the UK but, using data from 2000 to 2010, had drawn a graph. They thought there was a link between rhino horn re-export applications in the UK and poaching incidents in South Africa. The graph proved them correct.

By September 2010, new measures had been agreed and they were announced in December. The Minister for the Natural Environment and Fisheries stated his *'extreme concern'* about the rise in auction sales of rhino horns and the Government's belief that this was *'providing a financial incentive for poachers'* as well as encouraging the use of rhino horn in medicine'. Steps were to be taken 'to refuse future applications for the export of rhinoceros horn'.[26]

The Head of Animal Health's Wildlife Licensing Team explained that *'comparatively poor examples of taxidermy containing rhino horn have been selling for £40,000–£50,000 far exceeding their worth as art objects'*, and wild rhino populations needed protecting. So apart from a small number of notable exceptions, future applications for the export of rhino horn would be refused because there was evidence that *'such applications, if approved, could potentially fuel demand for rhino horn, which may lead directly to an increase in poaching'.*[27]

Animal Health then contacted all auction houses and major antique trade associations in the UK, highlighting the implications of the rule change and explaining this meant a refusal to grant most applications to export rhino

[23] Roland Arkell The Antiques Trade Gazette July/Aug 2010.

[24] Pers. comms. Esmond Martin and Lucy Vigne, March 2010.

[25] R. Arkell *Export ban expected for rhino trophies* Antique Trade Gazette, 23. Aug. 2010. See http://www.antiquestradegazette.com/news/2010/aug/23/export-ban-expected-for-rhir Accessed 27/02/2016.

[26] Animal Health *Animal Health to Prevent Export of Rhino Horn from UK*. See http://www.defra.gov.uk/animalhealth/cites/news/archived_news/200910-Prevent-Exp Accessed 28/02/2011.

[27] Ibid.

horn. All applications would be considered on a case-by-case basis, and to be granted, at least one of the following criteria had to be met:

* The potential item was of such artistic value that it exceeded its potential value on the illegal medicine market.
* The item was part of a genuine exchange of cultural goods between reputable institutions (museums).
* The item had not been sold and was a heirloom moving as part of a family relocation.
* The item was part of a bona fide research project.

The UK had changed its law. It was time for the EU to take action.

**The European Union:** The Single Market in the EU brought down trade barriers between the Member States, so European-wide action was required, and all Member States needed to comply. Animal Health made its case to the Committee on Trade in Wild Fauna and Flora, and the result was a guidance document. Prepared by the UK's CITES Scientific Authority, its annex included the graph, more information on the status of the species and the impact of poaching and illegal trade.

With the restrictions in place, suddenly UK auction houses were no longer the targets. The action switched to auction centres in other Member States.

Germany followed the UK, strengthening its own legislation in October 2010. After that, other Member States started getting applications for re-export of rhino horn or requests for information about how such applications would be handled. The obvious conclusion was that there were people out there trying to find a new 'easy way' of getting rhino horn out of Europe, now that the UK and Germany had tightened up their procedures.

The Committee was already aware of the epidemic of rhino horn thefts that had taken place across Europe, that Europol had recorded 50 thefts and 10 attempted thefts in 13 Member States, that 60 specimens had been stolen and that probably far more thefts had taken place but, for various reasons, these had not been recorded.[28] They were also aware of the fact that most of these had probably been stolen by a single organized crime group, the Rathkeale Rovers, a third of them certainly had been. No other group had been identified with the thefts.

---

[28] Guidance document: *Export, re-export and intra-Union trade of rhinoceros horns* March 2012.

They knew that the gang tended to sell their ill-gotten gains to intermediaries, who, in turn, applied for certificates and permits to trade the horns. This would account for the rise in applications. That the horns were destined for the Vietnamese and Chinese market and that this would further fuel demand for rhino horn in Southeast Asia.

It was this knowledge, together with the realization that *'such large demand for high-valued products represents a lucrative market which is very attractive to poachers and illegal traders'* could provide a very tempting driver to obtain even more horn, so placing even more rhinos in peril of their lives, that prompted the EU to take action. One expected/hoped-for result was that it would cause such difficulties to the gang they would end up taking *'desperate measures... thus leaving them more susceptible to law enforcement action'.*[29]

Because so many Member States had been targeted, it was obvious that a common approach was needed, and it was just the sort of situation to which the precautionary principle should be applied, so another guidance document was produced. It applied to mainland China, but not to Hong Kong, Macau or Taiwan, because their legislation authorizing trade in rhino horns was compliant with CITES rules.[30]

In February 2011, Defra/Animal Health (the UK Management Authority) released the new EU rules. They came into effect immediately and made it illegal to sell mounted but otherwise unaltered rhino horns in the UK.

Before that, under what is known as the '*antiques*' derogation, rhino horn(s) could be sold provided they fell into the category of '*worked items*'. These were horn(s) that had been prepared and then acquired in this condition before 3 March 1947 and not altered after that, and they included '*mounted rhino horns in their natural state*'.

The EU guidance document made it illegal to sell such a specimen unless it had been sufficiently and obviously altered to qualify under the *antiques* derogation. Mounted rhino horns had not so became *unworked* specimens, and since these were already banned from sale in the UK, it followed that it would no longer be possible legally to sell mounted rhino horn.

It was also made quite clear that, under the export restrictions that came into force in the UK in September 2010, CITES re-export certificates were unlikely to be granted in respect of exporting such items.[31]

---

[29] Ibid.

[30] Guidance on: *interpretation of EU rules on export and re-export of rhino horns: applications for permits under Article 5 of Council Regulation (EC) No 338/97.*

[31] Animal Health *New rules make it illegal to trade mounted rhino horns in the UK* 18 February 2011. See http://www.defra.gov.uk/animalhealth/news/180211-new-rules-rhino-horns-in-the-uk Accessed 28/02/2011.

# 20

# Organized Crime

Illegal trading in wildlife is highly lucrative, and the Rathkeale Rovers have shown just how difficult it can be to fight such gangs of organized criminals who do so much damage, so we must now examine the role they are playing in the destruction of our animals.

## 20.1 The United Nations Convention Against Transnational Organized Crime and the Protocols Thereto

This treaty was a major step forward in the battle against organized crime. It demonstrated:

- *'The political will to answer a global challenge with a global response'.*
- *'If crime crosses borders, so must law enforcement'.*
- *'If the rule of law is undermined not only in one country, but in many, then those who defend it cannot limit themselves to purely national means'.*[1]

The fact that it came into force in 2003, just 3 years after it had been adopted by the UN General Assembly, indicated that countries were demanding action. It sets out a number of measures including:

---

[1] United Nations Convention Against Transnational Organized Crime and the Protocols Thereof, *Foreward*, United Nations, New York, 2004. See https://www.unodc.org/documents/middleeastand-northafrica/organized-crime/UNI Accessed 9/03/2016.

© Springer Nature Switzerland AG 2019
B. Martin, *Survival or Extinction?*, https://doi.org/10.1007/978-3-030-13293-4_20

- The creation of domestic criminal offences, such as participation in an organized criminal group, money laundering, corruption and obstruction of justice
- The adoption of new and sweeping frameworks for extradition, mutual legal assistance and law enforcement cooperation
- The promotion of training and technical assistance for building or upgrading the necessary capacity of national authorities

The United Nations Office on Drugs and Crime (UNODC) operates with other authorities when dealing with matters concerning wildlife crime (designated as wildlife and forest crime (WLFC)), and in 2013, the CITES Secretary General was also asked to cooperate about possible links with poaching, illegal trading in wildlife and national security in some African countries.[2]

The Convention was drafted to ensure that it did not need to be amended every time a new type of crime emerged, so there is no precise definition of *transnational organized crime*. However an *organized criminal group* is defined as *a structured group of three or more persons, existing for a period of time and acting in concert with the aim of committing one or more serious crimes or offences established in accordance with this Convention, in order to obtain, directly or indirectly, a financial or other material benefit*. There is no doubt the Rathkeale Rovers fit this description.

By not providing a strict definition of transnational organized crime, it becomes possible to widen its scope to the extent that it includes *almost all profit-motivated serious criminal activities with international implications ... considers the global complexity of the issues involved and paves the way for international cooperation on the widest possible range of common concerns*.[3]

Wildlife (and forest) crime falls within the category of *serious crime*, which again has benefits. Not only is it punishable by at least 4 years in prison, to act as a deterrent, but greater efforts are made to bring the perpetrators to justice. Methods such as electronic surveillance and telephone taps can be employed. There is increased international cooperation between law enforcement agencies and with the judiciaries. Best of all though, is the *higher rate of convictions to motivate frontline law enforcers to investigate and pursue cases of wildlife and forest crime*.

Two important definitions are *wildlife and forest* which includes animals, birds and fish, timber and non-timber forest products; and *wildlife and forest*

---

[2] Report of the Secretary General on the activities of the United Nation's Regional Office for Central Africa and on the Lord's Resistance Army – affected areas. (S/2013/297).

[3] Rios 3013.

*crime'* which includes the taking, trading (supplying, selling or trafficking), importing, exporting, processing, possessing, obtaining and consumption of wild fauna and flora, including timber and other forest products, in contravention of national or international law. In other words, the illegal exploitation of the world's wild animals and plants.[4]

The gangs use the same routes and techniques as the smugglers of other illegal commodities. Gaps in national law enforcement and criminal justice systems are also exploited, particularly in developing countries where the governments themselves *'often lack the capacity to regulate the exploitation of their natural assets'*. And with very damaging results. *'Rather than promoting economic progress, poorly managed natural wealth can lead to bad governance, corruption or even violent conflict'.*[5]

Not only that some of *'the billions of dollars generated by this illegal business'* are used to finance terrorism, they contribute to instability and are also *'closely interlinked with money laundering, corruption, murder and extreme violence'.* They pose threats not just to biodiversity and endangered species but to the livelihood of people and have a severe impact on national security and social and economic development.[6]

A powerful summing up by a United Nations Office that is not given to exaggeration, making it obvious why wildlife crime falls within the jurisdiction of this Convention. But how to fight such powerful forces? One way is to set up a programme.

### 20.1.1   The UNODC Global Programme for Combating Wildlife and Forest Crime (GP)

Due to last 4 years, it aims to link regional efforts in a global system and to enhance capacity building and wildlife law enforcement networks at regional and subregional levels. The Programme acts both with and for the wildlife law enforcement community to ensure that wildlife and related crimes are treated as serious transnational organized crimes.

Bringing the Rathkeale Rovers to justice has already shown us that fighting such crime needs help, and, in essence, UNODC is there to provide it, particularly through the provision of specific technical assistance activities such as the Global Programme (GP) and the Sustainable Livelihoods Unit. They are

---

[4] United Nations Office on Drugs and Crime (UNODC) Everywhere, *Wildlife and Forest Crime Overview 2016*. See https://www.unodc.org/unodc/en/wildlife-and-forest-crime/overview Accessed 01/03/2016.
[5] Ibid.
[6] Ibid.

involved in capacity building activities in countries from southeast Asia, south Asia, east Africa and America.[7] The GP can provide equipment, legislative help and training to police, customs, border officials, forestry/wildlife officials and others, while the Sustainable Livelihoods Unit can provide strategic advice and assistance.

Once again, it is all about working together. So the GP/SLU team continues to work on developing partnerships both with other UN agencies and international organizations that are concerned with wildlife crime such as INTERPOL, the International Consortium on Combating Wildlife Crime, the World Bank, World Wide Fund for Nature and other NGOs. Also with widening their geographical range.[8]

Even with this relatively new addition to the fight against wildlife crime, it is still going to be hard work to bring these organized criminal gangs to justice. They tend to have a great capacity for survival as our next case study shows.

### Case Study 7: The Singapore Seizure

The Singapore Seizure of over 6 tonnes of elephant ivory took place in 2002 and led to the detection of a major ivory smuggling syndicate, with a network that, since the mid-1990s, had shipped enormous quantities of ivory from Southern Africa to China and Japan. Indeed, some members of the syndicate had been involved in ivory trading since the 1980s when it was still legal. But this trade had led eventually to the catastrophic decline in elephant populations, a decline Richard Leakey had been determined to end when, in 1989, he burnt Kenya's considerable ivory stockpile on a bonfire, the action that led directly to the complete ban on international sales in elephant ivory in 1990.

Once again we are indebted to the EIA who, working with investigators from the Zambia Wildlife Authority (ZAWA), Malawi's Anti-Corruption Bureau (ACB), the Lusaka Agreement Task Force and Singapore's Agri-Food and Veterinary Authority (AVA), carried out this undercover investigation. They followed the trail from Zambia, discovering the Lilongwe pipeline on their way. This is what they finally managed to unravel.

In the target factory in Lilongwe, they found a paper trail. Until he died in 1998, MacDonald John Gwedeza Zulu had run the family business, Allena Curios, and he was an enthusiastic record keeper. He kept the names of both

---

[7] United Nations Office on Drugs and Crime Global Programme 2016. See https://www.unodc.org/unodc/en/wildlife-and-forest-crime/global-p Accessed 01/03/2016.

[8] UNODC Partners 2016. See https://www.unodc.org/unodc/en/wildlife-and-forest-crime/partners.html Accessed 01/0 3/2016.

sellers and buyers, the amounts of ivory he had received, and their value (calculated on their weight), everything linked together.

However, it was the receipt books from 1994 to 1996 that, metaphorically speaking, yielded gold. They linked Sindikani Banda's illegal ivory (he featured in our poaching chapter) with buyers from southeast Asia. Certain names kept reappearing, Chow, Pang and 'Peter' Onn. Indeed, Pang and 'Peter', the main buyer, spent USD185,000 on ivory between February and September 1996. This represented almost 10 tonnes of ivory, a staggering amount.

The investigating team began accruing evidence. Telephone and fax contact numbers for Peter Onn in Singapore, Japan and Hong Kong SAR. They had been scribbled in a notebook. An actual receipt made out to Banda in 1996, which showed he had been paid USD7500 for 500 kg of ivory. Records in notebooks of various sums of money in US dollars, paid out to other members of the poaching gang.

Now we must return to Chipata, to the hub of this particular gang, to visit Banda's house and other places. Searches and enquiries revealed even more ivory being collected, that a container in Lilongwe was being prepared and that its destination was southeast Asia. A staggering amount of ivory was involved, about 13 tonnes of it. So 'a ZAWA investigator returned to Lilongwe to hunt for the rumoured container and its cache of poached ivory'.[9]

We already know that smugglers frequently disguise or hide their illegal cargoes, so the discovery of a currency declaration for the export of stone sculptures had to be deeply suspicious. Furthermore, it had been made out by someone who lived in Hong Kong. It was a consignment of six wooden boxes, from Seng Luck Trading Company (located in the same district as the Gwedeza family business) to a company in Singapore. However, the Seng Luck Trading Company did not exist, its contact details were false, and the Customs Declaration Form for the container listed Allena Transport and Trading as the exporter. They had found the missing/hidden ivory.

Contact was immediately made with Singapore. Docking was due in the next couple of days. No time was wasted. The Singapore Seizure followed.

The Malawian ACB could now look for further evidence. The first thing they did was search the premises of Gwedeza company, where they found documents showing a succession of suspicious shipments that had taken place since the middle of the 1990s. Records which linked to So Tat Wing, who lived in Hong Kong, with the currency declaration for the export of stone

---

[9] Jo Hastie, Julian Newman and Mary Rice *Back in Business: Elephant Poaching and the Ivory Black Markets and Asia*, EIA, 2002, p. 8.

sculptures. Geoffrey Kulupajili and 'Imoto' from Asia had organized the shipment's paperwork. Six boxes had been collected from a private address by a large truck that had been hired by Kulupajili. These were delivered to a container in the Manica yard. An 'Asian' man and his common-law wife, Kulupajili's sister, occupied the private address. When the ACB searched it, they found tools for carving and packaging, but the 'Asian' had fled.

The documents indicated that over an 8-year period, *'tens of tonnes'* of ivory had passed that same way through the Allena operation in Kawale.

Gift Gwedeza and Geoffrey Kulupajili were arrested over their involvement in the despatch of the container that had been seized in Singapore. It was destined for Delight Harvest (Singapore) Ltd., an electronics manufacturer and general trading company, which had already received four other consignments of stone sculptures or wooden curios from Allena Transport and Trading.

But Singapore was merely being used. There were loopholes in the regulations governing the lawful trade in ivory that the gang took advantage of. This made it an important staging post. The cargo, as on at least 15 other occasions, was merely passing through. Its final destination was Japan, which, in those days, was still a major consumer of ivory. And the powerhouse of this transnational organized crime gang was a group of ivory traders based in Hong Kong SAR.

The seized consignment's bill of lading had been sent by courier, by Kulupajili to Toh Yew Lye, who acted on behalf of 'Peter Wang', the owner of Delight Harvest. He admitted he had arranged for several containers from Lilongwe to be transhipped to Japan, but he thought they contained sculptures. However he was linked to Toh Brothers Timber, a company that had itself received six shipments from Lilongwe.

Delight Harvest also had some interesting linkage. Apparently its name had been used without their knowledge, and the directors claimed to know nothing about the seized ivory. This was despite the fact that one of them, Chan Chun Hung, was a founding director of Kyomi Handicraft and Trading Pte. Ltd., a Singapore firm that had been set up by a group of ivory craftsmen from Hong Kong SAR. The firm stated its business was *'ivory manufacturing and carving'*, but it too had received a consignment (of sculptures) from Allena.

Allena operated under yet another front in Lilongwe, this time as Sheng Luck, which sent nine shipments described as sawn timber, but suspected to be ivory, to Singapore. We've already seen that one of the names that kept reappearing in the receipt books found in the raid on Allena Curios was 'Peter' Onn. Shipping forms for some of the consignments of this sawn timber were sent by fax to Hong Kong and Japan. The numbers corresponded to 'Peter' Onn's contact details.

'Peter' Onn, one of a number of pseudonyms, appeared to be the kingpin, the mastermind behind the ivory smuggling syndicate. He had travelled to Lilongwe on a number of occasions to facilitate the shipment of ivory. And although he operated out of Hong Kong SAR, he made full use of the loopholes in Singapore's legislation. Nor was he the only one. A director of Delight Harvest and three founding directors of Kyomi Handicrafts and Trading Pte. Ltd. and Fung Ivory Manufacturing Pte. Ltd. also took advantage.

The evidence showed quite unambiguously that these three companies were all interconnected. It also showed that many of the business men connected with these companies had connections with the Poon Family, whose ivory empire stretched *from Europe to Africa and the Middle East to the far East* during the 1980s. Some of the companies, like Fung Ivory Manufacturing Pte. Ltd., were set up by some gang members and then run by other gang members. A powerful and destructive empire.

In 1998, for some reason the operation changed, and the ivory was smuggled not to Singapore but to southern China. Three containers of it.

In 1999, the first of the one-off sales of elephant ivory was held. All the ivory had to be sold to a single, reliable country. That country was Japan and they bought 50 tonnes of ivory. Yet the investigation carried out by the EIA showed some of the illegal ivory was clearly destined for Japan. Here are some of the evidence:

- The shipment included an enormous number of hankos, over 40,000. Hankos are traditional Japanese name seals.
- Some of the tusks were marked with the name of the Japanese port 'Yokohama'.
- Investigations in Singapore showed there had been a number of tranship-ments of ivory to Japan.

It left no room for doubt. The EIA concluded *'it is apparent that the syndicate has successfully moved tens of tonnes of ivory into the Japanese market, supposedly the most strictly regulated in the world'.*[10]

---

[10] Ibid, p. 12.

## 20.2  Organized Crime Gangs in East Africa

This is where organized crime gangs and corruption have been described as representing *'a true collusion of evils'*.[11] The problem, however, is to progress beyond intelligent suspicion to hard evidence. In Uganda, for example, there is suspicion but insufficient evidence to suggest the presence of a very powerful, transnational criminal syndicate controlling the collection, transportation and delivery of ivory and other high-end commodities.[12]

In September 2013, UNODC published its report *'Transnational Organized Crime in Eastern Africa: A Threat Assessment'* which highlighted the impact organized crime and corruption was having on East Africa. Kenya and Tanzania were two countries particularly in the frame, not only because they had important ports like Mombasa and Dar es Salaam where goods came and went from markets all over the world but also because their governments were weak and had a *'limited capacity to deter cross-border criminality'* and there was great deal of poverty, all possible contributory factors to corruption.[13]

The region was internationally recognized as a major hub for various kinds of trafficking, and corruption was rife. Furthermore it was *'widely accepted that both OCGs (Organized Crime Gangs) and corrupt officials are the main facilitators of ivory trafficking through Kenya and Tanzania to East Asian criminal-syndicate buyers'*.[14] In fact, it is hard to overestimate the important role they played, for they were involved at both low and high levels all along the chain. They organized the poaching gangs, consolidated the ivory the gangs sent to them and then shipped it out in containers. Ivory trafficking on an industrial scale.[15]

And CITES ETIS data and records of seizures clearly demonstrated the unpalatable truth in the number of large shipments (over 500 kg) seized since the year 2000. In fact, ivory for shipments of this size cannot be collected together without the involvement of organized crime gangs and an element of corruption. It is interesting to note that from 2012 to 2014, 61% of all the ivory confiscated worldwide came from large shipments.[16] And in 2013, about

---

[11] Tom Maguire and Cathy Haenlein *An Illusion of Complicity: Terrorism and the Illegal Ivory Trade in East Africa* Royal United Services Institute for Defence and Security Studies (RUSI), Occasional Paper, September 2015, p. 33.

[12] Rossi, A. (2018). *Uganda Wildlife Trafficking Assessment.* TRAFFIC International, Cambridge, United Kingdom, p. 90.

[13] UNODC *Transnational Organized Crime in Eastern Africa: A threat Assessment* 2013, p. 33.

[14] See Maguire and Haenlein (n.379), p. 34.

[15] Ibid.

[16] Ibid.

80% of all large seizures related to ports in Kenya, Tanzania and Uganda. Mombasa had a serious problem.

Another recurring theme in our story is the complex smuggling webs used to move both elephant ivory and rhino horn from **A** to **B**. We can see them again here. Tusks from Tanzania (**A**) were going to Kenya (**X**) via Uganda (**X**), although Entebbe Airport in Uganda (**X**) and the Airport at Addis Ababa, Ethiopia (**X**), seemed to be growing in importance in the direct transfer of ivory to east Asia (**B**) or to the Gulf of Guinea (**X**). We already know that shipments by air are small, but ivory from a series of small shipments to Guinea could be gradually collected until it filled a sizeable container, which could then be shipped out of one of Guinea's less regulated ports.[17]

This trade is sophisticated, and it will be hard to end. But despite the continuing gaps in knowledge and distraction of possible links with al-Shabaab (which we will be examining shortly), there are grounds for hope. At least the problem has been identified, its existence is supported by facts, and the international community is taking action. An international community comprised not just of nation states but also of enforcement bodies such as the Lusaka Agreement Task Force, INTERPOL, UNODC and the World Customs Organization. The USA has become actively involved. Recognizing the vital role local communities must play if these twin evils of organized crime and corruption are ever to be stamped out, it has 'combined forces' with Kenya's Northern Rangelands Trust (more of them later) and is supporting their development of a security community and an adjunct intelligence body. This is welcome news.

## 20.3   Organized Crime and Rhino Horn: South Africa to Viet Nam

Rhinos, particularly those in South Africa, have been hammered by poachers in order to supply the lucrative trade in their horns, a trade and market that focuses on Viet Nam. We start by looking at the 'trade chain'.

**The Five Levels of the Trade Chain:** These have been identified by South Africa's National Wildlife Crime Reaction Unit.[18]

---

[17] Ibid.

[18] These levels are similar to the poaching and trafficking chain for elephant ivory recently discussed at the Uganda Wildlife Trafficking Stakeholder Workshop (2016), except that an additional layer has been added. It reads: 1. Poacher; 2. Local middleman; 3. Transporter; 4. Urban middleman; 5. The exporters; 6. Kingpin. See Rossi (n.12), p. 80.

**Level 1:** Encompasses the illegal killing of rhinos, which takes place on both private land and the protected areas. The animals are poached by individuals or groups. The horns are removed and passed on up the chain.

**Level 2:** Poaching groups deal with buyers, and a national level courier receives the horn(s) and takes them to a middleman collection point.

**Level 3:** The middlemen, normally South Africans of Asian origin, are internal dealers. This time couriers, buyers and exporters are all involved.

**Level 4:** Now the trade chain becomes international. Once again the personnel involved are couriers, buyers and exporters.

**Level 5:** The top. Again the business here is international, this time involving buyers and consumers.

**The Rise and Rise of the Crime Syndicates:** The gangs involved in rhino crime are no different from those engaged in trafficking elephant ivory. They all operate multinationally and indulge in numerous kinds of serious international criminal activities. We shall see below that some gang members were South African, while others had been granted permanent residency status in South Africa. A few were even members of the Vietnamese diplomatic community; the evidence of their complicity was unambiguous.

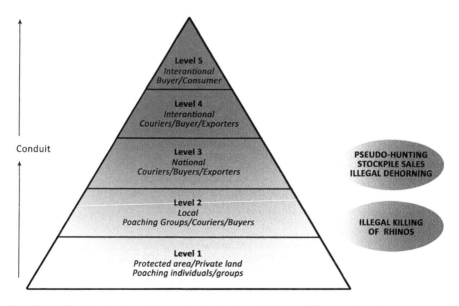

**Fig. 20.1** The five levels of the trade chain. Copyright T. Milliken and J. Shaw, TRAFFIC

**The Poaching Gangs:** In 2006, law enforcement officials realized they might have a major problem on their hands. There had always been a small amount of poaching particularly in Kruger National Park, but these poachers were local, opportunistic, merely trying to supplement their meagre incomes. Suddenly, the poachers seemed to become more efficient.[19] Their attacks were well organized and directed at specific targets. Perhaps they had cell phones. Cell phones are very handy kit for arranging the transport required in poaching operations.

By 2012, money was no object. The poachers could afford the best equipment. Numbers were no object either. The gangs wanted to secure their prey, and if it took as many as 16 men to carry out the task, they were there. They had been observed as *'often moving in para-military formations to provide protection against national security patrols',*[20] very dangerous for the men protecting the rhinos. And it gave the poachers an advantage. With so many foot soldiers, some of them could provide protection for the assassins carrying out the killing.[21]

These poachers were highly organized. They infiltrated themselves into local communities if the target rhinos were in private ranches or provincial parks. That way they learnt the location of the animals, saw how they were protected and could work out a minimum of two escape routes. They could spy out the land, plot and plan all at the same time. For the rhinos it was deadly. The poachers worked in the communities as well, for there were reports indicating the use of carpets and furniture to smuggle out horns.[22]

They were also professional, *'highly experienced in bush lore and ... known to use full moon periods when it is easier to see at night'.*[23] And they were a mix of people, as we can see in the following demographic analysis of 55 poachers arrested in Kruger National Park:

- All of them were male.
- Ninety-six % were black.
- Sixty % were south African citizens and 40% citizens of Mozambique.
- Forty-one % were aged between 20 and 29 years, another were 41% between 30 and 39 years, and only 18% were over 40 years.[24]

---

[19] R. Taylor, in litt. to TRAFFIC 2009 in Milliken, T. and Shaw, J. (2012) *The South Africa – Viet Nam Rhino Trade Nexus: A deadly combination of institutional lapses, corrupt wildlife industry professionals and Asian Crime syndicates* TRAFFIC, Johannesburg, South Africa.

[20] Swart 2012.

[21] See Milliken and Shaw (n.19).

[22] Swart 2012.

[23] See Milliken and Shaw (n.19), p. 79.

[24] Eloff 2012.

Nor were they confined by boundaries. They went where they could kill rhinos. And they got cross-boundary operational support from Mozambique and Zimbabwe.

The 'Pseudo-Hunts': At least five criminal syndicates had been identified by South African law enforcement officers as early as 2007. They, with the help of some corrupt local professional hunters, had organized and carried out some 203 separate rhino hunts in South Africa between 2005 and 2007: the number of hunts increasing exponentially each year with, seemingly, all the horns ending up in Viet Nam. And this was despite the best efforts of the South African government. Every attempt they made ended in failure. Finally, in April 2012, the Minister for Water and Environmental Affairs announced that the issuing of hunting licences to Vietnamese nationals had been suspended.[25]

So what were these *pseudo-hunts*? They were, as the name suggests, false hunts. Hunts, whose true purpose was to acquire rhino horns, not demonstrate hunting skills. Most of those involved were Vietnamese. But you can see what the government was up against. When, in desperation, it decided to restrict would-be hunters to only one hunt a year, the syndicates merely provided their Thai sex workers with another activity. Rhino hunting! The sex workers were issued permits despite the fact that they couldn't hunt. None of them could hunt. They were merely there to take away the horns of the dead rhinos. The rhinos allegedly killed by them.

**The Involvement of South African Officials, Wildlife Industry Professionals and the National Middleman Dealers:** These were a diverse set of individuals, some hooked-in by the pervasive effect of corruption.

**National and provincial conservation officials:** It is very sad when this happens, but according to Milliken and Shaw, '*serious concerns have … repeatedly been raised about irregular conduct by both some national and provincial government officials all the way up to senior levels, including management staff*'.[26]

Two examples will suffice. Both occurred in 2012. The first took place in Kruger National Park where 4 SANParks (South African National Parks) rangers were arrested in connection with rhino poaching activities.[27] The second example concerned the Reserve Manager for a Nature Reserve in

---

[25] Hubschle 2010; Swart 2012.
[26] See Milliken and Shaw (n.19), p. 79.
[27] Anon 2012b. See ibid.

Limpopo. 5 rhinos were killed. It was alleged that he and 3 Mozambique nationals were involved. He committed suicide.[28]

The fact is that although poachers will often turn gamekeepers, old habits can die hard, and poverty is frequently a contributory factor.

**Wildlife industry professionals:** By 2007, poaching was becoming a serious matter for private sector game ranches. There was a new breed of poacher, the white professional game industry insider, and this seemed to be the result of '*pure greed*'. '*The prospect of windfall profits apparently led some professional hunters and wildlife industry personnel (including landowners, wildlife veterinarians and game capture professionals) to begin illegally killing rhinos on poorly policed properties*'.[29] Most of them were Afrikaans.

This was a shocking development. These poachers could not plead hardship. And they brought invaluable skills into this deadly game. Pseudo-hunts and the trade in 'loose' horns (horns from rhinos who had died naturally) brought them into contact with the crime gangs. These corrupt individuals had almost certainly been responsible for killing their neighbours' rhinos as well, perhaps sometimes in retribution for their refusal to sell their loose horns to them.[30] It was a very disturbing and sinister state of affairs.

Returning to the trade chain, we are ready to move to level 3, while recognizing that there will always be a certain amount of overlap between the different levels, such as the syndicate boss from levels 1 and 2 acting as a buyer for level 3.

**The national middleman dealers:** These were normally South African business men of Asian origin. They employed 'runners' to pick up/procure the poached horns or obtain the 'loose' horns, which were then sold on to the Asian syndicates who disposed of them to end-use buyers.

There were also people involved in activities relating to illegally obtaining rhino horn, but who did not kill the rhinos. People involved in:

- '*Pseudo-hunts*' such as game farmers and professions hunters
- Rhino horn thefts
- Violations of TOPS permits, for example, illegally dehorning rhinos
- Illegal rhino horn sales and possession

---

[28] Gill 2012. See ibid.
[29] Ibid, p. 80.
[30] Ibid, p. 201.

Fortunately for the rhinos, rhino crime investigators considered that relatively few people in the private sector were involved.[31]

## 20.4   Africa-Based Asian Syndicate Dealers

Here, finally, in levels 4 and 5 are the 'kingpins', some of whom were now living permanently in South Africa. Curiously, some of the Vietnamese nationals granted permanent residency status were known to be involved in rhino horn crime.

We can see their position in the trade chain:

- They had direct involvement in organizing and financing the 'pseudo-hunts'.
- They were linked to the middlemen.
- To the shadowy trade channels which the illegal horn passed through on its way to the end-use markets.

They were also responsible for organizing the mules and couriers who transported the horns out of South Africa, mainly by air. Several Vietnamese students registered to study in South Africa were arrested for rhino horn trafficking at international airports on their way home, as were Vietnamese travellers at Maputo International Airport in Mozambique.

**Vietnamese Involvement:** Viet Nam was, and still is, the most important end market for rhino horn; therefore it should come as no surprise that even in the early years, up to 2006:

- *'Law enforcement officers believe that at least half of the 24 poached rhinos during 2006 had an overt Vietnamese connection.*
- *The majority of rhino horn cases occurring in South Africa during 2004 to 2006 that link to foreign nationals, involved suspects acting on behalf of Vietnamese citizens'.[32]*

Even more disturbing were the syndicate's quite blatant links with the diplomatic community. These included:

- *'Commercial alliances with diplomatic personnel*

---

[31] P. Jones pers. Comm. 2012. See ibid.
[32] R. Taylor in litt. to TRAFFIC 2009. See ibid.

• *Access to vehicles bearing diplomatic plates for the risk-free movement of rhino horn within the country; and, possibly, the use of a diplomatic pouch to move contraband to home markets'.*[33]

In fact, the Viet Nam connection started in 2003, with nontraditional hunters and organized crime syndicates soon becoming involved. In Asia, the trade for rhino horn revived and was partly supplied by horn from hunting trophies, particularly from the 'pseudo-hunts', trophies which the Vietnamese acquired by intentionally exploiting loopholes in the South African legislation. Poaching and 'loose' horn made up the remainder.

Deeply concerned, in 2009, the South African government placed a temporary moratorium, which was later challenged, on domestic sales of rhino horns.

## 20.5   Smuggling Out the Rhino Horn

Most horn leaves by air. Luckily for the smugglers, the largest airport in the whole of Southern Africa, O.R. Tambo International Airport, is conveniently situated in South Africa. It is big, busy and carries a heavy volume of traffic, all ideal factors for a would-be rhino horn trafficker. Fortunately the personnel are also aware of this, so are all geared up to prevent the smugglers succeeding.

The airport has a permanent enforcement unit of Special Investigations Officers, Environmental Management Inspectors (EMIs), who mainly inspect the cargo area and international mailing sections of the airport.[34] They also carry out random inspections of passengers and passenger luggage. They are responsible for CITES and for South Africa's domestic legislation implementing CITES.

And it is effective. In 2008, 4 white rhino horns were discovered in a suitcase due to be exported to either Viet Nam or China. They had been wrapped in newspaper which was sealed with tape. Customs officers found the horns. EMIs then carried out an investigation which resulted in the arrest of 2 Vietnamese nationals, and although one of them was released, the other was successfully prosecuted. The second one was eventually linked to an Asian organized crime syndicate that was operating in South Africa.[35]

---

[33] Ibid, p. 82.
[34] Du Toit and Craigie 2008 in ibid.
[35] Ibid.

In another case, this time involving a Thai national and taking place over a 2-year period between 2007 and 2008, a total of 300 kg of horn was smuggled out of the country. The perpetrator confessed. He had taken horn out on 15 separate occasions, normally using Thai International Airlines. The horns had been packed in his own personal check-in luggage and then, reportedly, routed to China by way of Thailand and Laos PDR.[36]

In February 2008, South Africa passed a legislation making it illegal to transport rhino horns as a trophy or personal effect within hand luggage. Needless to say, this did not remove the problem. Seizures are still made, and it is impossible to know how much horn still gets through because other airports are used as well. One thing may help though. Vietnamese buyers prefer purchasing whole horns and these are easier to detect.[37]

---

[36] R. Taylor in litt. to TRAFFIC, 2009. See ibid.
[37] Ibid.

# 21

# Rebel Militia and 'Blood Ivory'

Even more deadly than the involvement of organized crime in illegal wildlife trafficking are the far more sinister militia groups, some of whom have turned to ivory as the new 'blood diamonds' to fund their wars against humanity.

In May 2013, Ban Ki-moon, the Secretary General of the United Nations, delivered his report linking the Lord's Resistance Army with Central Africa and the illegal trade in elephant ivory.[1] It identified four countries whose elephants were at greatest risk of being poached:

- Cameroon
- Central African Republic
- Chad
- Gabon

In other words, countries where the critically endangered forest elephants live.

These elephant populations were already struggling, so it was very serious news indeed when they were identified as the targets of rebel militia who saw them as a highly lucrative source of funding. Nor was the Lord's Resistance Army the only group involved. Some of the weapons used for this poaching were believed to have come from Libya and were of the type to cause horrendous damage to the poor animals. The herds were being decimated, and Cameroon was just one of the countries that decided to bring in their national army to help hunt down poachers.

---

[1] Report of the Secretary General on the activities of the United Nations Regional Office for Central Africa and on the Lord's Resistance Army - affected areas (S/2013/297).

© Springer Nature Switzerland AG 2019
B. Martin, *Survival or Extinction?*, https://doi.org/10.1007/978-3-030-13293-4_21

John Scanlon, CITES Secretary General, also spoke out, for he was worried about rogue elements of the military.

The result was a fringe meeting held later that year at CoP16, a roundtable on transnational organized wildlife and forest crime and its impact on species, ecosystems, livelihoods, economies and national and regional security. It was organized by the International Consortium on Combating Wildlife Crime (ICCWC) with Ministers and high-level representatives attending.

As January 2014 drew to a close, the United Nations Security Council passed two Resolutions, the first applied to CAR and the second to DRC. Both related to the protection of elephants and were aimed specifically at armed groups who were financing their operations by trafficking in wildlife, namely, illegal ivory. Sanctions were to be applied. Individuals involved were to have their assets frozen and themselves be subject to travel bans. And the pre-existing arms embargo on various groups in DRC was to be renewed.[2]

## 21.1   A Historic UN Resolution

Almost 18 months later, in July 2015, the UN General Assembly adopted its historic Resolution: 'Tackling the Illicit Trafficking in Wildlife'. It had taken 3 years for the diplomats to achieve, and it marked a turning point. Finally the world had woken up to the plight of wildlife.

The Resolution committed every country to increase their collective efforts to end the global poaching crisis and to tackle the huge amount of illegal trade. The Director General of WWF International, Marco Lambertini, described it as a landmark Resolution, marking *a new phase in the fight against wildlife crime, which is threatening countless species with extinction while jeopardizing national security and sustainable development* and adding that it *proves that ending wildlife crime is no longer just an "environmental" issue and not just limited to a few countries: it has become a priority for every nation*.[3]

---

[2] BBC News *UN sanctions on DR Congo militias financed by poaching* 31 January 2014. See http://www.bbc.co.uk/news/world-africa-25978768 Accessed 22/03/2016.

[3] WWF *World unites to fight wildlife crime as UN adopts historic resolution* 30 July 2015. See http://wwf.panda.org/wwf_news/?250072/World-unites-to-fight-wildlife-crime-as-UN. Accessed 04/08/2015.

## 21.2 RUSI's Research into Possible Links Between Illegal Ivory Trafficking and Al-Shabaab

We have seen how difficult it is to disentangle the workings of organized crime, but this is straightforward compared to penetrating the dark and shadowy world of the armed militias. Here, rumours abound, and it is hard, without adequate data, to distinguish fact from fiction. So it is with relief that we can turn to an Occasional Paper published by RUSI, the Royal United Services Institute, and which was produced without anyone being put in danger.

The possibility, put forward in a now widely discredited report by the Elephant Action League, that one chilling terrorist group, the notorious al-Shabaab, had *'received up to 40% of its running costs through the illegal ivory trade alone'*, was investigated by two researchers, Tom Maguire and Cathy Haenlein, whose results were published in 2015, in the Paper *'An Illusion of Complicity, Terrorism and the Illegal Ivory Trade in East Africa'*. This is a thought-provoking and compelling document. They did not find an al-Shabaab-ivory nexus.

But if such terror groups are almost impenetrable, why should we believe these results? Because RUSI is *'the UK's leading independent think-tank on international defence and security. Its mission is to be an analytical, research-led global forum for informing, influencing and enhancing public debate on a safer and more stable world'*, and this widened the range of sources.

We have seen that countries like Kenya, who play a key role in our story, have some fairly intractable problems to overcome, so the presence of a terror group as well could make those problems seem insurmountable. These countries need to know what they are facing. This is why.

*National and transnational organised criminal networks are the most serious threat to East Africa's wildlife. Furthermore, many are concerned by the prospect that terrorists in the region could also be participating in this trade for financial gain – enhancing their ability to use violence to terrorise civilians and governments alike. This dangerous fusion presents us with an even greater potential threat. But it is important that we focus on the evidence available to us. A false assumption or mistaken conclusion could lead us to focus on policy areas and capabilities which do not have the impact on the illegal ivory trade that we so urgently need.*

These words in the foreword to the RUSI Paper were written by Lord Hague, a former Foreign Secretary of the British Government and a person with significant experience in this area. The Paper itself begins by setting out the evidence that led to the *widespread acceptance* of the nexus, a *narrative* that *'has been embraced by researchers, journalists and politicians and has captured the public's imagination'*. A narrative that *'is beginning to have an impact on policy making'*. A narrative with two main components:

- That *'Al Shabaab participates directly in poaching'*
- That it moves vast quantities of ivory down to Somalia and the ports there, now that the Kenyan authorities are tightening up in Kenyan ports.

It then examined the narrative and concluded it to be *deeply flawed*.
Here are some of its findings:

- Al-Shabaab's revenues came from *black gold* and sugar rather than ivory. In fact, UNEP and INTERPOL produced a report in 2014, the *Environmental Crime Crisis: Threats to Sustainable Development from Illegal Exploitation and Trade in Wildlife and Forest Resources*, which concluded that *'Al Shabaab's charcoal revenues likely dwarf those potentially deriving from ivory'*. Somalia has acacia forests, and the charcoal produced from these trees is highly prized, especially in the Arabian Peninsula where it sells for high prices. This was why organized crime gangs were involved in charcoal smuggling. And sugar smuggling, another profitable activity, had been carried out for decades. It could be described as an industry.
    Smuggling activities such as these produced high rewards for very little risk.
- Although elephant poachers did come from Somalia, the picture was unclear. The communities who live around elephants are not rich, and recent increases in the price of elephant ivory might tempt some of their members to become involved. This was backed by statements from Kenyan Wildlife Service personnel working on the frontline. A *diverse, localised picture* then.
- The data regarding the use of ports in Somalia rather than say Mombasa or Dar es Salaam was sparse and heavily dependent on a single source. However, charcoal and sugar smuggling was documented by the UN Monitoring Group on Somalia and Eritrea, and *this group has never found evidence of ivory crossing Somalia* although the routes that would be used fall within their jurisdiction.
- In financial terms, ivory smuggling would not make sense. In other words, the sums didn't add up. Al-Shabaab would be an additional middleman, and that would be to no one's advantage.

## 21.3    Gabon

While al-Shabaab might have been out of the picture, in 2013, Gabon's elephants were identified as being at high risk from poachers from the Lord's Resistance Army, and unfortunately it turned out that this was not the only terrorist group targeting them. In January 2018, the results of Operation Nzok revealed that they were also being poached by the notorious Boko Haram, the Islamist group operating in Nigeria. In an investigation that lasted 2 years, a large ivory trafficking network was uncovered in which a Chadian family was involved. They were arrested and the network dismantled.[4]

## 21.4    Bryan Christy's Experiment

Such in-depth analysis is rarely possible. Apart from the problem of funding, it is incredibly dangerous to carry out research into terror groups such as the Lord's Resistance Army, Janjaweed, the Sudanese Armed Forces and, in the Congo, the FDLR. The veteran reporter/document maker Bryan Christy, however, had a brain wave. He decided to track the progress of a pair of fake elephant tusks fitted with a GPS tracking device.

The first thing to do was to make the decoy tusks. This took three years, two years of them searching for the right person to carry out the task. The third year was spent making them. For the plan to succeed, they had to be perfect. The first six months were taken up with *trying out*, trying out various materials and resins, and then correctly balancing the tusks, for as we learnt earlier in our story, ivory has very special characteristics, even to the sound it makes when tapped. A GPS tracker had to be installed so that it could transmit the signals but at the same time be sufficiently protected. That took another six months. The final task was to paint them.

They had been made, but would they actually deceive anybody? The answer came swiftly. When he landed in Tanzania, Bryan Christy was immediately arrested as a suspected ivory trafficker and spent the night in prison. Nor did he find it easy to convince the experts the tusks were fakes. They had passed their first test with flying colours and could now be smuggled into the poachers' supply chain.[5]

---

[4] Dionne Searcey *Gabon Says It Busted a Major Ivory Smuggling Network The New York Times*, 18 January 2018. See https://www.nytimes.com/2018/01/18/world/africa/gabon-ivory-smuggling.html Accessed 23/01/2018.

[5] Warren Manger *Ivory poaching's direct links to the world's most evil terrorists, warlords and dictators is exposed* 18 September 2015. See http://www.mirror.co.uk/news/world-news/ivory-poachings-direct-l Accessed 19/04/2016.

In order to track their progress, the tracker sent out one signal a day. If it came from the same place, it meant that the tusks had not moved, but if it had changed, the new location would be noted, and the two dots would show the distance travelled. This was their route:

- They entered the smugglers' supply chain in Mboki, in DRC.
- They then travelled through CAR until they crossed the border into South Sudan.
- From there they moved into the disputed territory of Kafia Kingi which is controlled by Sudan.
- They moved another 74 miles in that territory and then stayed put for 3 weeks.
- On the move again, this time they travelled quickly into Sudan itself. The speed suggested a motorized form of transport.
- Then a further 32 miles to what was then, their last known location.

The 3-week stay was in Joseph Kony's camp. Kony leads the infamous Lord's Resistance Army. The presence of the fake tusks there showed not only a direct link with ivory trafficking but also the position of his camp, near the town of Songo. Thus it provided the most accurate position for some time.

In all, the tusks were tracked 592 miles. Then they stopped moving. The assumption was that they had been buried.

## 21.5   Michael's Story

This was what one small child experienced when he was pressed into becoming a member of Kony's Army.

Michael was only 9 years old when he was abducted by an armed group from the LRA, forcibly removed from family and friends, and then forced to poach elephants in Garamba National Park in DRC. The Army wanted to use their ill-gotten gains to bankroll their nefarious activities, so the ivory was to be traded with the Sudanese Army, exchanged for weapons and supplies.

Some seventeen years later, Michael defected, and he was able to show the Ugandan Army where a hoard of ivory had been hidden.[6] A sad and disturbing footnote.

---

[6] Brent Stirton, a photo story that won the 'Wildlife Photojournalist Award: Story' in the Natural History Museum's 2015 Wildlife Photographer of the Year competition, in action *WWF Magazine* Spring 2016, *Ivory Wars*.

# 22

# The Internet

'The internet has become the world's biggest marketplace, one that is open for business 24 hours a day, seven days a week. It is without boundaries, largely unregulated, free and mostly anonymous, and provides easy opportunities for criminal activity'.[1] It is also very difficult to police. This is due to the speed at which transactions occur and to the fact that these transactions are anonymous.

## 22.1  CITES

At CoP15, the problem of Internet trading finally 'came alive'. An earlier CITES Resolution Conference 11.3 (Rev CoP15) was amended and that, together with two decisions (15.57 and 15.58, which dealt with its administration), committed Parties to a number of actions. These included ensuring they had sufficient measures in place to control legal trade and investigate illegal trade. Appendix I items were to be given high priority. Those caught offending were to be punished.

The Resolution also recommended that a dedicated unit should be established at national level by the Parties. It would:

* Investigate 'wildlife crime linked to the internet' itself.
* Incorporate 'wildlife trade issues into existing units that investigate or monitor computer or cyber-crime'.
* There would be 'a mechanism to coordinate the monitoring of internet related wildlife trade'.

---

[1] IFAW *Wanted – Dead or Alive Exposing Online Wildlife Trade* 2014, p. 6.

© Springer Nature Switzerland AG 2019                                                    **241**
B. Martin, *Survival or Extinction?*, https://doi.org/10.1007/978-3-030-13293-4_22

- Provide for *'the timely sharing between designated contact points in CITES management and Enforcement authorities of information that results from these activities'*.

And the Parties and INTERPOL should undertake a number of measures, some of which included:

- Submitting certain information to the Secretariat.
- Making sure there was enough funding to investigate and target illegal internet-related trade in CITES-listed specimens.
- Regarding the use of certain data.
- Providing yet more money, this time to establish *'a full-time position, dedicated to e-commerce aspects of wildlife crime, within the General Secretariat of Interpol'*. That person would be responsible for *'ensuring that all information or intelligence regarding e-commerce is consistently collected and disseminated to the relevant Enforcement Authorities designated by Parties'*. Once again we see the emphasis on good data and sharing it.

Unfortunately, the two working groups, e-commerce and e-permitting, set up at the same time to explore the above issues have not been very active.[2]

## 22.2   Internet Sales

If we are to understand what is happening on the Internet, we need data, preferably going back a number of years. IFAW has been monitoring this illegal wildlife trade since 2004, so has it. We are going to look at the results of three surveys conducted in 2008, 2011 and 2014, respectively, to find out what it tells us about this illegal trade in elephant ivory and rhino horn.

### 22.2.1   The 2008, 2011 and 2014 Surveys

#### 22.2.1.1   The 2008 Survey

Carried out in 11 countries, IFAW found that products made from elephant ivory accounted for more than 70% of all wildlife products sold online.[3]

---

[2] Ibid, p.22.
[3] IFAW *Killing with Keystrokes: An Investigation of the Illegal Wildlife Trade on the World Wide Web* 2008.

That same year, 2008, the two largest e-commerce websites in China, Taobao.com and Alibaba.com, banned sales of elephant ivory. In order to make the ban effective, they installed screens that automatically filtered out any words for ivory. This caused problems for the illegal sellers who obviously had no intention of letting a little thing like a ban stop their trading. How to get round it? Use false labels and code words to avoid being detected. They were ingenious. The phoney words were essentially a play on *elephant tooth* which are the Chinese words for ivory. They included, for example, *tooth carving, tooth from the largest mammal on land* and even *XY, Xiang Ya,* which is the Chinese word for ivory. These were comparatively easy to crack, unlike the more obfuscating *patterned tooth*. Where code words bore no resemblance to elephant ivory, the potential buyer had to be attracted by detailed pictures.

## 22.2.1.2    The 2011 Survey

This took place in China, where some 13 websites, identified as having a large number of illegal wildlife listings, were monitored. They consisted of a mixture of online forums, auction and collection sites and e-commerce platforms.

As China's State Forestry Administration had not given its approval to any online shop to trade in elephant ivory, it followed that all sales in this commodity would be illegal. Unfortunately this did not appear to be discouraging the sellers, because in the one week, the sites were monitored; the researchers found 17,847 listings of ivory products for sale, 97% of which were on websites specializing in selling collectables and antiques.[4]

One of the collection websites was responsible for more than 12,000 of the illegal listings and one seller, based on the China/Viet Nam border, *'had posted 526 ivory products since the shop registered in September 2010'.* Worse still, rhino horn was amongst the other specimens of wildlife for sale. The researchers noted that *'most of the wildlife products listed had already been sold'*, which suggested a ready market.[5]

Although the online part of the 2011 survey was complicated by the use of code words and false labels, the investigators still managed to get some evidence, even some sellers' own admissions of their unlawful trading practices, after the product had been identified as ivory.

---

[4] Grace G. Gabriel, Ning Hua, and Juan Wang *Making a Killing: A 2011 Survey of Ivory Markets in China,* IFAW, p.12.
[5] Ibid.

### 22.2.1.3 The 2014 Survey

This survey was global. The IFAW researchers looked at 280 online market-places in some 16 countries simultaneously over a 6-week period.

There were 9482 advertisements, which between them were selling 33,006 endangered wildlife and wildlife parts and products, and although ivory and rhino horn only made up part of this, with these enormous numbers, even a small portion amounts to a lot of animals. And the estimated total value from the sales was at least USD10,708,137. Very big business indeed!

Ivory and 'suspected ivory' items featured in almost a third of the adverts, some 3047 of them. However, as we saw earlier, there are many different kinds of ivory, including 'plastic ivory', so an expert who could identify elephant ivory was a key member of the team. Even then, it was not always possible to be quite sure what kind of ivory it was. And once again there was the added difficulty of code words, although some sellers on some online marketplaces quite blatantly sold 'ivory'. Breaking the code required analysis not only of the words themselves but also of the accompanying pictures and prices of the items offered for sale.

**Some Results from the 2014 Survey:** These were mixed, and some countries must have been disappointed.

**China:** The researchers found 18,590 items being offered for sale in 2106 online advertisements over 21 websites. Although the self-imposed ban by sites such as eBay, Alibaba and Taobao had been having some effect, the figures still looked bad. But China was making progress. It had managed to clamp down on Internet crime and had constantly monitored 20 key trading websites since 2012, which had resulted in more effective regulation of online market websites and stronger enforcement action.

The illegal sale of rhino horn continued to be an intractable problem though. Rhino horn cannot be traded within China. There is no legal domestic market. And international trade is forbidden under CITES. But the researchers found the second most common item for sale was rhino horn or rhino horn products (including medicinal products). There were also 164 advertisements for rhinoceros carvings and white rhino horn beads.

China and IFAW have been working together on the problems of illegal sales on websites since 2005.

**Europe:** The results were discouraging.

**In Belgium:** Ivory and *suspected ivory* items accounted for half of all the online wildlife trade. During the survey, one seller posted 11 advertisements on 2 of the websites, offering to sell 'suspected ivory' bracelets, carvings and vases.[6]

**In Russia:** The number of ivory and 'suspected ivory' advertisements had more than doubled since the 2008 survey. The researchers identified 81 advertisements for jewellery, ornaments or antiques and noted that one vendor actually posted 23 adverts during the survey.

**In the UK:** The results were equally disappointing, with 38% of all the adverts being for ivory or *suspected ivory* pieces. Although most of the sellers claimed their ivory was antique, so could be legally sold, only six advertisements contained any supporting evidence to show the provenance of the item(s). The UK differed from the other countries in that some 62 of the sellers posting adverts were not only based overseas but completely outside the EU. They offered to ship items, mainly from China or the USA.

**In France:** The results were good. There had been a dramatic drop in adverts, down from 192 postings in 2008 to only 7. The suspect items were jewellery and statues, and the two sellers had used code words. The investigators considered this was because of the sales ban imposed by eBay.fr.

## 22.2.2   The USA and Canada

(The data on the USA was collected at the same time as the live auction survey we looked at earlier.)

Two major aggregator websites, LiveAuctioneers.com and AuctionZipLive.com, both open to the general public, were monitored in 18 sessions. Unfortunately, most of the data came from LiveAuctioneers.com because AuctionZipLive.com didn't offer information on lots that had been sold.

These companies are huge, AuctionZipLive being *the world's largest online auction marketplace* and *the nation's largest online bidding destination* with *thousands of new items* posted every week. LiveAuctioneers.com formed a marketing partnership with eBay in 2002. This meant auction houses all over the world could go online with their live sales, and their phone apps *'have opened*

---

[6] See IFAW (n.1), p.28.

*up a new mobile pipeline to bid anytime, from anywhere, with complete anonymity*.[7] By comparison, live auctions pale into near insignificance.

The investigators monitored 340 online auctions. The lots were posted by 223 auctioneers and galleries, and their online catalogues registered with either AuctionZipLive.com or LiveAuctioneers.com. On average, 465 lots of ivory were on offer for sale each week, an estimated 24,186 ivory lots for sale a year. Some auction houses held a number of auctions during the period they were surveyed. They advertised the sort of objects we have already seen in surveys, puzzle balls, figures, carvings and tusks (both raw and worked). In other words, objects made entirely or almost entirely of ivory. There were lots in which ivory was an important feature, in items such as walking sticks, statues, chess sets, billiard cues and balls and even page turners. And there were items that contained very little ivory. These included handles for knives, ivory inlay in furniture, ivory used in musical instruments and guns. The items sold for prices ranging from USD5 to USD90,000, the average being USD991.

The figures, coming only from the database of LiveAuctioneers.com, showed that during the 9-week survey period, 2277 ivory lots were sold, which would translate into 13,156 ivory lots sold per year (253 a week), raising over USD13 million a year (making a weekly average of USD250,933). When the buyer's premium, an extra 21%, was added on, the total estimated annual commission on ivory sales would be approximately USD2.75 million. An enormous sum of money and unfortunately *almost none of the businesses surveyed were able to provide any documentation on the provenance of their products*.[8]

## 22.3   TRAFFIC and Online Marketing in China

TRAFFIC has been concerned about the online sale of illegal wildlife products in China for many years, so once again we must look to the surveys, see what their data reveals and what happened as a result.

**The 2006 Survey:** Lasting 8 months, the results were disturbing. They showed 332 ivory products and 193 rhino horn products on offer for sale on a number of Internet sites in mainland China, Taiwan and Hong Kong SAR.[9]

---

[7] Ibid, p.27.
[8] Ibid, p.58.
[9] Xiao Yu and Wang Jia *Moving Targets: Tracking online sales of illegal wildlife products in China* TRAFFIC Briefing February 2015.

**2010–2011** TRAFFIC started the routine surveillance of online sites, counting the number of advertisements selling illegal wildlife products. During 2010 and 2011, they shared the results with law enforcement agencies in a series of workshops which were held in conjunction with China's Management Authority for CITES. It was obvious that the authorities still did not have enough knowledge if they were to detect and deter the trade.

Therefore a comprehensive programme of online monitoring in China began in late 2011. The results were shared not only with enforcement agencies such as the National Forest Police Bureau but also with the website and platform managers. They all needed to be brought into the frame if the illegal trade was to stand any chance of being controlled. Training was another 'must'. In fact, TRAFFIC has been regularly training e-commerce professionals to identify species and the latest code words used to hide illicit transactions and also alerting them to any changes in the law and other relevant matters.

Knowing and/or identifying the code words is one of the keys to defeating the illegal sellers. Once a code word has been linked to a particular product, it can be blocked, provided it doesn't also apply to a legal product. Related adverts can then be deleted.

**2012** In January, TRAFFIC started manually tracking 15 selected websites in depth, looking for 5 key illegal wildlife products, including elephant ivory and rhino horn and sea.rching for 12 code words used to disguise the products.

That was just the beginning. Illegal sellers are reluctant to give up. So as well as identifying the ever-changing code words, it was necessary to keep on expanding and adapting as new sites were used and new products came onto the market.

**2014** The end of September saw TRAFFIC routinely monitoring 25 e-commerce and antique-selling websites every month for 8 specific wildlife products and 64 code words. Eventually it became possible to use automated searching and screening methods. Another improvement, as they provided more detailed results which were then sent to the relevant enforcement agencies.

The problem of fake products was always present. The screeners had to carefully examine written descriptions of the product and any images posted online. Then hope for the best! Another problem that arose was that of duplicate information. This was where the dealer advertised the same product on several different websites. Protocols now cover this situation.

TRAFFIC continued to train and guide website managers and share information with them. The website managers made great efforts to block various code words and remove offending adverts. The result was *a steady and signifi-cant drop in the number of TWPA (total number of illegal wildlife product adver-tisements) in the first half of 2013*', a sharp fall to less than 10,000 adverts, where it remained.[10] This clearly showed 'working together' was paying divi-dends and was most encouraging.

The NWPA (number of new wildlife products advertisements) provided a more accurate picture, removing the influence of old, previously recorded adver-tisements and enabling new trade trends to be picked up more rapidly. The results were startling. From July 2012, the NWPA fell to around 1500 and, with slight variation, continued to stay there. Unfortunately, advertisements for ivory gener-ally made up over half the new adverts posted every month, and ivory continued to be the most frequently advertised of all the eight products monitored.[11]

## 22.4   Sales on Social Media

The move to social media was a blow. These sales have a distinct advantage over online selling, providing major benefits for illegal dealers who have better control. And all transactions are undertaken in private which makes them much more difficult to monitor.

TRAFFIC's monitoring of the social media in China has shown significant quantities of illegal wildlife products being sold by this method, and in March 2014, they began researching its increasing use.

Dealers must attract and then interact with potential customers. They do this by releasing photos and information about illegal wildlife products. Some use 'agents', which both increases the number of potential buyers and speeds up sales. Using Facebook in our example, it works like this:

- The agent reposts information about the products onto his site.
- One of his 'friends' reads the information and wants to buy the product.
- He, the agent, goes back to the dealer and buys it.
- Then resells it to his 'friend', at a higher price. After all, he has to make a profit too!

It also means that the dealers can control who remain their 'friends'. They can block any 'friend' who might appear suspicious, perhaps because the 'friend' asks too many questions or never actually buys anything. Like research-ers. So they are then blocked.

---

[10] Ibid.
[11] Ibid.

Another problem the researchers encountered was where a sale involved both a website and social media, sellers and buyers switching from one to the other during a transaction.

Nothing daunted, TRAFFIC continues to work on different ways to overcome these difficulties.

## 22.5    Methods of Payment

These can provide yet another means of detecting a sale.

An increasingly popular way of paying is by *guaranteed transaction* services. Let us see how this works.

- The intending buyer communicates with the seller that he wants to purchase a particular ivory statue. A sum of money is agreed.
- The buyer pays the money to a *guaranteed* intermediary.
- The seller sends the statue to the buyer.
- He checks it and then, if satisfied with it, instructs the intermediary to release the money to the seller.

As the companies that deliver the illegal products to the purchaser are often unaware of what they are doing, TRAFFIC has been working with a number of couriers on ways to avoid this.

All in all, although there is no room for complacency and definitely room for further improvement, China is one country making good progress in its attempt to control this illegal online trade. In June 2012, the declaration at the end of a workshop convened by the National Forest Police Bureau in collaboration with China's CITES Management Authority and TRAFFIC stated that *'they had a zero-tolerance policy towards their services being used to conduct illegal wildlife trading'*. It was signed by 15 of the leading e-commerce companies, including Alibaba, Taobao and Tencent.

## 22.6    Malaysia's Facebook Groups

Facebook, too, is causing problems. In March 2016, TRAFFIC published a disturbing report about the situation in Malaysia.[12]

It monitored 14 Facebook groups (most of them 'closed groups') in Peninsular Malaysia, for 30 min every day for 5 months. There were 68,000

---

[12] Krishnasamy, K. and Stoner, S. (2016) *Trading Faces: A Rapid Assessment on the use of Facebook to Trade Wildlife in Peninsular Malaysia* TRAFFIC. Petaling Jaya, Selangor, Malaysia.

active members. The results took them by surprise. The groups were *'turning Facebook into a wildlife marketplace, driving a roaring and often illegal trade in iconic and threatened animals'*. This was totally unexpected because open wildlife markets were not found there.

Their report recommended:

- Closer cooperation between Facebook and the enforcement agencies, making it easier to target offenders effectively
- Setting up a dedicated forum at both regional and global levels *'for governments, social networking bodies and NGO partners to find realistic solutions to the problem'*

TRAFFIC took their evidence to Facebook and Peninsular Malaysia's Department of Wildlife and National Parks. Both were concerned with this disturbing situation, and they all started working together on practical solutions.

Although neither ivory nor rhino horn were mentioned in this survey, there have been large seizures of ivory in Malaysia, and we have seen just how easy it can be to trade illegally on the Internet. So here we have another potential nightmare. *'Social media's ability to put traffickers in touch with many potential buyers quickly, cheaply and anonymously is of concern for threatened wildlife and enforcement agencies which demands nothing short of a concerted global response'.*[13]

## 22.7    Education

China is using education such as the *'New communications approaches to reduce demand for threatened wildlife and enhance public awareness'* workshop, to raise awareness of problems flowing from the growing trend in social marketing.

This was held in April 2017 and co-hosted by the Shanghai Wildlife Conservation Centre, Shanghai Zoo and TRAFFIC. Attended by wildlife conservation and law enforcement officers, NGOs and Wan Ziming, the Deputy Director of the Shanghai branch of China's CITES Management Authority, it brought the participants up-to-date *'with the latest social media tools and techniques to mobilize public support for wildlife conservation'.*[14]

---

[13] Ibid.

[14] Ziming (2017) in TRAFFIC *Shanghai workshop promotes use of social marketing to change illegal wildlife consumption* 13 April 2017. See http://www.traffic.org/home/2017/4/13/shanghai-workshop-promot Accessed 20/04/2017.

# 23

# Religious Ivory

The Lion Man is the earliest artefact we have of ivory carved into a precious religious object, and we are about to see some more manifestations of this 'cult of idolatry'.

## 23.1  Thailand

Thailand is a good starting point, in Phayuha Khiri, whose carving industry was founded by a monk. So it was hardly surprising to find so many shops selling life-size images of famous monks and small images of the Buddha, as well as large quantities of smaller religious artefacts, which only the orange-robed monks were buying.

Surin, the other major carving centre, was where the Elephant Monk could be found. Formerly the Scorpion Monk, he believed he was an elephant in a past life. A controversial character with followers in Singapore and Malaysia as well as his native Thailand, he told the investigator that *'ivory removes bad spirits'*, and he was wearing an ivory elephant head pendant suspended from ivory prayer beads which represented the 108 human passions. He gets his main income from amulets; ivory ones can sell for large sums of money. And he has been accused of cruelty to elephants.[1]

Amulets play an important role in Thai life. They bring good luck to their owners, as well as protect them from harm and black magic. Apparently all the soldiers in the Thai Army stationed along the border with Cambodia were issued with amulets *'to ward off Cambodia's black magic'*. And many

---

[1] Bryan Christy *Blood Ivory* National Geographic, October 2012, p. 6.

Thais own and wear at least one, often several of them. Bangkok even has an amulet market, which is *'huge, with countless vendors selling tens of thousands of small talismans'* made of various materials, including ivory and even compressed dust.

### 23.1.1  The Elephant Monk and Smuggling

According to the Elephant Monk, smuggling ivory into the country presented few problems. He explained in precise detail, just what the investigator should do, from cutting the ivory into pieces to fit into his suitcase to his arrival at Bangkok airport where the monk had (possibly still has) followers in immigration and what to say if anything should go wrong there (that he was bringing the ivory to the monk's temple).

Fortunately, Buddhist leaders in Thailand are now working with TRAFFIC to spread the word that using ivory products is not part of the Buddhist tradition.

## 23.2  The Philippines

In 2012, the *National Geographic Magazine* published the findings of an investigation in the Philippines by the veteran reporter Bryan Christy, into the role ivory played in religion there.

Before the investigation, the Philippines had not been recognized as an important consumer of ivory, although a number of large seizures had been made there over the years, including 7.7 tonnes in 2005 and 5.4 tonnes in 2009.[2] However, far from being merely a transit country, ivory passed through on its way to China; it was actually a destination country. Perhaps, though, some whisper of this came and passed un-noticed when, in 2006, 6.1 tonnes of ivory bound for the Philippines was seized in Taiwan. Seizures that represented the killing of some 1745 elephants.

Much of the ivory was connected with the Santo Niño de Cebu (Holy Child of Cebu), a small, elaborately dressed wooden statue and the country's most important religious icon. And to the efforts of one priest, a noted ivory collector. Every year, the little statue was undressed and ceremonially dipped into barrels of water. That transformed the water into holy water, all ready for whatever use was required. It was a Roman Catholic ceremony, in a country

---

[2] Ibid.

with the third largest number of Catholics in the world. Every January, the faithful walked in procession with the statue, each of them carrying their own miniature copy, which, if they were lucky, would be carved from ivory. This was because the more you could afford to spend on your icon, the more blessings you would gain in return. And every year, just outside Manila, there was a Santo Niño exhibition to celebrate the best of the parishioners' collections. Families even opened bank accounts in the name of their icons and named them in their wills. In fact, *'on Cebu the link between ivory and the church is so strong that the word for ivory, garing, has a second meaning: "religious statue"'*.[3]

Although there were many family heirlooms, the curator of the exhibition Father Jay was instructing people to buy new ivory items, so that the history of their artefact would start with them. Only too well aware that these new ivories came from Africa, he was completely unfazed that they had been smuggled in. Indeed, his words seemed to indicate that he positively revelled in the unlawful activity. *'It's like straightening up a crooked line: You buy the ivory, which came from a hazy origin, and you turn it into a spiritual item. See?'* he says, with a giggle. His voice lowers to a whisper. *'Because it's like buying a stolen item'*.[4]

Monsignor Garcia had an anteroom, a mini-museum, in his church. It housed a collection of religious figures whose heads and hands were made of ivory, as well as items made only of ivory, like the *'near life-size Mother of the Good Shepherd seated beside an ivory Jesus'*.[5] He also had connections with the Vatican. His Santo Niño had been blessed by Pope John Paul II. And like Father Jay, he knew all about smuggling. How should the investigator get his proposed purchase of an ivory Santo Niño into America? Smuggle it in. How? *'Wrap it in old, stinky underwear and pour ketchup on it so it looks shitty with blood. This is how it is done'*.[6]

And it was the same in Manila, the capital city. While it is easy to understand why Filipino priests living overseas might be customers, why would the dealer from Mexico *'gathering up new ivory crucifixes, Madonnas, and baby Jesuses in bulk'* to smuggle into Mexico in his luggage, go to so much trouble? Was it just about money? Did the profits far outweigh any risks?

---

[3] Ibid, pp. 3–4.
[4] Ibid.
[5] Ibid.
[6] Ibid.

## 23.2.1  Smuggling Ivory Into and Out of the Philippines

The investigator found Monsignor Garcia most helpful regarding his proposed purchase, providing invaluable information such as:

- The names of his favourite carvers.
- Whose wife overcharges.
- Who fails to meet deadlines.
- Phone numbers and places.
- The National Museum of the Philippines might provide a certificate to the effect that the artefact was an antique, if it was too large to be hidden in a suitcase.
- A carver could alter the carving date so that it predated the 1990 international ban on ivory sales or produce a paper declaring it was imitation ivory.

Everywhere he went, the investigator found someone who could suggest a way of smuggling ivory into the USA. Methods included:

- Painting the ivory brown to resemble wood, with watercolour paint of course, so that it could be washed off afterwards.
- Making identical resin statuettes and laundering in the ivory one. Should he be unlucky enough to get caught, he had to tell the Customs officer/s that it was resin.

It seemed that the smuggled ivory followed the ancient trade routes out of Africa, with Muslims from Mindanao bringing it into the country, its path smoothed by bribes.

## 23.3  The Vatican

Until recently, the Vatican had a long, and arguably dubious, connection with ivory artefacts. It owns some very beautiful examples, and it has received ivory gifts from Heads of State, like the ivory Santo Niño given, in 2007, to Pope Benedict XVI by the then President of the Philippines, Gloria Macapagal-Arroyo. And it has given ivory gifts, presenting the American President Reagan and his wife with an ivory Madonna which they then bought from the nation to own themselves. Even former Kenyan President Daniel arap Moi gave Pope

John Paul an elephant tusk. But that was before he made his magnificent gesture to elephant conservation by burning 13 tonnes of the country's stockpiled ivory. And Roman Catholics could buy their very own ivory artefacts in St. Peter's Square, Vatican City, just like the Philippines.

# 24

# Bribery and Corruption

Another key strand in this story is corruption, which can take many forms. It may be a failure to dig deeply enough into a cargo so that the ivory hidden underneath the timber remains undiscovered. Except that this failure was deliberate, a 'knowing' failure. Or it may be a quiet tip-off that officials are coming to check your paperwork. Sometimes, it is quite blatant though, as we saw when those corrupt but unsuspecting Chinese officials were willing to sell their illicit tonne of rhino horn to the investigators from the EIA. Or their offer of help of local police transport to take the haul to Hong Kong SAR. Sometimes, it is achieved by money laundering.

## 24.1 The United Nations Convention Against Corruption

This very important Convention, adopted by the UN General Assembly in October 2003, entered into force in December 2005. The nations of the world had finally accepted that the time had come to deal with this silently creeping evil, an evil that causes so many problems to the innocents, human and animal, in this world, for it is normally the poorest and weakest who suffer. Nor must we forget the income that governments are cheated of. No country can claim to be squeaky clean. In some countries, the problem is endemic, only one of a number of problems besetting them. Most countries are making valiant efforts to deal with the situation.

© Springer Nature Switzerland AG 2019
B. Martin, *Survival or Extinction?*, https://doi.org/10.1007/978-3-030-13293-4_24

The treaty's main provisions, all of which are matters relevant to our story, are concerned with:

- Prevention
- Criminalization
- International cooperation
- Asset recovery

Heading the list is *prevention*, but frequently the other three elements are needed to achieve this. Corruption should be made a *criminal offence* in every country and that requires *international co-operation*. Agreeing to include *asset recovery* was a major breakthrough. There is no ambiguity about it. This is a fundamental principle of the Convention. Corruption deprives many developing countries of their desperately needed income, so we must hope that asset recovery will see this wrong remedied.

The importance of the treaty in relation to the illegal trade in wildlife was reiterated in the Declaration agreed at the end of the 2014 London Conference on the Illegal Wildlife Trade (more of this later). The section *'Building on the existing international framework for action'* includes *'effective international co-operation demands the active participation of partners that support Governments in different sectors, in particular … the UN Convention against Corruption'.*

And *'Actions'*, B *'Ensuring Effective Legal Frameworks and Deterrents'* addresses *'the serious problem of corruption and money-laundering facilitating wildlife trafficking and related offences by adopting or amending legislation, as necessary, criminalising corruption and bribery facilitating poaching, wildlife trafficking, and related offences'.*

All governments are urged to sign up to the Convention and put measures into place to prevent corruption and money laundering. If/when this happens, it should provide a lot more protection for our beleaguered animals.

Many countries already have some anti-corruption legislation. Let us take the UK as an example. The Bribery Act was passed in 2010, and the government continues to amend and update its anti-bribery policy and legislation. It is safer to do business if there is no corruption, and bankers have a crucial role to play in uprooting this evil. This was made clear in the British Bankers Association's *Anti-Bribery and Corruption Guidance* which was published in 2014, the same year the UK passed its Bribery Act. In 2018, the National Crime Agency served its first Unexpected Wealth Orders (UWOs), which it might also be possible to use against suspected illegal wildlife traffickers.[1]

---

[1] Sean O'Neill *McMafia wife also owns golf club* The Times, 9 October 2018.

Corruption and its associated problems, such as money laundering, play a core role in our story. They were discussed at the Symposium on Combating Wildlife Crime, where one of the speakers swiftly demolished some myths/misconceptions:

- That it was caused by poverty *because wealth can also cause corruption.*
- That corruption goes if corrupt officials are removed from public office, because *even good and clean public officials are subject to corruption.*[2]

We already know it happens, that it causes severe damage to both animals and people, is hard to pin it down and hard to prevent. Unfortunately, much of the evidence tends to be anecdotal or hearsay, second hand though sometimes it is stronger. But winning prosecutions in a court of law requires first-hand evidence, evidence that will withstand an attack by the defence lawyers, evidence of what is happening on the ground and evidence strong enough to actually convict an offender charged with a corruption offence.

## 24.2   Elephant Ivory

It is easier to understand what can happen by looking at some examples.

### Case Study 8: Elephants in Zambia

(Once again, we are indebted to the EIA for our information.)

Benson Nkunika, a notorious ivory poacher, had been arrested by the ZAWA game scouts for illegally possessing firearms and an ivory trophy. He claimed he had committed this crime because he was poverty stricken, so they decided that instead of prosecuting him, they would employ him as a trainee scout who would act as an informant for them. Unfortunately, the 'poacher turned gamekeeper' soon reverted back to poacher.

He was rearrested, this time by RATZ, a private antipoaching unit, set up and paid for by private enterprises and concerned individuals to help the struggling ZAWA. While in custody in ZAWA's local headquarters *'Nkunika revealed the existence of widespread corruption among ZAWA staff, and alleged that he had been contracted by the then warden of South Luanga National Park*

---

[2] Thaung in *Symposium on Combating Wildlife Crime: Securing Enforcement, Ensuring Justice, and Upholding the Rule of Law. The Proceedings.* Editors Kala K. Mulqueeny and Francesse Joy J. Cordon, 2014, p.78.

*… To poach 100 elephants from the park and surrounding areas*.[3] He had already killed 58 when he was arrested. He also described how the ivory was moved to Lilongwe.

That warden was replaced and the new warden decided he needed to find out what exactly was going on and not just because so many elephants were being killed. One incident was particularly troubling. Three elephants had died close to ZAWA's headquarters, which suggested the poachers knew where the ZAWA scouts would be. The Regional Manager took charge.

His team of investigators was led by a Senior Investigator from Head Office and drawn from ZAWA staff outside the area. They discovered the Lilongwe Pipeline and the Chipata network used by Sindikani Banda to traffic the poached ivory from Chipata to Lilongwe. Not only that, they found the *fabled* factory that was rumoured to take deliveries of ivory from the Chipata network. And when it was raided in 2002 by ZAWA investigators and staff from Malawi's Anti-Corruption Bureau, they struck gold, finding documents that showed *'a massive and co-ordinated ivory trafficking operation'*.[4]

Although the Zambian authorities tried hard, they did not manage to stamp out corruption completely, and the EIA discovered some disturbing examples of bribery when they returned to Tanzania and Zambia early in 2010 to investigate reports that both poaching and ivory trafficking were beginning to increase again. The next CITES CoP was due to be held in March that year, and both countries had put in a request that they be allowed to downlist their elephant populations from Appendix I to Appendix II, which would then enable them to sell their legally held stockpiles of elephant ivory. It was essential therefore, to either substantiate or disprove the reports, and they did not have much time to do this. Unfortunately, they found not just poaching and ivory trafficking but bribery as well.

The investigators obtained their information from traders who had no qualms about speaking at length on how they operated. From, or having connections with Zambia, DRC, Angola, South Africa and Zimbabwe, the traders offered the investigators both worked and raw ivory and even rhino horn!

Here are some of the examples of bribery/corruption they gave:

- From Angolan refugees, using contacts from Angola: Ivory transported in place of medical supplies in Red Cross vehicles. Medical supplies in boxes were removed and replaced with ivory; *the boxes are never searched.* They

---

[3] Jo Hastie, Julian Newman and Mary Rice (2002). *Back in Business: Elephant Poaching and the Ivory Black Markets of Asia*, EIA.

[4] See Hastie, Newman and Rice (n.1), p.4.

even claimed *'they could supply one tonne of ivory immediately'*. Allegedly, rhino horn could be supplied in the same way.

- A trader from DRC who had *family ties to a high-ranking officer in the Congolese army*: claimed he used military vehicles to escort unworked ivory to the DRC/Zambia border, from whence it was taken to Lusaka. He further claimed he had been *'operating in the ivory business with impunity since 1995'*, he had 250 kg of ivory and *'could supply any amount on demand'*.
- One trader described: how he travelled into the bush in Angola and Zambia to collect ivory from small traders. He claimed he would then pay Zambian soldiers who transported the ivory over the border in Zambian military vehicles. This method of transport was also described by other traders.[5]

Although it may be clear from what is actually taking place on the ground that some sort of corruption/collusion is occurring, it is not always possible to follow anything up. Another example will serve to demonstrate. This time it comes from DNA analysis. Ivory from a seizure in China was found to have come from elephants in Tanzania, Mozambique and Malawi. In order to consolidate such large shipments, there would have to be networks capable of collecting the ivory from across many borders and that would suggest high levels of immunity. In other words, the poaching networks had been able by some nefarious means, to obtain *'the collusion of politicians, security forces and even wildlife rangers, as well as buy the local support needed to find and poach so many elephants so quickly'*.[6]

A priest in the Philippines explained to another investigator how bribery was enabling new ivory from Africa to enter that country. *'To the coast guards, for example ... Imagine from Africa to Europe and to the Philippines. How long is that kind of trip by boat? ... And you just keep on paying so many people so that it will enter your country'*.[7]

## 24.3   Rhino Horn

It is instructive to return to Case Study 3, to the EIA investigation into the illegal 6 tonne stockpile of rhino horn in China. Their researcher Rebecca Chan who was working undercover at the time, was told by one of the phar-

---

[5] Ibid.

[6] Virun Vira, Thomas Ewing and Jackson Miller *Out of Africa: Mapping the Global Trade in Illicit Elephant Ivory*. Commissioned by Born Free and carried out by C4ADS, published 2014, p.15.

[7] Bryan Christy *Blood Ivory* National Geographic, October 2012.

macists in Taipei, Taiwan, that they, the pharmacists, were warned by the police whenever a raid was planned. Furthermore, even if any of them (pharmacists) were found to be in possession of rhino horn, they were not prosecuted. That was in the early 1990s, and soon after that, mainly because of pressure from the American government, Taiwan started to take a proactive role in tackling the illegal sales of rhino horn and other wildlife products.

Taiwan might have reformed, but hard evidence confirmed just how widespread corruption was in the current poaching epidemic in South Africa. It became obvious that wildlife professionals, hunting outfitters, game capture operators, some of the professional hunters themselves and even some wildlife veterinarians were involved. Rampant corruption, rampant poaching. Deeply shocking and distressing news for most South African people and pockets of it still linger today.

An early indication that it was beginning to occur within the South African game industry was evidenced in 2010, when an alleged rhino poaching syndicate operating in the Limpopo province was found to include veterinarians. After that, the evidence steadily mounted. In 2012, 4 government rangers were arrested for killing rhinos in Kruger National Park, while the death of 5 more rhinos in the Atherstone Nature Reserve in Limpopo resulted in the apparent suicide of the Reserve manager. Further evidence was provided by the high-calibre weapons frequently used by poachers to kill the rhinos. They were *characteristically only used by wildlife industry professionals*.[8]

Provincial administrators ignored *pseudo-hunting*, and high-ranking wildlife officials were exposed in a number of irregularities. Some private sector owners, concerned that officials had provided intelligence to poaching gangs were reluctant to report mandated information on the number of live rhinos they owned and their stockpiles of rhino horn.[9]

There was also considerable evidence linking Vietnamese Embassy personnel in South Africa to the illegal trade in rhino horn. It first came to light in 2006 when the commercial attache was arrested. He had with him 2 rhino horns, diamonds and large sums of cash. When questioned, he allegedly claimed to have used a diplomatic bag on various occasions to move rhino horns from South Africa to Viet Nam. Although diplomatic immunity meant he escaped prosecution, he left South Africa under a cloud, having been *recalled and disciplined*,[10] and there were concerns that he might

---

[8] Milliken, T. and Shaw, J. (2012) *The South Africa – Viet Nam Rhino Trade Nexus: A deadly combination of institutional lapses, corrupt wildlife industry professionals and Asian crime syndicates* TRAFFIC, Johannesburg, South Africa.

[9] Ibid.

[10] Anon. 2008b, in ibid., p.82.

have re-entered the country *'on at least one occasion since his initial arrest'*, using a nondiplomatic passport.[11] Another Vietnamese diplomat was implicated when 18 kg of rhino horn was seized from a vehicle at a casino in Northern Cape province.[12]

However, the most notorious incident was filmed during an undercover investigation by the South African Broadcasting Corporation's *'50/50'* programme, shown on national television. The Embassy's First Secretary was seen handing rhino horns to a known smuggler and then *'heading back to the Embassy smiling'*.[13] At the same time, the car of another member of the Embassy, Counsellor Pham Cong Dzung, could be seen parked on the other side of the road, with a Vietnamese national standing next to it. This posed the question why, when most of the Embassy cars were parked in the Embassy,[14] should that car be there? The Counsellor had been implicated in the earlier 18 kg seizure. He might even have been involved in two rhino hunts.[15]

# 24.4   Diplomats

As well as the glaring examples above, there are claims that other Asian diplomats, particularly from China, have 'made use' of their diplomatic privileges, in the form of embassy pouches, to spirit rhino horn and ivory from both Zimbabwe and Mozambique.[16] The doctrine of diplomatic immunity can be exploited.

## 24.4.1   North Korea (DPRQ)

Enter the North Koreans. Although other commodities such as gold were known as being smuggled, ivory was unexpected as there was no evidence of an ivory carving industry in North Korea and, unlike Viet Nam, there was no rising middle class with plenty of money to spend on luxuries made from elephant ivory and rhino horn. No affluence but a lot of poverty.

---

[11] R. Taylor in litt. to TRAFFIC 2009, in ibid.

[12] Anon. 2008a, in ibid.

[13] Anon. 2008b, in ibid.

[14] Ibid.

[15] D. Newton pers. comm. 2010, in ibid.

[16] Vira, Ewing and Miller (n.6) p.26.

Sometimes smuggling even seemed to benefit the perpetrator. For example, one junior diplomat, Han Dae Song, was expelled from Zimbabwe after he had been accused of using a diplomatic bag to smuggle rhino horns out of the country. That was in 1992. By 2017, he had become *one of North Korea's most prominent diplomats, heading up the country's permanent mission to the United Nations in Geneva and serving as its ambassador to Switzerland*.[17]

In fact, the Global Initiative Against Transnational Organized Crime found that *North Koreans have been implicated in 18 of at least 29 detected rhino horn and ivory smuggling cases involving diplomats in Africa since 1986*.[18] And between 1996 and 2005, after repeated attempts at smuggling, both ivory and rhino horn out of various countries in Africa had failed, at least 6 North Korean diplomats were forced to leave their posts. Nor were the quantities small. In Kenya, 689 kg of ivory were involved.[19]

In 2012, a North Korean citizen was arrested while trying to smuggle 130 pieces of ivory out of Mozambique by air, using a flight to Korea going via South Africa.[20] Three years later, 2 men, one of whom turned out to be the political counsellor at an embassy in South Africa and the other a Taekwondo master suspected of spying for North Korea, were also detained in Mozambique. Their vehicle contained a large sum of money together with 4.5 kg of rhino horn. But they were released after the intervention of the North Korean ambassador to South Africa.[21]

Before he defected to South Korea, one North Korean was posted to China to set up an office to *make money through trade and financial business*. There, he met DPRQ diplomats from Africa bringing with them rhino horn, mainly from South Africa (where the DPRQ embassy was the main supplier) and Mozambique, but also from Ethiopia and Angola, and ivory, mainly from DRC, but some from Angola, all to be turned into hard currency. Rhino horn was more valuable than ivory. According to the defector, *'he regularly facilitated transactions between diplomats based in Africa and Chinese organized crime networks'*.[22]

---

[17] Koller Frederic 2017 in Julian Rademeyer *Diplomats and Deceit: North Korea's Criminal Activities in Africa* The Global Initiative Against Transnational Organized Crime, September 2017.
[18] Ibid.
[19] Asher 2005 in Varun Vira, Thomas Ewing and Jackson Miller *Out of Africa: Mapping the Global Trade in Illicit Elephant Ivory* Commissioned by Born Free and carried out by C4ADS, published 2014, p.26.
[20] AllAfrica 2012.
[21] See Rademeyer (n.17), p.12.
[22] Ibid, pp.15–16.

### 24.4.2   Zimbabwe

In March 2018, soon after the soft coup that removed President Robert Mugabe from office, news broke that his wife Grace, the former first lady, was being investigated by the Zimbabwean authorities over allegations that *'she ran a poaching syndicate that smuggled tons of ivory out of the country'* to buyers *'assumed to be organised criminal groups operating out of China and Malaysia'*.[23] Although Zimbabwe holds the second largest population of elephants in Africa, some parts of the country have suffered heavy losses due to poaching.

Much of the evidence, some in the form of undercover footage, was obtained by Adrian Steirn, an Australian photojournalist who spent 3 months undercover and managed to infiltrate the criminal networks by posing as a prospective buyer. Several human sources directly implicated Mrs. Mugabe, and there was further evidence from a number of key documents.[24] This led to a trap for suppliers allegedly working for Mrs. Mugabe, being set by police and whistleblowers.[25]

## 24.5   Reasons to Be Encouraged

Recently, there has been a crackdown on corruption because countries have begun to understand just how damaging it is to them. And this has made it more difficult for those who would be corrupt to carry out their damaging activities.

We have already seen how in Zambia, as soon as they became aware of the widespread corruption among ZAWA staff, the authorities replaced a suspect warden and sent in a team of outside investigators led by a Senior Investigator from Head Office. Eventually, and working in conjunction with staff from Malawi's Anti-Corruption Bureau, they managed to greatly reduce the level of corruption even though they were not able to eliminate it entirely. They acted very promptly once they were made aware of the problem.

In the case of the Vietnamese Embassy's First Secretary, uproar followed the television broadcast, and Mrs. Moc Anh was immediately recalled to Vietnam. The Vietnamese Ambassador was interviewed by a Vietnamese journalist.

---

[23] Peta Thornycroft, Roland Oliphant and Adrian Blomfield *Tusks, gold and diamonds flowed out of Zimbabwe* The Daily Telegraph, 24 March 2018.

[24] Ibid.

[25] Jan Raath (Harare) and Harriet Salem (Nairobi) *Grace Mugabe 'smuggled poached ivory'* The Times, 26 March 2018.

He was blunt in his condemnation of his former colleague, whose actions he described as *stemming from pure* greed. He also had the foresight to see that '(this was) not just a private act, it is one that has repercussion on the Embassy and the image of Viet Nam itself'. Counsellor Dzung was *'on leave in Vietnam. We have requested that the Ministry of Foreign Affairs summon him for explanation'.* Viet Nam's Foreign Ministry also *felt compelled to issue a statement'.* This indicated that it would *'strictly punish violation related to wildlife trafficking'.*[26]

## 24.5.1 'Eyes in the Courtroom'

There have long been suspicions about corruption in the legal systems in some African States. Most of the evidence however had been anecdotal until this pioneering scheme. *Eyes in the Courtroom* was set up in Kenya by the NGO WildlifeDirect holding *a watching brief on behalf of civil society, communities in Kenya that derive their livelihoods from wildlife – and elephants.* It uses legal interns to monitor court proceedings and the public is both kept informed of the trial's progress and alerted should irregularities occur.[27]

Their first Courtroom Monitoring Report published in 2013, provided *concrete evidence of systemic failings in the prosecution of wildlife crimes.* Five years of monitoring exposed the stark fact that the majority of suspects were *walking free from the courts.* This was because they were either pleading guilty and paying *a paltry fine* or pleading not guilty because they were confident their case *would be dismissed due to procedural irregularities.*

The Kenyan Government immediately took action. It established a Wildlife Crime Prosecution Unit, issued guidelines to prosecutors and magistrates who were also given training and passed legislation that dramatically increased penalties for those found guilty of committing offences against wildlife. In some cases, even life imprisonment. And this worked, for the second Report, published in 2016, showed better case management, more convictions even where defendants pleaded '*not guilty*', and tougher sentences imposed. Except for the kingpins. It concluded that *'no high level trafficker has yet been convicted and sentenced by Kenyan courts'.*[28] This all changed later that year, when the notorious Feisal Mohammed Ali was finally convicted of possessing illegal ivory, fined and sentenced to 20 years in prison (see later).

---

[26] Anon 2008, in Milliken and Shaw (n.439).

[27] Paula Kahumbu *Kenya jails ivory kingpin for 20 years* The Guardian, 23 July 2016. See https://www.theguardian.com/environment/africa-wild/2016/jul/23/ Accessed 04/10/2016.

[28] Ibid.

## 24.5.2    A Court Syndicate in KwaZulu-Natal, South Africa

Allegations have been made from a number of sources into possible corruption in parts of the judicial system in KwaZulu-Natal. The Nzimande Report, a confidential report for the Magistrates' Commission contains, among other matters, allegations of bribery, and some of the names in the report are the same as those provided by a whistle blower who claims *'he took money given to a lawyer from rhino-horn kingpins and paid it to people within the judiciary'*. One of the kingpins mentioned is Dumisani Gwala, who, as we shall see later, has managed to evade justice for a number of years.

As yet, no action has been taken on either the Nzimande Report or the results of the police investigations into the bribery allegations, which is of some concern as one of the reserves regularly targeted by poachers is Hluhluwe iMfolozi Park.[29]

---

[29] Alastair Leithead *South African rhino poaching: 'Bribes paid to court syndicate'* BBC News, 15 August 2018. See https://www.bbc.co.uk/news/world-africa-45089699 Accessed 04/09/2018.

# Part IX

## Enforcement

All illegal trade in wildlife poses an immediate global threat to the sustainable conservation of biodiversity, so anything that can in some way reduce it, is to be welcomed. Demand reduction is the key. No demand, no poaching. So any resolution to poaching and wildlife trafficking will require demand to be reduced to the absolute minimum. And, of course, there are the porous borders which enable effective and efficient trade routes to be set up by the professional criminals, the organized crime gangs. And there is the ever-present threat of corruption.

Enforcement, another key to resolution, was never going to be easy. While many of the range States and consumer countries have a substantial body of legislation in place to protect elephants and rhinos, the protection still tends to be rather ineffective. And there are often problems relating to enforcing the law. Where enforcement is successful it can make a big difference. We have some evidence for this. Whereas the number of white rhinos killed in South Africa had been rising steadily since 2006, in two provinces, KwaZulu-Natal and North West, it showed some decline, partly because of 'the increased co-ordination of security efforts and security operations, leading to a number of high profile arrests of poaching syndicate operatives'.[1]

More to add on recent situation Quazulu

[1] Milliken, T. and Shaw, J. (2012) *The South Africa – Viet Nam Rhino Trade Nexus: A deadly combination of institutional lapses, corrupt wildlife professionals and Asian crime syndicates* TRAFFIC, Johannesburg, South Africa

# 25

# The Enforcers

We'll start at the top, with INTERPOL, the international enforcement agency.

## 25.1 INTERPOL

Although we don't have an international police force as such, we do have INTERPOL, which is a network of police forces from all over the world, from in fact 190 countries. And its job is *'to help police in different countries work together to solve crimes that cross borders'*.[1] It is a network because each country has an INTERPOL office, its own National Central Bureau, which is connected to that country's own police force and to the other members of INTERPOL. This means they can both share information and have access to high-tech tools and resources.

Like CITES, INTERPOL has a strong administrative set-up, with its General Secretariat, Regional Bureaus and Liaison Offices. This structure helps make it effective.

Let us use the analogy of a spider's web, where INTERPOL is the spider. The central hub is the Command and Control Centre which has operation rooms in both France and Argentina and eventually in Singapore. These are manned 24 h a day, every day of the year by staff from all over the world. They send messages, issue notices and exchange information. Spiders sit in their webs waiting to catch their prey. Our spider, which of course, must be called

---

[1] *What is Interpol* INTERPOL. See http://www.interpol.int/ipsgapp/educational/what-is-interpol.html Accessed 17/09/2015.

© Springer Nature Switzerland AG 2019
B. Martin, *Survival or Extinction?*, https://doi.org/10.1007/978-3-030-13293-4_25

INTERPOL, has two spots in her web where she hangs out. But she doesn't just watch and wait, she is proactive as well. Her prey had better beware!

INTERPOL has a super secure network (in other words, its information remains within the law enforcement domain). It is called 1–24/7, and as well as linking all the member countries to each other, it enables them to search INTERPOL's international criminal databases. Their police can view INTERPOL's international criminal information and can share alerts with other member countries. It would seem that INTERPOL the spider has her own private Internet!

INTERPOL also keeps the police forces of its member countries informed of international crimes. It does this by issuing notices that are colour-coded, and are a form of alerts. They range from requests for information to both 'wanted' and 'warning' notices and work well as a means of communication with other jurisdictions. So how are they used? Let us take the case of rhino poacher Rajkumar Praja as an example. The year was 2013. Praja was wanted in Nepal, having been found guilty of both poaching rhinos in Chitwan National Park and trading internationally in rhino horns. He had been sentenced to prison for 15 years. INTERPOL issued a Red Notice after the Nepalese law enforcement authorities had dismantled a network of 13 poachers earlier that year. Another Red Notice, issued after a request by the Kenyan Government, was in respect of Feisal Mohamed Ali. We shall meet him again.

The police officers who work for INTERPOL are highly skilled, experienced people who bring with them their own expertise from the various countries they come from. They must also be linguists, familiar with at least one of INTERPOL's four official languages, English, Spanish, French and Arabic. And they work with civilians. However, they are only on loan and after a time will return to their own police forces taking with them the new expertise they have learned from working in a global/international environment.

Our spider has the potential to inflict a severe bite.

So how does INTERPOL protect our animals?

**Project Wisdom:** This was INTERPOL's response to the growing threat of the illegal trade in elephant ivory and rhino horn.

Its core objectives are to:

- Encourage communication, cooperation and collaboration with respect to intelligence exchange, cross-border investigations and training courses.
- Contribute to the apprehension of criminals and organised groups.
- Develop a global picture of the criminal activity hindering the ongoing conservation of elephants and rhinos.

- Reduce wildlife crime in Africa.
- Support and coordinate law enforcement operations.
- Provide case-oriented assistance.
- Emphasize deterrence as well as intelligence-led law enforcement.
- Contribute to broader civil objectives.

And they are being achieved by means of a number of Operations. We will look at five.

**Operation Tram (2010):** Eighteen countries targeted the illegal trade in traditional medicines containing wildlife products, and as a result, illegal products worth more than EUR10 million were seized.

**Operation Prey, Phase I (2012):** The object of Operation Prey, which consisted of three phases, was to target the illegal trade in Asian big cats and their derivatives. Phase I was conducted across Bhutan, China, India and Nepal, and nearly 40 people were arrested. 'The haul' of illegal products included both rhino horns and ivory.

**Operation Worthy (2012):** This Operation specifically targeted the illegal trade in elephant ivory and rhino horn and was coordinated across 14 African member countries. It worked well, resulting in nearly 2 tonnes of illegal elephant ivory, more than 20 kg of rhino horn and more than 30 illegal firearms being seized (as well as various other wildlife products).

**Operation Wendi (2013):** Another success story, this was aimed at combating the trafficking of elephant ivory in west and central African countries. Nearly 4000 ivory products were seized, including 50 elephant tusks and 88 firearms (again, with other products, and 222 live animals were released back into the wild).

**Operation Wildcat (2013)** This was a combined Operation in that it had two targets: to combat ivory trafficking and illegal logging across Southern and Eastern Africa. As a result, 660 people were arrested, 637 firearms and 44 vehicles were removed and 240 kg of ivory (and 856 timber logs) were recovered.

Why 'Wildcat'? Because the Operation was also supported by the Wildcat Foundation and the Norwegian Agency for Development Cooperation (Norad).

**Working Together in INTERPOL:** We have just seen one very successful example of working together, and although the organized crime gangs know how to exploit this to their advantage, the enforcers are even better at forging successful partnerships.

Formal intergovernmental networks, such as the Green Customs Initiative and the International Consortium on Combating Wildlife Crime (ICCWC) already exist, and in 2013, INTERPOL's member countries were invited to establish NESTS, National Environmental Security Task Forces. These promote a multidisciplinary approach, assessing a country's needs when tackling wildlife crime and promoting best practice.[2]

NESTS: *'gather, analyse and share information and intelligence about environmental crime and criminals'* with other INTERPOL member countries and participating agencies.

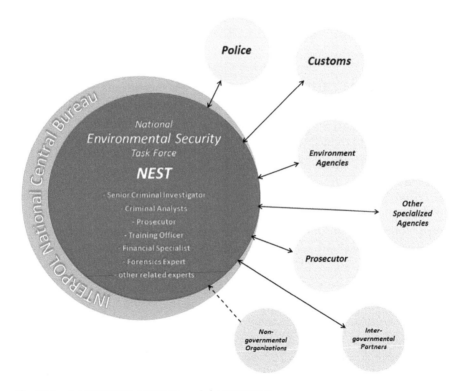

**Fig. 25.1** An INTERPOL NEST. Copyright INTERPOL

---

[2] INTERPOL Environmental Security Sub-directorate *Environmental Compliance and Enforcement Committee: Meeting and Events – Final Report* February 2014, p.6.

All 190 INTERPOL member countries are involved in the INTERPOL Environmental Compliance and Enforcement Committee whose object is to harness global support for environmental law enforcement. One of its three working groups is:

**The Wildlife Crime Working Group:** This *'initiates and leads a number of projects, led by different countries, to combat the poaching, trafficking, or possession of legally protected flora and fauna'.*[3] Academics and members of NGOs can also attend its meetings. As well as Project Wisdom, its Projects include:

**Wildlife Forensics Project:** Led by the USA, it has produced a wildlife forensics manual which is available on the INTERPOL website.

**Internet Crime Project:** Ongoing and led by Australia.

**Rhino Project:** This is *'currently focused on building links between Asia and Africa with regard to seizures made, to ensure both ends of the supply chain are appropriately covered off'.*[4]

A number of important issues were discussed at the 24th meeting of the Committee, which was held in Kenya in 2013. It dealt with Operations, including:

- African member States wanting help with an operation targeting rhino crime and illegal trade in rhino horn.
- The need for NGOs to be brought in to help reintroduces threatened species back into areas previously affected by poaching.
- Help needed to harmonize legislation for endangered species across the different range states, to make enforcement and prosecutions much more straightforward.

A number of additional special events also took place, including a workshop on *Environmental Crime Scene Investigation DNA and other Forensic Tools*, which recognized the serious threat of wildlife crime, the severe impact it was having on animals and that it posed *'an immediate global threat to the sustainable conservation of biodiversity'.*[5] Advances in DNA technology would help combat the threat, as would the INTERPOL Firearms Reference Table.

---

[3] Ibid, p.18.
[4] Ibid.
[5] Ibid, p.22.

This enables enforcers to identify a firearm used in a crime, while the INTERPOL Ballistic Information Network makes it possible for ballistics data to be shared and compared internationally.

At the meeting, the ICCWC was put forward as an excellent example of working together at both national and regional levels.

## 25.2 Working Together: The International Consortium on Combating Wildlife Crime (ICCWC) and Others

Five major players make up this important Consortium, whose object is to fight wildlife crime effectively. They are:

- The CITES Secretariat
- INTERPOL
- The United Nations Office on Drugs and Crime
- The World Customs Organization
- The World Bank

The ICCWC is there to catch and punish wildlife criminals, particularly those involved in serious offences, and it intends to use all the weapons at its disposal to achieve this.

Two other major players are ASEAN-WEN in Asia and the Lusaka Agreement Task Force in Africa.

**ASEAN-WEN:** The world's largest wildlife law enforcement network. It was formally launched in Bangkok in 2005 and is:

- A regional intergovernmental law enforcement network designed to combat the illegal wildlife trade
- A pro-active response to southeast Asia's alarming levels of wildlife trafficking and loss
- A mechanism by which countries can share information and learn from each other's best practices[6]

---

[6] *What is ASEAN-WEN?* See http://www.asean-wen.org/index.php/about-us/what-is-asean-wen Accessed 17/09/2015.

The police, customs and environment agencies of all 10 ASEAN countries, namely, Brunei, Cambodia, Indonesia, Laos, Malaysia, Myanmar, the Philippines, Singapore, Thailand and Viet Nam are involved. And there are links with CITES offices, INTERPOL, the USFWS, the US Department of Justice and other wildlife law enforcement groups. Once again, training workshops together with an annual meeting play a vital role.

ASEAN-WEN is administered by a permanent Secretariat, ('the Program Coordination Unit'), in Bangkok, Thailand, which also provides some financial support. Working with its partner organizations, one of which is TRAFFIC Southeast Asia, it formulates Action Plans, such as the ASEAN Regional Action Plan on Trade in CITES Wild Fauna and Flora (2011–2015). The USAID ARREST Program (Asia's Regional Response to Endangered Species Trafficking) is a 5-year programme which tackles wildlife trafficking in Asia in three ways:

* By reducing consumer demand
* By strengthening law enforcement
* By strengthening regional cooperation and the anti-trafficking network

Once again cooperation is the keystone. China and South Asian countries have been included, as well as the ASEAN-WEN States, NGOs and private sector organizations. With everyone working together, the prospects of curbing the destructive trade are greatly enhanced. Indeed, WEN representatives at the 2014 Symposium on Combating Wildlife Crime enthused about the need for WENs to develop links with others, links that could result in a global network.

Moving on to Africa, LATF plays an important role in fighting wildlife crime.

**The Lusaka Agreement Task Force on Cooperative Enforcement Operations Directed at Illegal Trade in Wild Fauna and Flora (LATF):** This came into force at the end of 1996 and is based in Nairobi and shares its headquarters with the Kenyan Wildlife Service. An intergovernmental organization dedicated to fighting wildlife crime; its main function is to facilitate cooperative activities in the network's Parties and signatory States.

There are six Parties: the Republic of Congo (Brazzaville), Kenya, the Kingdom of Lesotho, Tanzania, Uganda and Zambia. And three signatory States: Ethiopia, the Kingdom of Swaziland and the Republic of South Africa. The LATF is, like INTERPOL and all these networks, a permanent law enforcement institution, drawing its staff from law enforcement officers seconded from the Parties, with local people providing the support staff. It has National Bureaus in Liberia as well as all the Party States. And it has a Cooperation Agreement with INTERPOL.

The LATF investigates transnational wildlife crime, particularly where national laws relating to illegal trade in wildlife crime have been broken, and is also involved in capacity building, partnerships and meetings/conferences.[7]

TRAFFIC is linked to many of these regional bodies and networks, supporting their work particularly in the field of training, and sometimes providing them with technical assistance. It played an important part in the establishment of both ASEAN-WEN and the Lusaka Task Force Agreement.[8]

Both ASEAN-WEN and the LATF are involved in COBRA.

**COBRA:** Fighting transnational wildlife crime is difficult, but it becomes easier when different countries and regions are determined to work together to defeat the criminals. And this is where COBRA comes in. It is a multi-regional wildlife law enforcement operation, which both ASEAN-WEN and the LATF and SA-WEN (the South Asian Wildlife Enforcement Network), China, South Africa and the USA, were instrumental in initiating, with FREELAND Foundation for Human Rights and Wildlife providing financial and logistical assistance. In other words, it is a joint effort to fight the organized criminals and their networks. Network pitted against network. And it *bridges source, transit and destination countries*.[9]

**Operation Cobra II:** This demonstrates just how successful such an operation can be. Although it only lasted a month, more than 400 alleged wildlife poachers were arrested, and agents from 28 countries seized. More than 3 tonnes of ivory and 36 rhino horns were recovered, as well as other items. Furthermore, it marked the first successful partnership between Chinese and African agents. And it was truly productive partnership, for an ivory trafficking kingpin was arrested, a criminal who had been funnelling large numbers of elephants tusks out of Africa and into China.[10]

So when someone with the experience of Wan Ziming, the coordinator of the National Inter-Agency CITES Enforcement Coordination Group and

---

[7] See http://lusakaagreement.org/ Accessed 17/09/2015.

[8] TRAFFIC – Networks See http://www.traffic.org/networks/ Accessed 17/09/2015.

[9] Lusaka Agreement Task Force, Joint Press release *COBRA initiative enhances partnership in fighting transnational wildlife crime* 15 September 2015. See http://lusakaagreement.org/?p=857 Accessed 22/09/2015.

[10] USFWS 2014.

director of the Law Enforcement and Training Division of the CITES Management Authority of China suggests that enforcement actions such as these should continue yearly, we should all take notice. He was particularly impressed by the *'exceptional enforcement actions and interagency coordination exhibited by countries like the PRC, India, Kenya, Nepal and Singapore'.*[11]

He also suggested that all participating countries should:

- Properly allocate human, technical and financial resources.
- Provide information on the seizures that took place, to the organizers and countries involved.
- Coordinate enforcement actions.
- Ensure they keep their records confidential.
- Make better use of modern enforcement techniques and technologies, especially controlled delivery and sniffer dogs.
- Conduct more follow-up investigations.
- Endeavour to wipe out the transnational wildlife crime syndicates.

## 25.3 Customs and Border Controls

These are the frontline troops stationed on each county's borders.

**The World Customs Organization:** This represents 180 Customs administrations from different countries across the world. Run by a Secretariat, its governing body is the Council, and, like INTERPOL, its staff is drawn from a number of nationalities. It plays a vital role *'in stimulating the growth of legitimate international trade'* as well as being a forum for dialogue and providing technical services and training. It also promotes *'the emergence of an honest, transparent and predictable Customs environment'* and it 'champions' working together—the partnership approach.[12]

Because the World Customs Organization is a global organization like INTERPOL, the problem of illegal trade can be put into context, essential when so many countries are involved. Each State has its own system of border controls, its own personnel to patrol and protect its borders, to catch illegal traders who traffic not only endangered wildlife but drugs, arms and people.

---

[11] Ziming 2014.
[12] *World Customs Organization (WCO) in brief* See http://www.wcoomd/en/about-us/what-is-the-wco. aspx Accessed 30/08/2016

Customs officers have been able to build up large repositories of knowledge on who smugglers are and how they operate. They forge links with shipping companies and airlines, which can help intelligence gathering. And, as we saw at the very beginning of our story, they can catch smugglers, even those who disguise rhino horns in a seemingly innocent sculpture.

## 25.4   Europol

This is the European Union's law enforcement agency, and once again a number of countries are involved, not just the Member States but other partners as well, including Australia, Canada, Norway and the USA. Based in the Hague (in the Netherlands), Europol handles criminal intelligence, to prevent and fight all forms of serious international organized crime and terrorism. And we've already seen it at work, tracking down the infamous the Rathkeale Rovers.

Europol differs in some important aspects from the other enforcement organizations we have seen, as it uses a multiagency approach to recruiting the about 800 members of staff. So as well as the police, personnel are drawn from the border police, the customs and the security services. There are also 145 Europol Liaison Officers who are seconded to Europol by both the Member States and their non-European Union partners. These *'guarantee fast and effective cooperation based on personal contact and mutual trust'*.[13]

## 25.5   Enforcement in the UK and Ol Pejeta Conservancy

It is interesting to compare the process of enforcement in the UK, a consumer country with that in Ol Pejeta Conservancy, home to both rhinos and elephants.

**The UK:** A number of agencies are involved, including HM Customs and Excise, the Border Agency and the police. And where there are suspected crimes against wildlife, they are often assisted by experts from NGOs.

The Customs and Excise Management Act 1979 sets out offences of exporting and importing illegal goods, which of course includes trafficked elephant

---

[13] Europol *About Europol/Europol* 2015. See https://www.europol.europa.eu/content/page/about-europlo-17 Accessed 17/09/2017.

ivory and rhino horn. It also empowers enforcers to seize illegal goods, search any premises where it is suspected they may be and search vehicles and vessels if they are suspected of carrying illegal goods.

Education and training play an important role. An early government publication '*Wildlife Crime: A Guide to Wildlife Law Enforcement in the UK*'[14] set out in some detail not only the relevant legislation but also essential information such as what Chinese Traditional Medicines (known then as TEAMS, Traditional East Asian Medicines) were, what they contained and what they were used for. Rhino horn was '*used as an ingredient in a range of products, most often in the form of tablets and herbal teas*'.[15]

And again there are examples of the benefits obtained by working together. As well as PAW, the Partnership for Action Against Wildlife Crime, there is the National Wildlife Crime Unit (NWCU). Comprised of a number of experts, it is there to help prevent and detect wildlife crime. Because the NWCU obtains and disseminates information from a wide range of sources, works with PAW and assists the police with their investigations into wildlife crime, it is easier to bring successful prosecutions. While its input into *thousands of investigations and hundreds of prosecutions* in the UK is invaluable, it also has a global profile and regularly works with law enforcement agencies around the world.[16]

The police operation, 'Operation Charm', ongoing for many years, has been responsible for many successful prosecutions, including two of the first prosecutions in the world for offering for sale items/remedies containing parts of endangered species. The cases took place in 1995, and both defendants were fined. Punishments, which can either be a fine, a prison sentence or both, are set out in the Control of Trade in Endangered Species (Enforcement) Regulations, a number of which have been passed to keep the legislation up to date.

**Ol Pejeta:** There things are rather different. Because the Conservancy is now home to the largest population of black rhinos in east Africa, a 'Key 1' population, which is the highest category of continental importance,[17] as well as the last northern white rhinos on the planet, enforcement is of the utmost importance. Armed rangers protect these rhinos 24/7. Drones monitor wildlife to pick up potential threats, and there is a response unit of the (Kenyan) National Police Reservist Rangers who have powers to arrest poachers. Dogs also play a

---

[14] Department of the Environment and Partnership for Action Against Wildlife Crime *Wildlife Crime: A Guide to Wildlife Law Enforcement in the UK* The Stationary Office, 1996.

[15] Ibid, p.118.

[16] RSPB *10 year celebrations of NWCU* Legal Eagle, February 2017, No 81.

[17] Kristi Foster *Closer look: Ol Pejeta Conservancy* See http://www.fauna-flora.org/closerlook/olpejeta/ Accessed 27/08/2016.

key role. In 2013, an anti-poaching dog unit was developed, their K-9 Unit, and the following year, 'multi-role' dogs were trained. In 2017 alone, the canine unit responded to over 100 incidents, both in and outside the Conservancy.

By 2017, Ol Pejeta was employing more than 150 rangers, including over 40 highly trained National Police Reservists to protect its rhinos. But even these, working closely with government authorities and other conservancies, and with their exceptional K-9 Unit, were unable to prevent 3 rhinos being poached, although in each case the poachers were caught.[18]

## 25.6   The Rangers

These are the troops on the ground in the producer countries, and many of them have been killed or terribly injured carrying out their duties.

One of their most important tasks is to regularly patrol the areas for which they are responsible, their presence essential to check on the animals, thwart potential attacks by poachers and deal with the dreadful aftermath of a successful attack. Patrols are normally carried out on foot and in vehicles, both methods potentially hazardous, but providing them with the opportunity of carrying out another vital role, data collection.

Myanmar shows how this can be achieved. There, the rangers work in Elephant Protection Units, in partnership with decommissioned logging elephants which carry them deep into the jungle. Working elephants helping to protect their wild cousins who need protecting. In 2014 alone, some 230 incidents were reported and many arrests made. And the rangers' station was burnt down in an arson attack. In 2015, the teams spent a month in Alaungdaw Kathapa National Park, a known hotspot for illegal activity, and collected data that was then entered into the SMART software database (more of SMART later).[19]

Data collection has been the subject of recent research in Queen Elizabeth National Park, Uganda. There, all surveillance patrols, whether on foot or in vehicles, use handheld GPS units to collect various data.[20]

They record their exact location:

* When they see an animal
* See signs of illegal activity
* At 30 min intervals after the last sighting or recorded position

---

[18] Ol Pejeta Annual Report 2017.
[19] Elephant Family, 2016.
[20] Critchlow et al. 2015.

This information is supplemented with *'supporting information'* that provides additional details such as the type of illegal activity. When it is all collected together, it provides a better idea of what is happening in the Park.

Snares can cause major problems to both elephants and rhinos, although most snares are set to catch bushmeat, food, an activity that has been carried out for centuries. Rangers look out for and dismantle snares.

Sean Willmore, President of the International Ranger Federation, paints a moving picture of the day-to-day lives of rangers. Of how, when on duty, they live in tents. How they cannot afford to let down their guard because there may be poachers about, so two always remain on duty, armed and prepared. And of the *affinity* between rangers and their charges, for animals often approach and stay near the camping sites during the night. The rangers are convinced they come there because that is where they feel safe.

Poachers, however, pose a very real danger. The very worst that can happen is that rangers are killed. But what if they are 'merely' injured? If the incident is serious, as with any soldier it may well result in post-traumatic stress syndrome. While terrible wounds, often inflicted by a gun do heal, the underlying emotional distress, perhaps brought on by the fear they might die next time, or because they have lost a friend or colleague, lingers on, festers away, gnawing at their confidence. And even minor wounds can sometimes prove fatal if they aren't or cannot be treated correctly at the time.

Even more dispiriting for the rangers can be the feeble sentences meted out to the poachers they have only recently caught and brought to justice. Poachers released from prison a month or two after their detention. Poachers released immediately to return to poaching, armed, perhaps with the very same guns that inflicted the previous damage, guns that killed and maimed some of their colleagues. And then there are the death threats. Sometimes directed at them, sometimes at their families. Threats uttered by people who would seemingly have no scruples about carrying them out.

Rangers are heroes!

So what happens to those rangers who suffer severe injuries but survive or to the families of those who die in or as the result of action?

**A Safety Net of Sorts:** Some of them are helped by the International Ranger Federation (IRF). Although this is only a small organization, it packs some punch. Here are a few of the things it does:

- Some rangers do not get paid enough money to fully support their families, so the IRF will 'top-up' the amount to help pay for food or other essentials.
- It may be possible to provide prosthetics for some rangers who lose limbs while in action.

- It has created farm jobs and subsistence livelihoods for some of the widows.
- It helps children of rangers who have lost their lives, to get back to school – which costs money of course.
- It can assist with training and equipment for essentials such as mosquito nets and disinfectants to clean wounds thoroughly.

And more help has arrived, this time in the form of war veterans from countries such as the USA and the UK. They have much to contribute in this ongoing war against the poachers.

One group is the International Anti-poaching Foundation (IAPF). Founded by Damien Mander, who had been a member of the Australian Army's Special Forces, its aim is *'to reduce poaching with direct action'*. It is currently working in Mozambique, where poaching is rife. And it is making a difference, a considerable difference, for there has been a massive drop in the numbers of white rhino poached in Kruger National Park. A difference recognized by Mozambique's National Agency for Conservation, which has engaged IAPF to help train the rangers in the country's National Parks.

Another group, Veterans for Wildlife, is from the UK. It has three objectives:

- To place highly skilled and knowledgeable volunteers into strategic positions throughout Africa to train, mentor and support antipoaching rangers and staff.
- To raise awareness about the wildlife poaching epidemic by running educational initiatives and programmes, sharing volunteer's stories and engaging with larger audiences.
- To fundraise, not just *'to place volunteers and run programmes, but also to support other anti-poaching initiatives'*. This includes *community uplift projects*, like building schools, creating jobs, lifting people out of the poverty that may well be driving them to poaching in the first place and educating them about the damage it is causing.

**Their Way of Working:** In early 2016, a BBC correspondent vividly recounted his experiences in the Borana nature reserve in Laikipia, Kenya. The rangers had benefitted from training by the veterans and were *'part of an SAS-trained private army'* protecting rhinos from poachers. There, they shoot to kill because too many rhinos were dying.

Hand-picked rangers, working in teams of four, one signaller, two riflemen and a sniper, were equipped for night vision and carried digital radios to call in air and ground support if necessary. They were all enrolled as Kenyan police reservists, as that gave them the authority to shoot to kill, and they operated in both Lewa and Borana Conservancies. Although 19 poachers had died, in 2015 no rhinos were lost to poaching in Lewa and only 1 in Borana.[21]

In August 2016, another BBC correspondent described his experiences alongside rangers also trained by the veterans, as they carried out their duties in a private game reserve in Quazulu Natal, South Africa. Their work was more dangerous though, because they could only shoot when shot at, unlike the National Parks, where again rangers can shoot poachers on sight.[22]

Teams of rangers are made up of men, or both men and women, but not the Black Mambas. All of them are women.

**The Black Mambas:** Formed in 2013, there were originally only 26 of these brave women, who were the first all-female antipoaching unit in the world. They patrolled with their specially trained dogs, every day and every night, in the 50,000 hectare Balule Nature Reserve, part of South Africa's Greater Kruger National Park, risking their lives to care for the rhinos, elephants and big cats that are found there.

And their duties included:

* Looking for dead rhinos
* Freeing any animals they find caught in a snare
* Removing snares

Clad in their military fatigues, they looked like soldiers, and the idea was that their mere presence would act as a deterrent to potential poachers. However, unlike poachers, they did not carry guns, which would have made them very vulnerable were it not for their backup. Their dogs, who are with them of course, and not far away helicopter patrols and armed rangers. Should they encounter a poacher, they must immediately radio-in their position to the armed rapid response unit, who will then come to their aid.

With their presence, poaching started to decrease. In 2007, 13 rhinos were poached in Balule Nature Reserve, the numbers soaring to over 1215 in 2014, but by the end of 2015, the number had been dramatically reduced.[23]

[21] Chris Haslam *The snipers trained to protect rhinos* BBC News Magazine, 7 February 2016. See http://www.bbc.co.uk/news/magazine-35503077 Accessed 27/09/2016.

[22] BBC Radio4 *From Our Own Correspondent* broadcast on 24 September2016.

[23] Jay Akbar *The real lionesses of Africa: Stunning 'Black Mambas' are first all-female anti-poaching unit risking their lives to protect big cats, rhinos and elephants from men with guns* MailOnline, 22 February 2016. See http://www.dailymail.co.uk/news/article-3458085/The-REAL-lionesses_Africa-Stunning- Accessed 1/03/2016.

In October 2016, some volunteers from Veterans 4 Wildlife spent 10 days with these exceptional rangers.

By 2018, their numbers had grown to 32, plus two men. They now undertake paramilitary training and have taken on additional duties including VHF rhino tracking. And in 2019, a special report in the Daily Telegraph revealed that a group of female veterans from Britain, in particular the Sandhurst Sisterhood, have been supporting them since 2016, involved in *'everything from tactical training and finances to running the operations room'*.[24]

## 25.7    Sniffer Dogs and Tracker Dogs

These are the canine members of the team, last but not least. *'Sniffer dogs may look cute, but their message is clear: they can't be fooled. They are on a canine mission to stop wildlife crime'*.[25]

These enforcement agents, working with their handlers, are playing an increasingly important role in the fight against illegal wildlife trade. Logistically, they can make a significant difference, not only because together dog and person make a formidable team but also because the dogs can do what people can't, partly through lack of time, scan cargo, luggage and packets for, in our case, ivory and/or rhino horn, without everything having to be opened and examined. And they are similarly efficient when tracking poachers in the Parks and reserves.

Early in 2012, after a flood of incidents involving elephant poachers, Virunga National Park, in DRC, got its first team of bloodhounds. They swiftly discovered a cache of illegal weapons. A few days later, rangers spotted a poached elephant whose tusks had been hacked off. A potentially more deadly game commenced. Two of the dogs and a specially trained ranger unit were transported by helicopter to the body, where the dogs picked up the scent and tracked the poachers to a small fishing village. After opening fire, the poachers dropped their weapons and fled. The dogs had tracked them some 7 km (four miles).

---

[24] Helping Rhinos *Black Mamba Anti-poaching Unit* 2018.See https://www.helpingrhinos.org/black-mambas/ Accessed 05/06/2018. See also Hayley Dixon *The British women leading the war on poaching* The Daily Telegraph, 5 January 2019 for a fascinating account of how the Black Mambas and the veterans are working together.

[25] Sarah Ruggiero *Canines and Contraband* WWF, 2014. See http://www.worldwildlife.org/stories/canines-and-contraband Accessed 31/03/2016.

A specialized Swiss centre and volunteers from the German police were involved in this *'bloodhound programme'*.[26]

By 2014, dogs were being widely used in South Africa, in reserves to track poachers and in airports and parks to detect smuggled ivory and rhino horn. Tracker dogs and sniffer dogs. Ol Pejeta has a team of 20 and uses them in both offensive and deterrent roles.[27]

But sniffer dogs can be used at more places than airports and ports and sniff out more than ivory and rhino horn. As their noses are *almost one thousand times more sensitive than a human's*, they can pick up a scent up to 300 h old.[28] And the reason they make such effective members of enforcement teams is because of the close bonds they form with their handlers. There must be trust and understanding between them because they must rely on each other, *'the sniffer dog for commands and directions and the handler for cues and expressions'*.[29] This time, it is human and dog working together, and if they are to be effective and make successful seizures, both handler and dog must be finely attuned to each other.

So how do the dogs communicate with their handlers? They are trained to identify the illegal goods, and when they find them, they learn, among other things, to sit or lie down beside the suspect container or package. Their handlers watch them intently for any sign of clues, such as sudden alertness and other expressions.

Enforcement is expensive, and sniffer dogs add even more expense, thus increasing the financial burden on countries some of which can ill afford it. So, many dogs are there as the result of NGO partnerships. In South Africa, one NGO partner has been the Endangered Wildlife Trust, which itself had partners, the African Consultants for Transport Security (ACTS), a cargo clearing company and Bidvest Magnum (Endangered Wildlife Trust 2014). The dogs, Rico, Renaldo, Heddi and Condor, worked in the cargo sections of OR Tambo International Airport, sniffing out smuggled ivory and rhino horn, and at Lanseria International Airport, where the sniffing was to be extended into passenger areas.

Antipoaching dogs can be trained:

- To sniff out arms and ammunition
- To detect bullets at crime scenes
- To track people

---

[26] *DR Congo employs dogs to tackle elephant poaching* BBC News Africa, 5 March 2012. See http://www.bbc.co.uk/news/world-africa-17256894 Accessed 22/03/2016.

[27] Ol Pejeta 2016.

[28] See Ruggiero (n.486).

[29] *How Dogs are Tackling Wildlife Crime* WWF Tx2. See http://tigers.panda.org/illegal-wildlife-trade/sniffer-dogs/ Accessed 07/01/2016.

This achieves results that might not otherwise have happened. Even brave and competent handlers are mere humans with limited senses. They find it hard to work at night. Not so dogs, and if they can also find stashes of weapons that the poachers intend to use on their next visit to the reserve, or spent bullets that can be linked to a poacher's gun and a dead rhino, they are doing a fantastic job. This is the kind of evidence that when used in a prosecution can result in convictions.

Zimbabwe also started to use dogs to protect its precious population of black rhinos, with funding provided by the International Rhino Foundation in partnership with GreenDogs.[30]

In summer 2015, a group of eight dogs which included a Spaniel, a German Shepherd dog, a Belgian Malinois and a German short-haired Pointer started work with Kenya Wildlife Service and the Wildlife Division of Tanzania. Trained by the African Wildlife Foundation, their job was to detect smuggled ivory hidden in shipping containers in the Tanzanian Port of Dar es Salaam. Other sniffer dogs were to be employed at two Tanzanian airports, Julius Nyerere International Airport and Kilimanjaro International Airport.

In Kenya, they joined an existing canine unit, one of the first in the region. There, their jurisdiction included border crossings through which ivory is smuggled.[31] Lewa Conservancy has its own K-9 unit. Two of its bloodhounds, Tripper and Tony, are regularly sent to Northern Rangelands Trust to help out. The latest member of the team is Zack, a 17-month-old Belgian Malinois. Together with Jack and Maxo, they are trying out the special boots they have to wear to protect their paws from the spiky thorns of the African bush and learning to bond with their new handlers. Belgian Malinois are particularly good at detaining prisoners on the run.[32]

In the Mkomazi National Park in Tanzania, where Chester Zoo is a partner organization, it contributes to the sniffer dog programme. These dogs are Belgian Malinois, and they help to track poachers with the antipoaching patrols in the Mkomazi Rhino Sanctuary.

Meanwhile, in 2015, in Nepal, on the other side of the world, two sniffer dogs, Murray and Sears, together with their handlers, started work in Chitwan National Park. Their mission, to sniff out evidence of poaching and track routes used by poachers, then uses this knowledge to apprehend and deter poachers in and around the Park. Trained in India, they are named after the

---

[30] Endangered Wildlife Trust Rhino Conservation Project *The Use of Dogs* 2014. See https://www.ewt.org.za/RHINO/dog.html Accessed 07/01/2016.

[31] *Ivory: Sniffer Dogs at African Ports Now on Smugglers' Tails: News…* Nature World News, 27 July 2015. See http://www.natureworldnews.com/articles/15837/20150727/ivory-p Accessed 07/01/2016.

[32] The Northern Rangelands Trust, Wild Frontiers, May 2016 update.

famous tennis star and WWF global ambassador Andy Murray and Kim Sears his wife.[33]

TRAFFIC and WWF started their work with sniffer dogs in Russia, China and India, collaborating with governments and law enforcement agencies in the fight to protect tigers by seeking out illegally traded products. In Nepal and India, the dogs were trained to combat poaching. China boasts the largest sniffer dog training centre in Asia, the Ruili Sniffer Dogs Training Centre, located in Yunnan province. It is an official partner of TRAFFIC.[34] The first two dogs were employed at Guangzhou Baiyun International Airport. Jin Kai and Jin Li successfully passed their exams and were able to detect wildlife contraband in a package centre, a shipping container and an airport luggage conveyor belt.[35]

Because of the heavy volume of container traffic at the ports used by the smugglers, the dogs could only investigate a fraction of them. The odds were still in favour of the smugglers. But in 2018, a possible breakthrough came in the form of Rasco, the Remote Air Sampling for Canine Olfaction. Air can now be sucked out of suspect containers, or even a random sample of containers, and passed through filters which are then presented to specially trained dogs, who will sit if they smell anything suspicious.

The device is being tested in Kenya, Mombasa, the port where so many ivory seizures have been made, and with as many as 2000 containers passing through daily even the keenest dogs could do with some additional help. WWF, TRAFFIC and KWS are working with them to provide this.[36]

[33] World Wildlife Fund – WWF – Worldwildlife.org *Sniffer Dogs Join Fight Against Wildlife Crime* Akash Shrestha / WWF – Nepal See http://www.worldwildlife.org/stories/sniffer-dogs-join-fight-against Accessed 07/01/2016.
[34] WWF *How Are Sniffer Dogs Trained?* See http://tigers.panda.org/illegal-wildlife-trade/sniffer-dogs/ Accessed 07/01/2016.
[35] See Ruggiero (n.486).
[36] *Meet an ivory trafficker's 'worst nightmare'* BBC News, 26 August 2018.

# 26

# Seizures and Arrests

Part of the enforcement process, seizures give some indication of the amount of illegal wildlife products that are destined to be laundered into the markets. Not only that, big seizures of ivory, more than 500 kg, *are indicative of the existence of a criminal organization driving the poaching, as only with such organization can the whole set of arrangements, planning, and intelligence be financially sustained*.[1]

Most seizures, which are of ivory rather than rhino horn, can and do take place all over the world, and although they are successful, they are not successful enough, despite the fact that enforcement authorities work steadfastly to try to prevent the illegal goods from either leaving their country of origin or reaching their final destination. The problem is, as we have already seen, that the smugglers are very good at their job. They have a multiplicity of routes to choose from and everything to play for because the rewards are financially so high.

Unfortunately, not many arrests are made because although the smuggled goods are there, the smugglers are somewhere else. A selection of examples provides a snapshot of some successes in preventing the illegal trade and explaining this apparent anomaly.

---

[1] Milliken et al., 2012 in Rossi, A. (2018). *Uganda Wildlife Trafficking Assessment*. TRAFFIC International, Cambridge, UK.

© Springer Nature Switzerland AG 2019
B. Martin, *Survival or Extinction?*, https://doi.org/10.1007/978-3-030-13293-4_26

**2010 South Africa:** At OR Tambo Airport, just before the 2010 FIFA World Cup opening ceremony was due to start, 2 Vietnamese traffickers were arrested trying to smuggle out 18 rhino horns.[2]

### 2010–2014 The European Union (the figures come from ETIS)

**UK:** There were 154 seizures with a total of 345.5 kg in raw ivory equivalent weight (RIE), an average of 30.8 seizures a year. Compare this with the previous 5 years when the average number of seizures was 15 and the total RIE was 134 kg. Most ivory was seized in 2012, when there were 45 seizures totalling 84 kg RIE, including one 15 kg seizure of hundreds of carved items which included bangles, pendants, beads and name seals. This consignment had been smuggled out of Nigeria and was destined for Hong Kong SAR and Guangzhou in China.[3]

**Germany:** Reported 236 seizures weighing 753.03 kg RIE.

**France:** Reported 268 seizures weighing 1561.38 kg RIE.[4]

**2011 Hong Kong SAR:** A major seizure from a shipping container from Cape Town, South Africa, recovered 33 rhino horns and items of carved ivory hidden in 63 packages of *scrap plastic*. The horns had been '*wrapped in aluminium foil, newspaper and plastic*'.[5]

**2012 Hong Kong SAR:** The law enforcers who seized elephant tusks from two separate containers must have been astonished by the size of their find, which consisted of over 1200 tusks, worth about USD3.4 million, and was one of the largest seizures of ivory ever made. To collect so many tusks suggested they must have come from a number of sources and that organized crime gangs must have stored them in various locations, mainly in Central Africa, until they were loaded into the two containers then shipped through Kenya and Tanzania to Asia.

---

[2] Many sources, see Moneron, S., Okes, N. and Rademeyer J. (2017) Pendants, Powder and Pathways TRAFFIC, East/Southern Africa Regional Office, Hatfield, Pretoria, South Africa, p.10.

[3] Border Force 2012 in Lau, W., Crook, V., Musing, L., Guan, J. and Xu, L. (2016) *A rapid survey of UK ivory markets* TRAFFIC, Cambridge, UK.

[4] Ibid.

[5] Anon 2011 in Moneron, S., Okes, N. and Rademeyer J. (2017) *Pendants, Powder and Pathways* TRAFFIC, East/Southern Africa Regional Office, Hatfield, Pretoria, South Africa, p.10.

Some of them may have come from the 400 elephants slaughtered in Cameroon by poachers from the Lord's Resistance Army. Fortunately, the President had reacted with determination, immediately sending an elite army to the area.

**2013–March 2016 Uganda:** During this period, Uganda Wildlife Authority recorded seven big seizures of ivory (each above 500 kg) on its Wildlife Trade Seizure Database, six of which took place at Entebbe International Airport. A record of 1478 kg of ivory from Uganda was seized in Kenya in 2013. This ivory had been disguised as fish maws. The suspects arrested in Kenya were then prosecuted, while those arrested in Uganda still appear to be waiting to be tried.[6]

**2013 Czech Republic:** The 2 rhino horns seized at Vaclav Havel Airport, Prague, were found hidden in *'electronic equipment'*, having been *'cast in resin and sealed inside an electric coil'*.[7]

**2014 Cameroon:** One of the final strongholds of the critically endangered forest elephants, the threat to the elephants became manifestly clear when in November, the eco-guards in the Dja Biosphere Reserve made a grim discovery. A total of 39 tusks, many of them small, so obviously from young elephants.

The seizure was the result of a tip-off from an intelligence network supported by the Zoological Society of London (ZSL). The tusks had been carefully hidden under cocoa bags in a truck, which was intercepted in the south of the country. Had this not happened, its owner, a local businessman, would have made a great deal of money from his ill-gotten gains, for the ivory, which weighed 91 kg, was one of the largest single seizures made in Cameroon. It was valued at more than USD190,000. The businessman was arrested.

The Ministry of Forestry and Wildlife Conservator stressed how important prosecution was to deterrence, so penalties are high, and a defendant could end up in prison for 3 years merely for trafficking a single elephant tusk. Unfortunately though, traffickers are rarely prosecuted.

---

[6] Rossi, A. (2018). *Uganda Wildlife Trafficking Assessment* TRAFFIC International, Cambridge, United Kingdom.

[7] Czech Republic CITES MA 2016 in Moneron, Okes and Rademeyer (n.2), p.10.

**2015 UK:** There was a massive seizure of 110 kg of ivory at Heathrow Airport. Found in abandoned luggage, it came from Angola, and its reported destination was Germany. The haul included tusks, carved bangles and beads,[8] but its contents were unusual. Normally only small amounts of ivory for personal use are smuggled into the EU,[9] whereas *this single seizure had more ivory, in terms of total volume, than the annual seizure totals in the previous 10 years*.[10] Once again, the UK was acting as a transit country.

That same year, the UK Border Force seized over 150 postal parcels containing ivory items en route to China. Most of the items were antiques, with one parcel contained 22 kg ivory carvings. The ivory was only discovered as the result of a specific operation the Border Force was carrying out, targeting such parcels, so this could have been happening for many years.[11]

**2015 South Africa:** The offender, a diplomat from North Korea, was arrested while driving a car in Mozambique. With him in the car, which carried a South African diplomatic registration, was almost 10 lb. of rhino horn and USD99,300 in cash. The South African government expelled him.[12]

**2015 Mozambique:** An enormous seizure of both elephant tusks and rhino horn was taken from a house in Matola (Maputo); the 340 tusks alone, weighing about 1.1 tonnes, were worth a fortune. Shockingly, there were also about 65 rhino horns, in itself a record seizure.[13] In total, *at least 170 elephants and more than 30 rhinos*' had been killed.[14] With its *fresh blood spatter and the rank smell of decay*' indicating *that some of the horns were from recent kills*', this was the stuff of nightmares. The police arrested a Chinese national, one of the occupants of the house, and the next day, another Chinese man, this time for corruption. He had *offered police investigators a $34,000 bribe to drop the case*'.[15]

---

[8] BBC 2015 in Lau et al (n.3) .

[9] Mundy 2014 in ibid.

[10] Ibid.

[11] TRAFFIC 2016.

[12] The Times *Diplomat had rhino horn in car* 24 December 2015.

[13] Wildlife Extra News *Record rhino horn seizure reported in Mozambique*. See http://www.wildlifeextra.com/do/ecco.py/view_item?listid=1&listcatid=1&listitemid Accessed 04/06/2015.

[14] Julian Rademeyer *Beyond Borders: Crime, conservation and criminal networks on the illicit rhino horn trade, Part 2 of a 2-part investigation into rhino horn trafficking in Southern Africa* The Global Initiative Against Transnational Organized Crime, 2016.

[15] Ibid.

Unfortunately, it did not end well. Not only did 12 of the horns vanish from supposedly safe custody at the police's provincial command headquarters, so too did the suspects. They had been released on bail!

The Head of the Criminal Investigation Division in Matola was arrested, together with three other policemen and two other people, all 6 charged with the theft of the rhino horns. But the case went nowhere, and potential evidence was deliberately destroyed by the government who decided on a public burning of the entire seizure. Although the intention was good, *'to send a signal that the country will not tolerate poachers, traffickers and the organized criminals which employ and pay them to kill our wildlife and threaten our communities'*, the South African authorities, who store DNA samples on a special database (more later), only had time to take samples from about 18 of the 65 horns.

**2015 Tanzania:** A kingpin was arrested. Or rather, a queen-pin, for the lady in question, a Chinese grandmother, was known as the *ivory queen*. Conservationists have described her as *'the most important trafficker ever busted'*.[16]

A Swahili translator for Chinese railway workers, Yang Feng Glan, moved to Tanzania in the 1970s and then stayed on, leading a double life. While she appeared to be a respectable businesswoman who owned a popular Chinese restaurant, she was also a highly successful smuggler, managing to spirit nearly 2 tonnes of ivory, worth an estimated £1.6 million, out of the country between 2000 and 2015, when she was arrested. After a high-speed car chase through Dar es Salaam, her reign ended. Ms. Yang was arrested, *charged with smuggling more than 700 elephant tusks out of Tanzania* and was facing up to 30 years in jail.

**2015 Singapore:** There was a massive seizure of some 2000 elephant tusks, which had been cunningly hidden in a cargo of tea leaves from Kenya.[17]

**2015 Vietnam:** Over 155 kg of elephant ivory and 9 kg of rhino horn were seized by customs officers at Noi Bai Airport, a *'significant highway for illegally traded wildlife parts'*, while at Da Nang city, the amounts were even larger, 3600 kg of ivory and 142 kg of rhino horn.[18]

---

[16] Ruth Maclean *Ivory queen who netted 700 tusks is arrested* The Times, 10 October 2015.

[17] News 24 *Singapore destroys tons of illegal ivory* 13 June 2016. See http://www.news24.com/Green/News/singapore-destroys-tons-of-illegal-ivory-2016061 Accessed 11/08/2016.

[18] TRAFFIC *Chi on display at Noi Bai Airport* TRAFFIC Wildlife Trade News, 4 February 2016. See http://www.traffic.org/home/2016/2/4/chi-on-display-at-noi-bai-airport Accessed 03/03/2016.

**2015 Vietnam:** 2.2 tonnes of ivory from Mozambique was seized at Hai Phong port.[19]

**2015 Vietnam:** At Tan Son Nhat Airport, Ho Chi Minh City, 1.39 kg of rhino horn was seized. It *'had been cut into small pieces and hidden in lobster heads, which were then kept in a refrigerated container'*. A Vietnamese citizen was arrested.[20]

**2016 Vietnam:** October was a particularly disturbing month in that four large seizures of ivory took place, two in the same week. The amounts included 3.5 tonnes, a tonne, half a tonne and 300 kg, and all of them had been carefully hidden, the largest haul in crates of wood, the tonne was hidden among boxes of fruit on their way to Hanoi, the 500 kg in two crates of timber from Mozambique, while the 300 kg, which came from Nigeria, had been deliberately mislabelled as glass. They were all bound for different destinations.[21]

**2016 Côte d'Ivoire:** This seizure came as something of a shock. It was the largest haul of illegal ivory in a decade, consisting of 150 kg of elephant tusks that had come from Nigeria, where things had seemed to have quietened down, and had reached the border town of Bonoua, near Ghana.[22]

**2016 South Sudan:** More than a tonne of elephant ivory was seized at Juba, the capital's airport. It had arrived on an Ethiopian Airlines flight from Uganda, hidden in a consignment marked 'food', and was bound for Cairo, Egypt. Once there, it was to be transferred to an EgyptAir flight going to Malaysia and, eventually, Asia. Two people were arrested, one from Uganda and the other from South Sudan.[23]

**2016 South Africa:** Junior rangers in Kruger National Park arrested one of their comrades and a veterinary technician, both of whom had worked for the Park Authorities for many years. In other words, 2 of their colleagues. It must have been a harrowing experience. This is what happened.

---

[19] *Three tonnes of ivory seized in Vietnam* Phys.org, 9 July 2017. See https://phys.org/news/2017-07-tonnes-ivory-seized-vietnam Accessed 06/03/2018.

[20] Huy 2015 in ibid., p.11.

[21] See (n.19).

[22] News 24 *I Coast seizes 150 kg of smuggled tusks* 10 June 2016. See http://news24.com/Green/News/i--coast-seizes-150kg-of-smuggled-elephant-tusk Accessed 11/08/2016.

[23] See    http://www.news24.com/Green/News/ssudan-more-than-a-ton-of-asian-bound-ivory-seized Accessed 11/08/2016.

Making their way towards gunshots, the rangers found a white rhino that had obviously just been killed and two men running towards a pickup truck. They managed to block the vehicle and arrest the poachers who turned out to be Mr. Landela, to whom they would normally be reporting and his veterinary accomplice. The evidence was incontrovertible: *'two rhino horns, a high calibre hunting rifle, blood-stained shoes and other poaching equipment'.*[24]

What made it even more tragic was the fact that Mr. Landela had won an award during his 15 long years on the frontline.

**2016 Zimbabwe:** Anomalies amounting to 228 kg were found in the country's legally held stockpile of rhino horn. An audit was carried out, and the Director General of Zimbabwe's Wildlife Management Authority went on leave so that the disappearance of the horn could be investigated, but there was insufficient evidence to make an arrest.[25]

**2016 South Africa:** A Chinese national was arrested at OR Tambo International Airport allegedly leaving the country and taking with him both money and rhino horn. As well as R1.5 million, there were 10 horns, 84 ornaments and 41 bangles. Quite a seizure! The suspect was charged with trafficking in rhino horn.[26]

**2016 India:** The Government Railway Police seized rhino horns from Shankar Rao, when he got off the train at Jaipur. Rao worked for Himalaya Export Company. While they were investigating the case, they recovered carved elephant tusks and antique firearms from his employer, Anurag Tiwari, now on the run and suspected of being involved in an international gang of wildlife smugglers.

The Himalaya Export Company, Tiwari's antique export business, appeared to have been providing cover. Searches uncovered various items including incriminating documents, and police departments in Kolkata, Goa and Delhi were working with the police in Jaipur to make a number of arrests.[27]

---

[24] Peta Thornycroft *Kruger Park ranger arrested over killing of white rhino* The Daily Telegraph, 30 July 2016.

[25] News 24 *Zimbabwe wildlife boss put on leave over 220 kg missing rhino horn* 28 June 2016. See http://www.news24.com/Green/News/zimbabwe-wildlife-boss-put-on-leave-over-220k Accessed 11/08/2016.

[26] Jeff Wicks *Chinese national nabbed with 10 rhino horns at OR. Tambo* News 24, 19 August 2016. See www.news24.com/SouthAfrica/News/chinese-national-nabbed-with-10-rhino-horns-at-or-tambo-20160819 Accessed 29/09/2016.

[27] The Times of India *Rhino horn seizure leads cops to int'l smuggling gang* 6 March 2016. See http://timesofindia.indiatimes.com/city/jaipur/Rhino-horn-seizure-leads-cops-to-intl-s Accessed 08/03/2016.

**2016 India:** There have been problems in Kaziranga National Park in Assam Northeast India, which contains the world's largest population of greater one-horned rhinos, some 2500 of them. This part of India is restless. It is not too far from the border with Myanmar, and there are many rebel groups who are known to be involved in the illegal trade in rhino horn, trade which can pass through Myanmar and into China.

Most of the local people love their rhinos. They are a major tourist attraction, and they provide work/income. They do all they can to protect them, providing vital information about possible attacks by poachers. They work as rangers, often risking, sometimes losing their lives to do this, even though their pay is low and their equipment is poor. The poachers have night vision goggles and AK 47 s. There is, however, a 'shoot on sight' policy in place, which has resulted in the deaths of dozens of poachers.

For 10 years poaching was minimal, but everything changed at the start of 2016. At least 12 rhinos were poached between January and June, one of them on the day that Prince William and his wife visited the park. So what had gone so terribly wrong? Perhaps the steep hike in the price of rhino horn. It had skyrocketed. For some people the temptation must be overwhelming.

When the mutilated, hornless body of a rhino was found partly buried in the park, two forest guards were arrested on suspicion of covering up poaching. Interestingly, the spokesperson for WWF India thought that it was a different sort of cover-up, not of a poaching incident, but rather because they were under so much pressure. WWF is working with the Assam government to set up a DNA database.

And there is the ever-present suspicion of corruption. In June 2016, the park's director was suspended by Assam's Forest Minister for failing to inform her that a rhino had been killed.[28]

**2016 Cambodia:** In August, 600 kg of ivory from African elephants and hidden in a cargo of corn were seized from a container that had been abandoned 2 years earlier, while in December, the authorities in Phnom Penh discovered and seized 1.5 tonnes of mainly ivory and pangolin parts in containers originating in Mozambique.[29]

**2017 Cambodia:** A year later, nearly a tonne of ivory (280 full or parts of elephant tusk) was seized in the port of Sihanoukville. The ivory, some of it

---

[28] News24 2016.

[29] *Cambodia seizes shipment of ivory hidden in hollow logs* Phys.org, 6 December 2017. See https://phys.org/news/2017-12-cambodia-seizes-shipment-ivory-hidden.html Accessed 06/03/2018.

hidden in logs, was found in an abandoned shipping container owned by a company based in Mozambique having left Africa in December 2016, from Côte d'Ivoire.[30]

**2017 Hong Kong SAR:** A man and two women were arrested after 7.2 tonnes of ivory were seized, hidden under a cargo of frozen fish from Malaysia.[31]

**2017 Malaya:** Working together as part of 1Malaysia Biodiversity Enforcement Operation Network, enforcement officers from the Department of Wildlife and National Parks (Perhilitan) and the Malaysian Armed Forces stopped two vehicles while carrying out road checks at the national park. Inside the vehicles, they discovered a hoard of weapons including various types of gun, ammunition, machetes, an axe, 27 firecrackers and 3 units of plastic explosives plus detonator. The weapons indicated their target prey were large mammals, including elephants.

They arrested the men, 7 Malaysians, *notorious poachers*, who had managed to evade capture since 2013. They also raided several premises finding more weapons, 2 elephant tusks and elephant meat.

The men were suspected of being part of a major poaching syndicate, supplying wildlife parts to neighbouring countries. They face prosecution under the Wildlife Conservation Act 2010, heavy fines, possibly up to 3 years in prison, or both.[32]

**2017 South Africa:** 12 suspected rhino poachers were arrested in Kruger National Park by officers from Operation Rhino 7, a multidisciplinary task force, and will appear in court.[33]

**2017 Uganda:** Enforcement officers from the Uganda Wildlife Authority seized a tonne of illegal ivory from an *opulent, expensive house* in Kampala, Uganda's capital city, and arrested the 2 West Africans they had been monitoring for several weeks. They suspected them of planning to ship the ivory abroad. As there are porous borders in Africa's Great Lakes region, the ivory

---

[30] Ibid.

[31] *Hong Kong seizes 7.2 tonnes of ivory* Phys.org 6 July 2017. See https://phys.org/news/2017-07-hong-kong-seizes-tonnes-ivory.html Accessed 06/03/2018.

[32] Nurbaiti Hamdan *Major poaching ring crippled* The Star Online, 14 February 2017. See http://www.thestar.com.my/news/nation/2017/02/14/major-poaching-ring-crippled-pe Accessed 16/02/2017.

[33] Simon Bloch *Several suspected rhino poachers nabbed* Times LIVE, 2 September 2017. See http://www.timeslive.co.za/local/2017/02/09/Several-suspected-rhino-poachers-nabbed Accessed 14/02/2017.

probably came from Tanzania and Congo. Uganda has recently set up a canine unit.[34]

**2017 Thailand:** A routine inspection of cargo at Suvarnabhumi Airport, Thailand, led to the discovery of rhino horns in suitcases, which, in turn, led the Thai police to Boonchai Bach, who was arrested at his operational base in Nakhon Phanom, by the river Mekong.

Steve Galster, founder of the NGO Freeland, which has been investigating the two Bach brothers since 2003, described it as *'like catching one of the Corleones'* (the fictional Mafia family). The brothers qualify as kingpins, for it is alleged that they are responsible for smuggling huge quantities of endangered wildlife including *'thousands of tonnes of elephant tusks and rhino horns'* from Africa to Asia.[35] They are associates of Chumlong Lemtongthai, another kingpin.

**Mammoth tusks:** There have also been seizures of mammoth tusks. We've already seen that there is a thriving market in ivory artefacts made from mammoth ivory, which can be laundered as elephant ivory. One seizure comprised a massive three tonnes and was confiscated from smugglers in Russia. There were 64 complete and 14 reconstructed tusks, which came from Siberia.

**Non-seizures:** These are seizures that should have but never actually happened, such as the case of the cache of cash, the 4.3 kg of rhino horn and the 2 North Koreans, one a diplomat, the other a Taekwondo master and possible spy, who were arrested in Mozambique but released after the intervention of the North Korean ambassador to South Africa (which we saw earlier).[36]

[34] Rodney Muhumuza *Uganda seizes ton of ivory, arrests two West African suspects* See http://www.680news.com/2017/02/18/uga Accessed 19/02/2017.

[35] Oliver Holmes *Thai police arrest notorious wildlife trafficking suspect* The Guardian 20 January 2018. See https://www.theguardian.com/environment/2018/jan/20/thai-police-arrest-notorious-w... Accessed 23/01/2018.

[36] Julian Rademeyer *Diplomats and Deceit: North Korea's Criminal Activities in Africa* The Global Initiative Against Transnational Organized Crime, September 2017, p.12.

# 27

## Evidence

Arrests are normally followed by prosecutions provided there is sufficient evidence to establish the defendant's guilt, so what sort of evidence might be used in a trial?

## 27.1 Permits

Some of the best evidence comes from paper and e-trails. Its permit system is one of the things that has helped make CITES into a very successful treaty and provided invaluable data over the years. However, these same permits can also be misused by traffickers who make good use of forgeries.

America provides an interesting example. Although it has tried to prevent illegal trade in elephant ivory, one result of its complex legislation has been a fairly widespread misuse of its CITES permits. This is how it's done.

There are two kinds of permit, one using the 'O' Source Code, the other the 'T' Purpose Code (commercial). The traffickers claim on the permit they are using that the ivory pieces they are importing are 'antique', except that they aren't. They are modern. In fact, between 2010 and 2012, of the 3242 ivory pieces that entered the USA for commercial trade as preconvention ivory, using the 'O' Source Code permit, 146 of them (4.5%) were seized as illegal or fakes, and this was probably an underestimate, as the customs agents were unlikely to spot all the fakes.[1]

---

[1] Daniel Stiles *Ivory Trafficking in California, USA* 2014.

© Springer Nature Switzerland AG 2019
B. Martin, *Survival or Extinction?*, https://doi.org/10.1007/978-3-030-13293-4_27

## 27.2   Fingerprints

Although fingerprints are a classic way of solving crimes, both in real life and detective stories, until very recently they could not be used to catch poachers. The problem with both elephant tusks and rhino horn has been the make-up of their exterior and now that has been solved for elephant tusks.

Even though like human skin they have ridges, they are also highly porous, quickly absorbing any grease and sweat which would be present on a normal fingerprint and to which the traditional powders stick. So no matter how many fingerprints the poachers left behind, they could not be identified. A different type of powder was needed, and this has now been developed by a specialist in the Forensic Department of the London Metropolitan Police working in his own time with King's College London. Consisting of magnetic particles, not only does this powder expose the previously hidden fingerprints, which can now be recovered up to a month later, it can also be reused.

Trials in Kenya where IFAW paid for the equipment produced very encouraging results. According to the Times, during 2017:

* It was used in a minimum of 4 cases;
* 15 people were arrested including 5 allegedly corrupt police officers; and
* 11 tusks and 50 worked ivory pieces were seized.[2]

The equipment works. It is easy to get hold of, is inexpensive, and can be used in the field, so it is fitting that IFAW gave Mark Moseley, who invented it, a special award.

## 27.3   DNA Evidence

DNA testing was a breakthrough. It has been successfully used to establish the origins of ivory, and it is hard to overestimate the important role it plays in the collection of evidence. This applies to rhinos as well.

### Elephants: DNA Testing

There was an urgent need to be able to distinguish both the age and origin of African elephant ivory so as to check that any ivory being offered for sale was, in fact, legal. MIKE and ETIS did not do this, and although the obvious method was to use DNA testing, in the early days, there were still some limitations.

---

[2] John Simpson *Police forensics kit snares poachers* The Times, 29 December 2017.

So it was encouraging when, in November 2010, at the Third African elephant meeting in Kenya, Germany's CITES Management Authority presented a paper on a project that would enable far more facts to be elicited from the data available.[3]

Until the mid-2000s, the use of forensics in the detection of the origin of illegal ivory was limited by the fact that typically, the information that could be provided only began at the point of shipment, and DNA analysis could only determine product origin if it could compare that DNA (collected at the point of shipment) with reference samples from different locations. The problem, a dearth of reference data, was overcome when new DNA assignment methods were applied that could *'determine the geographic origin(s) of wildlife products from anywhere within its range'.*[4]

Wasser and his team of distinguished international scientists came from:

- In the USA: the Centre for Conservation Biology, the University of Washington and the Department of Human Genetics and Statistics, the University of Chicago
- INTERPOL: their Working Group on Wildlife Crime and Department of Law Enforcement
- Israel: their Nature and Parks Authority
- Cameroon
- The Last Great Ape Organisation
- Kenya
- The Lusaka Agreement Task Force
- Tanzania: the Faculty of Veterinary medicine, Sokoine University of Agriculture

Being convinced that *'the most effective way to contain this illegal trade is to determine where the wildlife is being removed'*, they were only too well aware of the urgent need for this test to succeed. Once it was possible to work out which was the target population, the appropriate authorities could then direct intensive law enforcement to prevent this happening. It would also prevent countries *'from denying their poaching problems at home'* and *'thwart trade before it entered into an increasingly complex web of international criminal activity'.*[5]

---

[3] The project, *Determination of Age and Geographical Origin of African Elephant Ivory*, ran from 1 July 2010 to 31 December 2016. The final report for the project *Geographical Origin* was prepared by Stefan Ziegler, WWF West Germany, and published in March 2017.

[4] S. K. Wasser, W. J. Clark, O. Drori, E. S. Kisamo, C. Maitland, B. Mutayoba, and M. Stephens, *Combatting the Illegal Trade in African Elephant Ivory with DNA Forensics'*, Conservation Biology, Volume 22, No. 4, p. 1065. Published 2008.

[5] Ibid.

The new methods were used to determine the geographic origins of two large seizures of ivory, one of which was the Singapore Seizure (Case Study 7), where the ivory had been shipped from Malawi via Mozambique and South Africa. The second seizure took place in 2006, first in Hong Kong SAR and then, 1 month after that, in Cameroon, when two similar containers owned by the same company returned from Hong Kong SAR empty apart from a few chips of ivory. DNA was extracted from all the ivory specimens that made up the seizures, namely, large tusks, hankos, hanko shells and ivory chips.

The DNA samples were compared with *'over 600 reference DNA samples extracted from tissue and scat samples collected from elephants in known savannah and forest habitat locations across sub-Saharan Africa'*. Article 3 of the *Action Plan for the control of trade in elephant ivory* makes it clear that *'all elephant range States are recommended to cooperate with relevant research projects studying the identification of ivory, especially by supplying relevant samples for DNA and other forensic science profiling'*. On the basis of its DNA, each of the samples of unknown origin was classed as either a forest or a savannah elephant,[6] and then SCAT software, the *'smoothed continuous assignment technique'*,[7] was used to further refine its geographic origin.

The results indicated that all the ivory from the Singapore and Malawi seizures was from savannah elephants. The tusks came from *a relatively restricted part of southern Africa, centred on Zambia*. The origins of the hankos and the hanko shells closely overlapped each other and the tusks except that the hankos came from a rather more northeasterly area. In direct contrast, the Hong Kong ivory derived from forest elephants, the tusks coming from *a tightly clustered area centred on southern Gabon, near the Congo- Brazzaville border to the east*. The DNA data from the ivory chips found in the Cameroon seizure was consistent with that of the tusks, suggesting that they shared a common origin.[8]

All the ivory was destined for southeast Asia.

It is encouraging to know that since this early triumph by Wasser and his team, more DNA markers have been discovered, making it possible for the ivory to be identified even more accurately. So when the South Korean police confiscated ivory seals, D-loop sequences were used[9] to establish whether the

---

[6] Wasser et al. 2004, see (n.4).

[7] Ibid, p. 1068.

[8] Jo Hastis, Julian Newman and Mary Rice *Back in Business: Elephant Poaching and the Ivory Black Markets of Asia EIA, 2002*; and Wasser et al. 2007, see (n.4), p.1069.

[9] Lee et al. 2013, in Iyengar, A. *Forensic DNA analysis for animal protection and biodiversity conservation: A review* Journal for Nature Conservation (2014). See https://doi.org/10.1016/j.jnc.2013.12.001 Accessed 27/02/2014.

ivory had come from savannah elephants, forest elephants or even Asian elephants. The same technique used on an ivory idol, the subject of a suspected illegal trade, showed it was carved from the tusk of an Asian elephant.[10] It is becoming steadily easier to link seized ivory with its place of origin (but see later for a major breakthrough).

## 27.4   Radiocarbon/Bomb-Curve Dating

DNA testing, however, could not provide the date of the ivory, the answer to the all-important question: has this elephant died recently? The Third African elephant meeting decided on a new project, using radiocarbon dating to develop a database that would also include other endangered animals such as rhinos.

Because it takes about 5700 years for half of carbon-14 to decay radioactively, radiocarbon dating has not been an exact science, producing a reliable estimate rather than a precise date. What changed everything, however, were the tests carried out on nuclear weapons between 1952 and 1962. Carbon-14 comes from nitrogen. During the tests, neutrons bombarded nitrogen in the atmosphere, converting some of it into carbon-14, which was and still is being absorbed by plants and animals. Because these tests were carried out above ground, the amount of radiocarbon in the atmosphere nearly doubled. However, after 1962, testing took place underground, and the levels of radiocarbon have declined steadily ever since. This has produced 'the bomb curve'.[11] And the potential for more accurate dating of ivory.

Using an accelerator mass spectrometer to analyse small amounts of ivory taken from different points along the length of an elephant's tusk, it became possible to date even the exact year of the animal's death. Pre-1955 ivory could be identified by its low, pre-nuclear test levels of carbon-14. The most accurate results come from ivory closest to the 10-year nuclear test period. The dating is less precise the farther away the ivory is from this period, and it therefore becomes more difficult to give a precise date for the elephant's death.

It is easier to understand if we look at two examples. If an elephant is poached in 2003, the test result is so accurate that it will show that the animal died in 2003. However, if the elephant was poached in 2012, the test result

---

[10] Gupta, Thangaraj et al. 2011 in ibid., pp. 6 and 2.
[11] Uno et al. 2013.

can only establish that it died somewhere between 2010 and 2013, because the date is further away from *the bomb curve*.

It is immediately obvious just how important these forensic techniques are. We have already seen how, in some shops in China, it was not always clear whether the ivory object being sold really had been carved from the legally held stocks of African elephant ivory. But using DNA testing to pinpoint its origin and radiocarbon/bomb-curve dating to estimate when the elephant was killed will provide INTERPOL and other enforcing agencies with powerful investigative tools once radiocarbon/bomb testing becomes more widely available.[12]

## 27.5   DNA Evidence and Rhinos

Rhino horn is often found in powdered form, either on its own or mixed with other powders and made up into prescriptions, or as artefacts like daggers or beautifully carved vessels. But not all of this is genuine. By 2003 however, it was possible to identify the counterfeit from the real thing. Mitochondrial DNA could be used not only to identify rhino products but also the actual species of rhino.[13]

Another breakthrough came when the Rhino DNA Index System (RhODIS) was established. This was developed by Dr. Cindy Harper and the Veterinary Genetics Laboratory (VGL) of the South African University of Pretoria and provided another important step forward in the protection of rhinos. RhODIS was a new technique which has now become *a very powerful forensic tool for rhino horn management, monitoring and law enforcement*. Its main function is its use as a forensic tool to help investigate rhino poaching, and its importance lies in the fact that unlike microchips, DNA cannot be removed, changed or destroyed.

DNA samples collected from as many rhinos as possible, including those legally translocated and dehorned, from horn seizures and registered stockpiles and from animals that have been illegally hunted and killed are stored on the database.

So what is RhODIS and what can it do? It is a combination of DNA reference data and a DNA database for individualisation, a *central standard secure database of rhinoceros DNA profiles for forensic purposes*, and it is an important

---

[12] Daniel Stiles *Ivory Trafficking in California, USA*. 2014.

[13] Hsing-Mei Hsieh, Li-Hung Huang, Li-Chin Tsai, Yi-Chen Kuo, Hsien-Huei Meng, Adrian Linacre, James Chun-I Lee, *Species identification of rhinoceros horns using the cytochrome b gene*, Forensic Science International 136 (2003) 1–11. See www.elsevier.com/locate/forsciint Accessed 07/05/2015.

traceability tool. RhODIS: Rhino DNA Index System/CODIS – Combined DNA Index System. It can:

- Take DNA profiles directly from rhino horn.
- *Match recovered horn to individual rhinos* which enables it to *link poachers, traffickers, (and) horn from consumer countries.*
- Identify horns in stockpiles, individually.
- Enable recovered profiled horns to be returned to their owners.
- Ensure that in any legal sale, the horns can be identified as legal.

The DNA profile itself can provide a considerable amount of information:

- Whether the horn, even in powdered form, is rhino horn
- Whether it came from a black or white rhino
- Whether the rhino was male or female

And because it can be fed into population and pedigree analysis, it can provide a traceability system of individual rhinos and rhino horns, tracking the animal's movements from its source population.

It can also be used to ensure compliance with hunt permits. As each hunt and intervention requires a permit, this can be granted subject to the condition that DNA is collected from the rhino.

RhODIS was launched at the first international rhino DNA sampling training workshop, which was held in South Africa in November 2013, with law enforcement officers from each of the 11 African rhino range states, the South African National Parks, China, Thailand and Vietnam attending.[14]

By the end of 2013, there were already over 10,000 samples from black and white rhinos from Africa, with the early database including:

- *DNA profiles of rhinos from provincial and national parks and private owners in South Africa*
- *DNA profiles from all poaching cases*
- *DNA profiles from recovered horns and stockpile horns*
- *DNA profiles from rhinos from Zimbabwe, Botswana and Namibia*[15]

---

[14] *Forensics to support the fight against wildlife crime* Department: Environmental Affairs, Republic of South Africa, CITES. See http://www.cites.org/eng/news/pr/2013/20131106_forensics.php Accessed 07/05/2015.

[15] Dr. Cindy Harper. For a clear explanation of the database in its early days, see Dr. Cindy Harper *RhODIS profiling and a DNA database as a tool to protect the Rhino* published by Peppin L. et al in 2009, financial support coming from WWF, Stop Rhino Poaching, SAB, SYMCO and Rhino Alive.

By 2016, DNA from some 20,000 rhinos was on the system. Namibia, Botswana, Zimbabwe, Malawi, Zambia, Uganda and Kenya had all submitted data.

The 2016 RhODIS Scientific Workshop agreed that this was *'the first step towards an international rhino DNA profiling system'*, and that once the system had been refined, new techniques developed and existing methods validated, it could be used as *a global rhino DNA testing system*. Investigators would even be able to distinguish between a horn from a poached rhino or a horn stolen from a museum.[16] Participants included DNA forensic scientists, law enforcement officers and investigators from Australia, Botswana, Czech Republic, Hong Kong, India, Indonesia, Kenya, Malaysia, Namibia, the Netherlands, South Africa, South Korea, Thailand, Vietnam and the UK.

And now there is a rhino database for museum specimens, the Wildlife DNA Forensics—Diagnostics and Molecular Section of SASA, which was set up in Scotland by SASA (Science and Advice for Scottish Agriculture) and TRACE. It can be used to link recovered horns with specific thefts.

In 2018, there was yet another breakthrough. Scientists have been genetically analysing the DNA samples on RhODIS so that now, even small amounts of material from a dead rhino will enable that specific animal to be identified. The importance of this in terms of more effective law enforcement cannot be exaggerated.[17]

## 27.6   TRACE Wildlife Forensics Network

This is a network that operates as an information exchange for forensic scientists and enforcement agencies. An international NGO, whose global partners are PAW Forensic Working Group (in the UK), the Society for Wildlife Forensic Sciences and ASEAN-WEN, its aim is to promote the use of forensic science in the investigation of wildlife crime and biodiversity conservation.

Although it is involved in a number of projects, it is also there to educate, delivering training and capacity building in wildlife DNA forensics. For example, in 2018, TRACE and TRAFFIC will work with other partners to

---

[16] Dr. Lucy Webster *Global forensics to tackle rhinoceros poaching* Legal Eagle, November 2016, No. 80; and Hanti Schrader *Scientists join fight against rhino poaching* News 24 16//07/2016. See http://www.news24.com/SouthAfrica/News/scientists-join-fight-against-poachi Accessed 11/08/2016.

[17] The research paper, the result of a collaboration between scientists from USA, South Africa and Russia, was published in Current Biology in January 2018, see *Saving Rhinos Using DNA Smithsonian's National Zoo & Conservation Biology Institute* 9 February 2018, https://nationalzoo.si.edu/center-for-species-survival/news/saving-rhinos-using-dna Accessed 13/03/2018.

provide wildlife officials in Malawi, Zambia and Zimbabwe with the kind of specialist training they need to carry out wildlife forensic investigations, including crime scene investigation and DNA analysis. In other words, how to gather vital evidence if prosecutions are to succeed.[18]

## 27.7   Faking It

We have already seen the problems relating to seizures, and the passing off of illegal ivory as new legal ivory in China, now we look at the manufacture of fake antique ivory artefacts, artefacts that are modern, or comparatively modern, certainly not 100 or more years old.

Faking antiques goes a long way back in time. It was certainly not invented to get round CITES regulations, although it can prove very useful when applied to achieving that end. Antiques are faked because they are valuable, sometimes incredibly valuable. Even experts can be taken in. But proven fakes are good evidence in a prosecution.

Most traffickers, however, merely require items good enough to pass the scrutiny of potential customers. They want to make a sale. They're not too worried about enforcers. They know that antiques fetch more money than recent artefacts. They know that it is comparatively easier to sell fakes on the Internet. They also know that the CITES ivory trade ban in 1990 made international sales of ivory (with certain exceptions) illegal. One of the exceptions was antiques. So, all in all, selling fakes can quite profitable.

An ivory object can be made to look older than it really is by:

- Staining: which can be achieved by using a variety of materials such as tea or coffee, henna or even permanganate
- Cracking
- Wear marks: produced *by rubbing an ivory piece for long periods with a soft abrasive and/or by chipping a piece and rubbing the chipped area until it becomes smooth*[19]

Another way of tricking buyers into purchasing a supposedly valuable antique is by weaving a wonderful tale about it. Stiles recounts such an experience in CAR. When visiting a small shop in the art market in Bangui in 1999,

---

[18] TRAFFIC *Wildlife forensics: a new weapon for Southern African countries to help bring poachers to book* 16 March 2018. See http://www.traffic.org/home/2018/3/16/wildlife-forensics-a-new-weapon-for-southern Accessed 27/03/2018.
[19] Daniel Stiles 2015 p. 10.

**Figs. 27.1 and 27.2**   Faking it. *'What appears to be an antique Virgin Mary with the Christ Child. Note the longitudinal crack down the right side of the face and the off-white, dirt-smudged ivory surface to give the appearance of age. The figurine was actually carved in Paris in the late 1990s by a French master carver the investigator interviewed in 2004. The aging was accomplished by burying the carving in the ground for three weeks and then leaving it out in the sun to dry and crack'.* Copyright Daniel Stiles

he was taken into a back room and shown some carvings that were hidden there. *'The vendor said in hushed tones that the carvings were very valuable antiques that had been looted from Mobutu Sese Seko's Gbadolite palace after his death in 1997'.*[20]

And if you are the faker, it's a good idea to make sure there is only one copy around of this truly unique artefact or at least that the other copies are nowhere near each other!

---

[20] Ibid.

# 28

# Prosecutions

Now we arrive at the court hearing, the prosecution, the final step in this journey towards justice.

If we are to understand what happens to a person once they have been arrested and accused of carrying out a criminal offence such as poaching a rhino or being in possession of a pair of elephant's tusks, we need to look at the procedure in a court that hears and decides on criminal offences. However, different countries have different criminal justice systems and different court procedures, so we will turn to the UK, to see what happens there.

## 28.1 A Court Case in the UK

We begin with *the presumption of innocence*. In the UK, every person accused of a criminal offence is deemed to be innocent until they are found guilty in a court of law. This influences the course of a trial.

A crime has been committed. The police investigate it. They find evidence linking it to person D, so they arrest D and question him. If it appears to them there is sufficient evidence to make it highly likely that D did carry out the offence, they can charge him with it. If further evidence is needed, the police continue to look for it until the case comes to trial. Sometimes, if a very serious offence has been committed, D will be remanded (sent to) in prison until the trial. More often D will be released on bail, with various conditions laid down to ensure he returns for the trial. By this stage, D will be known as the accused or the defendant.

© Springer Nature Switzerland AG 2019
B. Martin, *Survival or Extinction?*, https://doi.org/10.1007/978-3-030-13293-4_28

In England, almost all criminal offences are tried in Magistrates' Courts. Only the very serious ones go to the Crown Court. It is an adversarial system.

In court, at the start of the proceedings, the defendant will be asked to plead (to say whether he is) guilty or not guilty. If he has been caught red-handed, he will normally plead guilty. The court can then proceed to sentence him, taking any mitigating circumstances (circumstances that might provide him with a reason for not being sentenced too severely) into account. However, if he pleads not guilty, this is what happens.

The prosecution starts by setting out their case that the accused, the defendant, D has committed the offence. They do this by producing evidence. Witnesses may be called, for example, to state that they found an elephant's tusk in the storeroom attached to the defendant's shop. The prosecution can question/examine each witness, to make sure the evidence is quite clear and reliable. The lawyer(s) for the defendant can cross-examine each witness, testing for flaws in the evidence, for they must try to discredit it. They can produce their own evidence to do this.

If the court is to find the defendant guilty, the prosecution must produce sufficient evidence to show that the defendant is *guilty beyond reasonable doubt*. This is known as *the burden of proof*. If there is not enough evidence to show that the defendant is guilty beyond reasonable doubt, in other words to satisfy the burden of proof, then, because of the presumption of innocence, the defendant must be found not guilty and acquitted.

If the defendant is found guilty, he will then be sentenced, normally to a fine or a term of imprisonment, sometimes to both.

## 28.2   Some Prosecutions

The problems surrounding the case of the Rathkeale Rovers provide a vivid example of the sort of obstacles to be overcome if there is to be any chance of a successful prosecution. And this is often the case.

In the UK in June 2008, Michael Elliot, a wealthy antiques dealer, pleaded guilty to seven offences concerned with illegal trading in endangered species. As part of an international network, he had masterminded the successful smuggling of ivory and whale tooth. The ivory, mainly hippo teeth, came from Africa and went to China for processing and then back through Eastern Europe for sale in the West. Scotland Yard's Wildlife Crime Unit, working with the American National Oceanic and Atmospheric Administration, investigated for 3 years, and when, in 2005, they raided Elliot's home in Kent, as well as ivory artefacts, they also found 18 elephant tusks. These, Elliot claimed, did not belong to him. However, they also found documents and a photograph which directly linked him to the illegal trade. The court sentenced him to 2 years in prison, the sentence suspended for 2 years, and then he was off to America to be prosecuted there.[1]

Some prosecutions are comparatively straightforward.

**2011 Russia:** A massive 3 tonnes of Siberian mammoth ivory, including 64 complete and 14 reconstructed tusks, were seized and then confiscated (into safe keeping) by the court. At the hearing, it was revealed that the gang had been smuggling mammoth tusks and bones across Russia's borders since 2004.[2]

**2012 South Africa:** The notorious wildlife trafficker Chumlong Lemtongthai, a kingpin, was sentenced to 40 years in prison, which was reduced to 13 years on appeal. His associate, Bach Van Limh, equally notorious, tried, ineffectively, to bribe an official to release him. The amount of the bribe? USD600,000! Chumlong later claimed it would have been accepted had the money been paid into an offshore account, which his associate refused to do, afraid that as there would be no record of this payment, the official would renege on the deal.[3]

**2015 USA:** The defendant, Yiwei Zheng, was an eminent Professor of Philosophy at St. Cloud State University in Minnesota. Arrested in March 2015 by the USFWS, he was accused of selling, between 2005 and 2011, illegal products made from rhino horn and elephant ivory to buyers in both the UK and China. Together with two *co-conspirators*, he used his own website and online auction sites to carry out large numbers of transactions of mostly small but valuable items.

We have already seen that although it is illegal to sell most ivory and rhino horn, on those occasions where it is possible, no transaction can take place without the appropriate CITES permit(s) having first been obtained, thus rendering the transaction lawful. In Zheng's case, there were no permits. Furthermore, he had used false labels to ship items through the postal service.

---

[1] David Brown *Antique dealer ran smuggling ring to supply illicit trade in ivory ornaments* The Times, 14 June 2008.

[2] *Mammoth haul seized* The Times, 21 January 2011.

[3] Nick Davies and Oliver Holme *The crime family at the centre of Asia's animal trafficking network* The Guardian, 26 September 2016. See https://www.theguardian.com/environment/2016/sep/26/bach-brothers-elephant-ivory Accessed 27/09/2016.

At the end of the hearing, Zheng was released on a USD25,000 bond, but he had to surrender his passport. A disappointing result, as he had been monitored for years.

**2013 Kenya:** A Chinese woman on a flight to Hong Kong was caught with 6.9 kg of ivory, disguised as bags of macadamia nuts. She was sentenced to 30 months in prison.[4]

**2015 Zimbabwe:** There had been a number of poisoning incidents in Hwange National Park. At least 62 elephants were killed by eating oranges laced with cyanide, and 22 were found dead at 2 watering holes, poisoned by cyanide. Then 7 men were arrested, at two different locations. While four of them were still on trial and one had been released for lack of evidence, two were found guilty. Both were given custodial sentences, but not for poisoning. One was sentenced to 10 years in prison, having been found in possession of elephants' tusks, and the other to 7 months, for possessing unlicensed ammunition.

It is hard to achieve justice for elephants and rhinos. Fortunately this did happen in a recent poisoning case although these are normally some of the most difficult cases to prosecute. This is because it is hard to find sufficient evidence, such as who did put down the poison, to obtain a conviction.

**2015 Tanzania:** The new President had pledged to root out both poaching and corruption. 4 Chinese men were convicted of smuggling 11 rhino horns out of Malawi and into Tanzania. Each of them was sentenced to 20 years in prison, and each of them was fined £3,000,000, ten times the value of the horns.[5]

Where there are convictions, some countries impose very heavy sentences. Punitive sentences to punish the guilty defendant and deter others who might be similarly tempted.

**2015 China:** The defendant, found guilty of illegally transporting wild animal products worth more than 260,000 yuan, products including 2.2 kg of ivory and 0.69 fragments of rhino horn, was sentenced to 10 years in prison and fined 50,000 yuan (USD7740), a sentence upheld on appeal to a higher court.[6]

On occasion, a punitive sentence may be tempered with mercy if there is a chance of the defendant reforming.

---

[4] *Ivory smuggler jailed* The Times, 23 August 2013.
[5] *Rhino horn smugglers given 20 years in jail The Daily Telegraph*, 19 December 2015.
[6] Xinhua April 13, 2016, Online Edition see www.shanghaidaily.com/national/Man-gets-10-years-for-illegally-transporting-wild-animal-products/shdaily.shtml Accessed 16/04/2016.

**2016 Zimbabwe:** A defendant convicted of poaching 2 rhinos was sentenced to 35 years in prison. However, this would be reduced to 20 years if, within the next 5 years, the defendant paid back USD480,000, the estimated value of the rhino horns. The Magistrate was quoted as saying: *'You deserve a deterrent sentence so that you can be a reformed person when you come out of prison'.* The poacher had not been working on his own. There was an alleged ringleader, an assistant officer in Zimbabwe's Secret Service, and two other gang members were supposedly on the run.[7]

Sometimes it may take years to bring the perpetrator to justice.

Zimbabwe provides yet another example. Although its legislation is good, its application is patchy, and *'police, prosecutors and magistrates are easily bribed to "make cases disappear"'.* The case involved the notorious poaching ringleader, Dumisani Moyo, who *'repeatedly evaded prosecution'.*[8]

Moyo was also identified as a supplier of guns and ammunition. Repeatedly arrested, he would then be released on bail or from custody and promptly abscond/disappear. INTERPOL, at Botswana's request, issued a Red Notice for his arrest, and about 4 months later, he was arrested and then released from custody to vanish yet again.

By 2017, he was wanted on several rhino poaching charges, not just in Botswana but also in Mozambique, Zambia and Zimbabwe. In January, the Zimbabwe Parks and Wildlife Management Authority named him as *'one of the suspected cross-border poachers wreaking havoc in the country's Rhino Intensive Protection Zones'*, and in August he was arrested in Zimbabwe. Together with a Zimbabwean Central Intelligence operative, among other offences, he was charged with four counts of killing rhinos and selling the horns to buyers in Zambia.

He appeared before magistrates in the Harare Magistrates' Court and was remanded in custody to appear before the Masvingo Magistrates' Court before being extradited to Botswana in December, with two of his alleged accomplices.[9] There he was again charged with possession of a rhino horn (the original 2012 charge) and, at the prosecution's application, remanded in custody, because he was a 'flight' risk.

---

[7] News 24 correspondent *35 years in jail for Zim poacher* News 24 Special Reports Zimbabwe, 1 January 2016. See http://www.news24/SouthAfrica/News/35-years-in-jail-for-Zim-poacher Accessed 17/08/2016.

[8] Julian Rademeyer *Beyond Borders: Crime, conservation and criminal networks on the illicit rhino horn trade (Part 2 of a 2-part investigation into rhino horn trafficking in Southern Africa)* Global Initiative against Transnational Organized Crime, July 2016.

[9] *NewsDay* Zimbabwe *Cross-border poacher faces extradition* 8 August 2017. See https://www.newsday.co.zw/2017/08/08/cross-border-poacher-faces-extradition Accessed 17/08/2017.

By July 2018, because the prosecution had still not produced reasons for his continuing detention, the Magistrate, acting on the presumption of innocence, granted him conditional bail. He had to pay P(ulas)1000, provide two sureties each paying P500, surrender his passport, report fortnightly to the CID offices in Francistown and attend court as and when required to do so. He was ordered to appear in court on 3 July.[10]

**2016 Kenya:** Convicted its first kingpin under its new legislation. Feisal Mohammed Ali, who, for years, had seemed untouchable. Then came a series of events that led to his downfall:

- **June 2014.** There was a massive seizure of 2 tonnes of ivory, consisting of 228 tusks, and some 74 pieces of unworked ivory were recovered from Fuji motors car yard in Mombasa (Kenya). The ivory was worth a minimum of £700,000.[11]
- Soon afterwards, a warrant was issued for his arrest, but somehow he managed to evade capture.
- **October 2014.** INTERPOL issued a red notice, at the request of the Kenyan government.
- **December 2014.** He was arrested in Tanzania by officers from INTERPOL and remanded in custody because *'he posed a serious flight risk'*.[12]
- **July 2016.** Finally Feisal Mohammed Ali was found guilty of illegally possessing ivory valued at 44 million shillings (USD440,000), fined 20 million shillings, and sentenced to 20 years in prison.

The presiding judge made some important comments:

- That elephants are *'part of Kenya's national heritage and a source of pride'*.
- That *'poaching is a menace in Kenya'*.
- That *'if nothing is done to stop it children in the future would only know elephants from what they read in books'*.
- (She) *'urged people not to wear ivory ornaments'*.
- She concluded by saying that in this case, more than 150 elephants had been killed to supply the ivory involved, *'the court had to put away the people*

---

[10] Lebogang Mosikare *Alleged poaching kingpin granted bail* Monegi, 26 June 2018. See http://www.mmegi.bw/index.php?aid=764 Accessed 05/07/2018.

[11] Jerome Starkey *Ivory kingpin caught after months on run* The Times, 24 December 2014.

[12] Jerome Starkey *Ivory baron remanded in Kenyan jail* The Times, 31 December 2014.

*who committed these crimes, as an example for those behind the poaching menace in the country'.*[13]

Catching a kingpin was never going to be straightforward. Once again it required a lot of working together, and Paula Kahumbu, Head of WildlifeDirect, with her *Eyes in the Courtroom* project, played a significant role in ensuring the trial proceeded in a manner that would bring justice to all, including the elephants.

**2017 Malawi:** Malawi's National Parks and Wildlife (Amendment) Bill 2017 enabled a chief magistrate to sentence three offenders to 36 years in prison for poaching a black rhino in Liwonde National Park.[14]

### 2010–2018 South Africa:

**September 2010:** Dawie Groenewald, his wife, a helicopter pilot, two veterinarians and professional hunters were arrested after a major investigation lasting 15 months. Charged with numerous offences relating to rhino poaching over 4 years, Groenewald was alleged to be the kingpin of a rhino horn syndicate. The accused were granted bail.

**2013:** About 26 dead rhinos minus their horns were discovered buried on Groenewald's farm.

**2017:** Three things happened:

- South Africa's constitutional court confirmed that the government must lift its 2009 moratorium on domestic trade in rhino horn, with the result that some 60 charges were dropped, including all the charges against his wife.
- Some American hunters claimed they had been misled into thinking they were legally hunting rhinos on the Groenewald's farm, and in June, INTERPOL executed an international arrest warrant for Groenewald and his brother. They were again released on bail.
- In December, the accused were served with an amended charge sheet, now reduced to some 1600 charges, which related to illegal hunting of rhino, dealing in rhino horn, racketeering and money laundering.

---

[13] Paula Kahumbu *Kenya jails ivory kingpin for 20 years* The Guardian, 23 July 2016. See https://www.theguardian.com/environment/africa-wild/2016/jul/23/ Accessed 04/10/2016

[14] *UK aid to crack down on criminal gangs driving the illegal wildlife trade* UK Government. See https://www.gov.uk/government/news/uk-aid-to-crack-down-on-criminal-gangs-driving Accessed 11/10/2018.

**January 2018:** The accused again appeared in court only to have their trial postponed to March so that some technical glitches could be sorted out.[15]

Although many cases involving illegal trading in wildlife never get as far as a prosecution, sometimes the most unexpected people are caught.

**2017/2018 Zimbabwe and the USA:** This is the case of the big game hunter and the American trophy hunter, neither of whom were ashamed to become poachers (we met them earlier). Three years earlier, the trophy hunter, a former vice president of the Dallas Safari Club, had paid £29,000 to the big game hunter to organize an illegal hunting trip. He admitted this in court and pleaded guilty to violating the Endangered Species Act. He was *fined £18,500, banned from hunting for 4 years and ordered to return his trophies to Zimbabwe.* The USFWS also investigated the hunter, who has now been indicted and a warrant issued for his arrest.[16]

**2018 Zambia:** Since the Zambia Wildlife Act of 2015 has been passed, several poachers and traffickers have been convicted. In December 2017, 3 foreigners and 2 Zambians were[17]sentenced to 5 years in prison with hard labour. They appealed, but at the hearing in the Lusaka High Court in July 2018, their conviction was upheld.

Forfeiture hearings are a rare occurrence however, so the following case is exceptional:

**2016 South Africa:** It involved yet another kingpin, the KwaZulu-Natal rhino horn syndicate dealer, Dumisani Gwala. According to the arresting officers, he had boasted to undercover policemen that he had buyers in both Johannesburg and Mozambique and could make up to R13 million during a single trip to sell horns. The application for the forfeiture of assets was held in the Durban High Court.

---

[15] Zelda Venter *Technical glitches delay rhino poaching trial* Iol, 29 January 2018. See https://www.iol.co.za/news/south-africa/gauteng/technical-glitches-delay-rhino-poach Accessed 06/02/2018.

[16] Jane Flanagan *Big game hunter faces arrest after offering poaching trip* The Times, 23 May 2018.

[17] See (n.14).

# 29

# Technology to the Rescue

While there is no doubt that vigorous enforcement can be a very effective deterrent, better still would be to end poaching, and the best way of achieving this would be by eliminating demand. But until this happens, we must do all we can to prevent this killing spree. We've already seen the vital role the rangers play in achieving this, but in this deadly game of cat and mouse between elephants and rhinos, the rangers who protect them and the poachers, can anything else be done to raise the odds against the poachers?

## 29.1   Planes, Helicopters, Autogyros and Aerial Surveys

Although these have been used for years, both aircraft and helicopters continue to be invaluable tools; indeed, when Mkomazi National Park was being reclaimed, one of the earliest tasks was to clear airstrips (more of this later). Some of the many purposes these aircraft can be used for include monitoring, dropping personnel into remote areas, rescuing rangers when they have been attacked by poachers and occasionally translocating rhinos.

   Much of the data collected for the Great Elephant Census came from aerial surveys using fixed-wing aircraft, although helicopters were used in the Kruger National Park, as their superior visibility in the rough ground and vegetation made it easier to see the animals. To ensure consistency, the areas to be surveyed were divided into strips calibrated by repeated flights over a marked-out runway. Block counts were used in mountainous areas. And larger herds of

© Springer Nature Switzerland AG 2019                                                **319**
B. Martin, *Survival or Extinction?*, https://doi.org/10.1007/978-3-030-13293-4_29

elephants were photographed, this being a better way to make an accurate estimate of numbers.[1]

Kelly Landen of Elephants Without Borders, who spent hundreds of hours counting the animals, described how it was done. Two wands were fitted a fixed distance apart to the plane's wing, and every elephant between the wands was counted. Not an easy task, so to ensure accuracy, there was also a camera on each side of the plane, taking photographs. A double-check and recount followed, to make sure no calves had been missed.[2]

And now Dragons have come to protect rhinos, more accurately, the Dragon GBT 1170 aircraft, an autogyro. As a result of yet another collaboration, Born Free with Chimera Aviation, Dragons will be used for ranger patrols in South Africa, in the Shamwari Private Game Reserve, and in the DRC, in Garamba National Park (more of this park later). Carrying a pilot, passenger and/or observer as well as technical equipment, these light aircraft can fly safely at low speeds, and they are quiet. They can also cover more ground.[3]

## 29.2   Camera Trapping

Rather surprisingly, camera trapping was first attempted in 1877. Now, transformed by technology, one of its many uses is in the fight against poaching.

## 29.3   Drones

These unmanned aerial vehicles can carry out many activities, but they are expensive to use, and it was almost inevitable that such innovative technology would come with problems. It did, when they were initially used in the war against poaching.

An early experiment was carried out by Air Shepherd (run by the Lindbergh Foundation), in South Africa's Hluhluwe Imfolozi Park. These were specialized drones, originally designed for use in the theatre of war to predict where

---

[1] Michael J. Chase, Scott Schlossberg, Curtice R. Griffin, Philippe J. C. Bouche, Sintayehu W. Djene, Paul W. Elkan, Sam Ferreira, Falk Grossman, Edward Mtarima Kohi, Kelly Landen, Patrick Omondi, Alexis Peltier, S.A. Jeanetta Selier, Robert Sutcliffe Continent-wide survey reveals massive decline in African savannah elephants PeerJ, 31 August 2016, See https://peerj.com/articles/2345 Accessed 01/09/2016.

[2] Alastair Leithead *Why elephants are seeking refuge in Botswana* BBC News, 31 August 2016. See https://www.bbc.co.uk/news/world-africa-37230700 Accessed 04/09/2018.

[3] Born Free *Enter the Dragon – Help fight wildlife crime* email, 24 September 2018.

IEDs (improvised explosive devices) were hidden, waiting to blow up human targets, American soldiers who moved along particular tracks at certain times, just like rhinos and elephants and poachers.

Carrying thermal-imaging computers, they could pick out both target animals and poachers, thus enabling the supercomputer that guided them to predict where the poachers would appear. The rangers could then be waiting for them, ready to catch them before they could attack the rhinos.

An algorithm was developed at the University of Maryland, USA, where the supercomputer was located. It generated flight plans that predicted the movements of both rhinos and poachers, where the animals would be at any particular time and where the poachers would be most likely to strike. These were uploaded into the drones before they went 'on patrol'.

The aim was to work out where a poaching incident was likely to occur and thus prevent it happening. The location was determined by linking together a number of factors:

- Historical data from rhino radio collars
- Where poaching attacks had previously taken place
- The season
- The weather
- What time of day it was

The algorithm managed to predict with considerable accuracy just where the protagonists were likely to be in Hluhluwe Imfolozi, which is a very large park, and the information from the drones narrowed down possible attacks to areas of about 2 mi². When a drone identified a poacher, the ground-control crew alerted a ranger team already in position, which could be immediately deployed to stop the poachers before they harmed the rhinos.[4]

Unfortunately though, as the terrain was difficult, the drones often suffered damage when they landed, so a ground crew always had to be on hand with, among other things, a 3-D printer and some epoxy glue, ready to fix whatever needed fixing, including broken wings.[5]

Namibia and Tanzania also experimented with drones. So did South Africa, but when, in March 2015, they were first used in South Africa's Kruger National Park, the drones found it difficult to distinguish moving objects from moving people, possible poachers, and it was not until April 2016 that

---

[4] Hill, Taylor *Supercomputer-Powered Drones Shut Down Rhino Poaching in This Park-Can They Save Africa's Elephants Too?* http://www.takepart.com/article/2015/03/09/drones-shut-down-rhino-elephant-poaching Accessed 23/07/2015.

[5] Hannaford 2015, see Hill, Taylor (n.574).

they got their first result. Three people were spotted acting suspiciously, and although they were not caught, a firearm was recovered.[6]

This vast tract of land shares some of its border with Mozambique where many poachers come from, and the senior investigator for the African National Parks' environmental crimes unit estimated that as many as ten armed groups a day were entering the park to poach rhinos. That was why it was decided to try drones. One problem seemed to be the high temperatures in the park; they might be interfering with the drones' infrared cameras. A ranger suggested making a base map that could take all the variables into account, including high temperatures, animal movements and extreme weather.

## 29.4 The Raspberry Pi and the Instant Wild Project

When the Raspberry Pi, a tiny computer about the size of a credit card, is plugged into a computer monitor or a television set, using a standard keyboard and mouse, it can do everything a normal computer can do, and more.[7] It is this 'more' that is being used to help protect rhinos in Kenya.

The Instant Wild Project uses citizen scientists from all over the world to save wildlife in a programme designed by ZSL to track endangered species. It uses camera trapping, but as this produces huge volumes of data, a special 'army' of scientists was required to help analyse the data, citizen scientists, who obtain the photographs on their smartphones using the Instant Wild mobile app.

The first cameras, off-the-rail cameras, were not up to the job. Although they were strategically positioned so they could monitor the wildlife, they required good cellular coverage to transmit the pictures they obtained, and frequently this coverage was inadequate. Raspberry Pi came to the rescue.

A different system was designed for them by Cambridge Consultants, and this used the reliable Iridium satellite network. There were two cameras in a single box, one to photograph images during the day and the other, an infrared camera, at night, both triggered by motion sensors. The images were then transmitted to a Raspberry Pi single-board computer that monitored and

---

[6]Yol Groenewald *Kruger drones struggle to take off* Oxpeckers Investigative Environmental Journalism, 5 May 2016. See Accessed 10/05/2016.
[7] *What is a Raspberry Pi?* See Accessed 20/08/2015.

uploaded them via the Iridium satellite network,[8] thus making them available to anyone with the requisite smartphone app.[9]

The system works well, it uses very little power, and it runs on the sort of cheap batteries that can be bought anywhere, even in remote areas. The cameras are in use in Kenya, in the huge Tsavo National Park whose population of black rhinos was being devastated by poachers. Before Raspberry Pi, monitoring the entire park presented considerable difficulties. But these have been eased by strategically placing cameras by water holes and on trails used by the rhinos, and therefore, potentially poachers as well, because movement by any of these will trigger the sensors. Now widely in use, in 2018 they were introduced to Lewa Wildlife Conservancy.

The system, the result of a partnership between conservationists and industry, was described as a cyber *safety net* by ZSL's then Head of Conservation, Professor Jonathan Baillie, intended as a lynch pin to significantly reduce poaching in the park.[10] Google donated £500,000 to the project.

## 29.4.1   Instant Detect 2.0

ZSL's Conservation Technology Team is working with partners on Instant Detect 2.0 which is designed to make data collection easier in more inaccessible areas. The new system is designed to connect camera traps and smart sensors to the *base station*, a single central location, usually buried and camouflaged, that gathers and stores the collected data, which can then be transmitted using satellites *to a user-friendly Internet-based interface anywhere in the world*. And because it is usually buried and camouflaged, unlike cameras, it can be kept safe from curious animals and the many other trials and tribulations that can happen to cameras in the field.

It can also aid in the war against poachers. Rangers place cameras and smart sensors designed to trigger when a poacher is present. Instant Detect 2.0 monitors the situation, and when a poacher is detected, a threat alert is immediately sent to the rangers who can then deal with the danger.[11]

---

[8] Steve Rogerson *M2M helps zoologists protect black rhinos and penguins.* (2015) See http://www.m2mzone.com/ddvrhino Accessed 20/08/2015.

[9] *Raspberry Pi aids cyber 'safety net' for African rhino* BBC News, 9 September 2013. See http://www.bbc.co.uk/news/technology-24014926 Accessed 20/08/2015.

[10] Ibid.

[11] *Instant Detect 2.0: A Connected Future for Conservation* See https://www.wildlabs.net/resources/case-studies/instant-detect-20-connected-future-co Accessed 15/02/2018.

## 29.5    GPS Tracking Systems and the Protect RAPID

The GPS (Global Positioning System) is a tracking device comprised of a satellite system connected to a cell phone system. Any animal fitted with a GPS device can be regularly monitored, which is why some elephants wear collars. Rhinos, however, tend to have the devices inserted into their horns.

Although elephants have been tracked by this method for many years, it was not until 2010, when the epidemic of rhino poaching was reaching disastrous proportions, that five rhinos in Mafikeng Game Reserve in North West province, South Africa, were fitted with GPS devices. The chips, which included alarm systems, provided quite detailed information, keeping the game wardens informed about where the animals were and alerting them to any unusual movements, such as a rhino running rather than ambling along, even that the rhino had left the park. And an alarm was programmed to go off if the rhino was apparently oversleeping (sleeping longer than its normal 6 h). Any of these circumstances could indicate that something was wrong, very wrong, so there was a reaction team on hand in the park, ready to track and quickly reach the animal in possible distress.[12]

But GPS systems were only devised to be tracking devices, so if they stopped working, it was not possible to know what was happening. This led to the development of the Protect RAPID.

The RAPID in Protect RAPID stands for Real-time Anti-Poaching Intelligence Device. It was invented to deal with situations where even the best antipoaching forces were finding it hard to operate successfully. Rhinos and elephants live in a landscape where huge tracts of land make it well-nigh impossible to monitor what exactly is going on and where and when poaching incidents are happening. Not many arrests are made; conviction rates are low. So poachers can operate without worrying too much about getting caught. There was little to deter them until Protect RAPID.

Designed by scientists working in the field with rhinos, scientists horrified by the growing carnage and driven to do something to prevent this slaughter, the improved system added not only a heart rate monitor but also a video camera implanted into the rhino's horn, to the GPS satellite collar, which broadcasts real-time information, 24/7, to a control centre.

So what happens when poachers attack a rhino wearing one of these devices? Immediately its heart rate rises and then falls. The heart rate monitor instantly

---

[12] Victoria Gill *Rhino horn GPS used to deter poachers* BBC Earthnews, 21 October 2010.

triggers the alarm, and the video camera confirms why this is happening. The control centre can see the poachers, work out where this is happening from the information taken from the GPS collar and then alert the rangers. Using helicopters and/or trucks, they can be on the scene within minutes, leaving no time for the poachers either to cut off the animal's horns or escape.

In other words, *'the Protect RAPID renders poaching a pointless exercise'*.[13]

In the Maasai Mara Game Reserve in Kenya, 25 rhinos were fitted with secure radio transmitters inside their horns, and 5 elephants were fitted with GPS collars round their necks to track their movements for elephants don't stay within the boundaries of the reserve and may even cross into Tanzania as they look for food and water. The data obtained can improve the management of the land and enable a more rapid response to be made if the animals are attacked.

In Tsavo, as part of the Rhino Impact Investment Project, black rhinos are being tagged with transmitters and having their ears notched, for the animals are difficult to spot in the park's thick savannah bush and woodlands. A device that enables regular tracking to be carried out provides better security and protection.[14]

These low-cost wildlife tracking devices are currently being modified.[15]

But in 2017, the journal *Conservation Biology* published an alarming article explaining that it would not be too difficult for poachers to hack into the tags on the animals, because security standards were not always of a sufficiently high standard. Professor Steven Cooke, one of the authors, provided a number of examples of how the system was already being abused, including the chilling incident of eight tagged grey wolves in Yellowstone Park, killed by hunters who had intercepted the signals.

## 29.6   Computers, Data and Statistics

Is it possible to predict the approximate location where poachers might attack? A team of researchers has examined spatiotemporal trends of illegal activities in a Ugandan national park, the Queen Elizabeth Conservation Area. The main aim of the research was to analyse extensive data in order to improve

---

[13] O'Donoghue 2015 *Research Article* 31 July 2015. See http://www.idtechex.com/research/articles/internet-of-things-can-pro Accessed 10/09/2015.

[14] ZSL Conservation > *Tracking black rhinos in Tsavo* ZSL, 4 January 2018. See https://www.zsl.org/blogs/conservation/tracking-black-rhinos-in-tsavo Accessed 09/01/2018.

[15] ZSL *Low-Cost Wildlife Tracking Devices* See https://www.zsl.org/conservation/conservation-initiatives/conservat Accessed 28/06/2018.

the effectiveness of law enforcement in the park, thus stemming the loss of biodiversity resulting from various illegal activities, including poaching animals for high-value commodities (including ivory). The data consisted of that commonly collected by ranger patrols in many protected areas.

The park contained a small population of elephants but no rhinos.

The results that concern us are those relating to the poaching of high-value animals (elephants). This is what they found:

- The driver was the density of the target animals. In other words, more animals, more poaching (which is what one would expect, possibly confirming what seemed to be happening in parks such as the Selous and the Kruger).[16]
- *'Travel cost from villages did not significantly affect any class of illegal activity'.*[17]

Within the park, there was no overall temporal trend in either commercial or noncommercial (bushmeat) poaching between 1999 and 2012, despite recent rises in ivory poaching elsewhere.[18]

These encouraging results indicated that rangers continued to fulfil a vital function and that traditional law enforcement activities were effective at protecting local sites; they did prevent increases in poaching. The findings were backed up by other results. For example, in southern Africa, apart from rhino poaching, other illegal activities were rare in areas that were heavily patrolled.[19]

So can ranger patrols be used to better effect, given:

- The size of the areas that must be covered
- The numbers of the rangers themselves
- Other limited resources

Let us see.

- Patrols need to be intelligence-driven.
- Although the past does seem to be the best predictor of the future, poachers can change their ways, and these changes could be missed by spatial and/or temporal analysis.

---

[16] Maingi et al. 2012.

[17] Critchlow R.; Plumptre A.J.; Driciru, M.; Rwetsiba A.; Stokes E.J.; Tumwesigye C.; Wanyama F.; and Beale C.M. *Spatiotemporal trends of illegal activities from ranger-collected data in a Ugandan national park* Conservation Biology, Volume 20, No. 5, 1458–1470.

[18] Burn et al. 2011; Maisels et al. 2015 in ibid.

[19] Beale et al. 2013a in ibid.

• Although ranger patrols can be concentrated on an area likely to come under attack, it is essential to make sure there are still sufficient patrols in other areas, not only to be present should an incident occur but also to collect vital data. After all, poaching 'hotspots' do change.

The ideal of course is deterrence, a state of affairs that forces the poacher to work out/identify the boundary between when he might well get caught, so doesn't go poaching, and when he probably won't get caught, so decides it is safe, worth the risk anyway, to carry out his unlawful activities.

What was so important about this research was the new statistical model it employed, and the hope that it could be used in other parks and game reserves, used to predict where poachers might strike next so that the rangers could be there. Waiting for them!

## 29.7   SMART Technology

SMART, or Spatial Monitoring and Reporting Tool, has been developed by ZSL and other leading conservation organizations to help rangers and those whose job is to protect wildlife from threats such as poachers. It, too, is about data collection and it has added yet another tool to the box.

It enables rangers to record essential information such as patrol routes, signs of wildlife and signs of illegal activity *with pinpoint accuracy* and provides managers with accurate conservation information that can assist with the day-to-day running of a reserve. It also stores information/data which it can analyse, thus making it easier to plan and target protection. It can even create maps. In other words, it makes both patrols and protection efforts more effective.

The number of partners who had a hand in developing SMART with ZSL is impressive. They include:

• CITES-MIKE
• The Frankfurt Zoological Society
• The North Carolina Zoo
• Wildlife Conservation Society
• World Wildlife Fund

They have now been joined by:

• Panthera
• The Peace Parks Foundation
• Global Wildlife Conservation

Already operating on sites in 31 countries, SMART is helping to protect forest elephants in the Dja Biosphere Reserve in Cameroon and savannah elephants in Kenya.

## 29.8  Connected Conservation

The brainchild of two companies, Dimension Data, a South African IT services provider, and Cisco, the American technology giant, Connected Conservation, was set up in 2015, in a private game park in South Africa. The system, based on a Wi-Fi network, is proactive and can be used to detect people unlawfully entering a reserve and then initiate an emergency response.

Security at the park depends on four tiers:

- Intelligence
- The fence line
- Patrols
- Armed response teams, including a helicopter

And it is dependent on data collection:

- CCTV
- Biometric scanning of visitors
- Seismic sensors in the ground
- Thermal imaging and drones in the sky

Inbound vehicles are tagged to track their movements in the park.[20]

The 62,000 hectare park is surrounded by a 72 km fence, and installed along its length are cameras, which include thermal cameras and CCTV cameras, biometric scanners and sensors all beaming data to the central control centre, which also receives data from rangers on patrol. Wi-Fi is provided by solar-powered local area network towers, housed in each of the park's four gates. Linkage to a national data base makes it possible to cross-check for known poachers/criminals and stolen vehicles.

Despite the challenging terrain, searing heat, wind, sand and floods, the system proved a great success, with a 96% reduction in poaching and no

---

[20] Endri Steyn, head of park security in Scott Carey's article *How technology helped foil the rhino poachers* Computerworld 9 May 2018. See https://www.computerworld.com.au/article/640959/how-technology-helped-foil-rhin0 Accessed 15/05/2018.

rhinos killed in 2017. The immediate access to data had cut response times from 30 min to 7, so that rangers could now intercept poachers before they started killing their targets.

By May 2018, phase two was ready to get started. This would entail an upgrade of the network to a LoRA protocol, *'plans to add sensors to vehicles, place magnetic sensors on the ground to detect weapons entering the park, and acoustic fibre to the fence line to better detect fences being cut'.*[21]

Other countries were also interested. As a result of government requests, work has already begun in Zambia and will soon start in Mozambique, as well as in a private reserve in Kenya.[22]

The slightly smaller Welgevonden Game Reserve is also using data collected from the Internet of Things (IoT) to tackle rhino poaching, although its approach is unexpectedly different. It isn't the rhinos that are collared but the prey animals, zebra, wildebeest, eland and impala. Then IBM's IoT platform is used to monitor and collect the data from these animals as they respond to disturbances, all of which is being used to create algorithms built on the animals' response to perceived threats, poachers, tourists and employees.

Connected Conservation works well because, as the name implies, the data is integrated. But the system is expensive.[23]

## 29.9    Possible Future Developments

There are some unusual developments in the pipeline.

### 29.9.1    Elephants and Seismologists

This work in progress is based on the results of research into the seismic activity generated by the elephants of Samburu, research carried out by Save the Elephants and Oxford University.

In addition to vocalising, elephants use their feet to communicate with each other, and a herd of elephants thundering over the ground generates seismic waves, vibrations, that can be felt/picked up in the ground, kilometres away. So the idea is that sensors in the ground will enable herds to be monitored, their position and activity triangulated and any unexpected running picked up. Then develop this into an alarm system to warn of poaching.

---

[21] Ibid.

[22] Ibid.

[23] Ibid.

## 29.9.2   Google App Imaging Software and Artificial Intelligence (AI)

Although this imaging system was originally developed to scan peoples' faces, ZSL and Google have developed the software to '*get to know*' elephants, rhinos and other very large mammals. Hidden automatic cameras, triggered by heat and motion, will capture and record each animal whenever it passes by, and the resulting data stored. This will enable profiles of the animals to be built up. They will also look for potential poachers and send out warnings if they seem to be suspicious.

Already trialled with other animals, such as giraffes in Kenya, a modified and more user-friendly system was released early in 2018. Dispensing with specialists, it will enable organizations to build their own machine-learning programmes, which will then release critical manpower back into the field, because humans will no longer have to spend long hours working their way through a mountain of images.

And in time, it is hoped the system will be able to recognize injured animals as well.[24]

## 29.9.3   A 'Gold Standard' Wildlife Protection Technological System

The result of yet another collaboration, the pilot for this new technology, will be used to protect black rhinos in Kenya's Ngulia Rhino Sanctuary which is home to about 10% of the country's black rhino population. A smartphone-based software platform will act as a first surveillance system, providing rangers and officers with a means of finding out what is happening in the Sanctuary, while in the second phase, sensors will be connected to the platform to provide perimeter control, intrusion detection and wildlife monitoring. In the third and final phase, advanced network and radar technologies will not only be able to detect large objects but extend the area that can be monitored. Drones may also be used for all-day surveillance, providing video and thermal imaging, detecting intruders and carrying out a census of the rhinos.

Why Ngulia? Although its rangers have military training, they need more sophisticated technology and training.[25]

---

[24] Charlie Parker *Facial recognition to save elephants from poachers* The Times, 24 January 2018.

[25] *Technology and Innovation to End Poaching and Wildlife Crime* See https://devex.com/impact/partner-ships/technology-and-innovation-to-end-poaching-and-wildlife-crime-748 Accessed 02/10/2018.

# Part X

## Important Developments

# 30

# The London Conference and What Followed

We begin with a series of conferences which resulted in many key countries committing to a series of measures that should help ensure the future of elephants and rhinos.

## 30.1   The Clarence House Conference

The first, the Clarence House Conference, took place in 2013 and was a precursor to the rest. Its hosts were HRH Prince Charles the Prince of Wales in his role as President of WWF-UK and the UK Government, and its aim was *'To build a cross government international approach to tackling the illegal wildlife trade'*. 22 countries attended.

## 30.2   The London Conference on 'The Illegal Wildlife Trade' and the London Declaration

It was followed in 2014 by the London Conference.[1] This time 41 countries and the EU were represented, as were multinational organizations and NGOs. The Prince of Wales and his sons Princes William and Harry were also present.

---

[1] See   https://www.gov.uk/government/groups/tackling-illegal-wildlife-trafficking-inter-m.   Accessed 28/09/2016.

© Springer Nature Switzerland AG 2019
B. Martin, *Survival or Extinction?*, https://doi.org/10.1007/978-3-030-13293-4_30

The resulting declaration set out five actions to be taken which were based on:

- The scale and consequences of the illegal trade in wildlife: in which elephants and rhinos received special attention. There was *a particularly dramatic escalation* in the rate at which they were being poached, which posed *a severe threat* not just to them but also to regional security and sustainable development.
- The need to build on the existing international framework for action.

They also recognized *'the importance of engaging communities living with wildlife as active partners in conservation, by reducing human-wildlife conflict and supporting community efforts to advance their rights and capacity to manage and benefit from wildlife and their habitats'*.

Although the representatives committed themselves to taking the actions, if they were to succeed, the international community would need to cooperate as well.

The five actions are:

- **Eradicating the market for illegal wildlife products:** This would require support and effectively targeted actions.
- **Ensuring Effective Legal Frameworks and Deterrents:** Here, the idea was to provide a better deterrent for the 'kingpins' using the UN Convention against Transnational Organized Crime and the UN Convention against Corruption and applying a zero-tolerance policy on corruption associated with the illegal wildlife trade.
- **Strengthening law enforcement throughout the chain:** A number of measures would have to be taken, including:

  - Capacity building to protect key populations of species threatened by poaching. This would include increasing *the number of well-equipped and well-trained law enforcement officers at key sites*.
  - Establishing and maintaining national cross-agency mechanisms.
  - International cooperation, sharing of expertise, including, but not limited to, *'criminal intelligence; controlled deliveries; traceability systems; risk profiling; detector dogs; ballistic analysis and the use of existing forensic technology including the further development of such technologies'*.
  - Strengthening of cross-border and regional cooperation.

- **Sustainable livelihoods and economic development:** Local communities should be helped to pursue sustainable livelihood opportunities and measures taken to eradicate poverty by developing measures such as *'community conservancies, public-private partnerships, sustainable tourism, revenue-sharing agreements and other income sources such as sustainable agriculture'*.[2] They should also keep the benefits for the conservation and sustainable management of wildlife.

- **The Way Forward:** This final action highlighted the need for more resources and further research into and improved understanding of the illegal trade, together with the impact of measures taken to combat and prevent it. And more working together. The parties welcomed the *Group of Friends* against illegal wildlife trafficking, within the UN, as well as the suggestion to establish a Special Representative of the Secretary General, to further the fight against the trade.

A number of other high-level events and commitments followed. They were designed to keep up the momentum, raise the profile of illegal wildlife trade and secure political support for these efforts. The next conference would be held in Kasane, Botswana, in 2015, where progress was to be reported and reviewed.

## 30.3   The Kasane Conference on 'The Illegal Wildlife Trade' and the Kasane Statement

The Kasane Conference aimed both at reaffirming the London Declaration and taking it forward. Once again, the representatives, most of whom had attended the London Conference, acknowledged that illegal wildlife trafficking was causing severe problems in many areas, including economic, environmental and social. It was also implicated in threats to security. The Kasane Statement (2015) recognized *'the need for decisive and urgent action to tackle the trafficking of endangered fauna and flora remains greater than ever'* because *'despite efforts to date for many species, the illegal trade, and the poaching which fuels it, is an ongoing and growing problem'*.

All the participants committed *'to provide the political leadership and practical support needed to find a lasting solution to the illegal wildlife trade'*, and they decided to:

- Undertake a number of actions.
- Ensure legal frameworks and deterrents were effective.
- Strengthen law enforcement.

---

[2] The *London Declaration*.

**The Actions:** As before, these were aimed at *eradicating the market for illegal wildlife products.* The conference delegates were to commit themselves and call upon the international community to:

- Strengthen partnerships with business and others to reduce demand and supply.
- Conduct and/or support research to improve understanding of the factors driving the market.
- Strengthen and, if necessary, establish partnerships among source, transit and destination countries to combat the illegal wildlife trade along its entire chain.

**Ensure Effective Legal Frameworks and Deterrents:** This concentrated on corruption, the representatives reaffirming their attitude of *zero tolerance.* Two actions were required:

- National laws should be reviewed and changed if necessary, to make offences connected to the illegal wildlife trade *predicate offences*, as defined in the UN Convention against Transnational Organized Crime, which would turn them into domestic money laundering offences. They could then be pursued accordingly. It would also bring in the Financial Action Task Force and its regional bodies, other multilateral organizations and asset recovery networks.
- The relevant legal personnel should *have the resources, knowledge and capacity effectively to investigate and prosecute financial crimes associated with wildlife crime.*

*Strengthen law enforcement:* Four actions were prescribed, all of them involving working together:

- The ICCWC to be strengthened so that it could take *'a leading role in providing coordinated global support to the law enforcement community, including enhancing enforcement capacities at national, regional and international levels'.* Working together would *promote increased sharing of best practice and lessons learned, facilitate the exchange of information and intelligence and foster cooperation, including…, through regional wildlife enforcement networks.*
- Regional wildlife enforcement networks (RWENS) to be established and/or strengthened.
- Awareness raised within the transport industries, which were to be supported as they worked out and implemented measures to eliminate the illegal wildlife trade.

- Concerned global and regional networks of prosecutors, some already in existence, some being established, should, because they are *'responsible for prosecuting offences involving organized crime as it relates to the illegal wildlife trade'*, foster international and regional cooperation both to raise awareness and strengthen the investigation and prosecution of these offences.

The Kasane Statement re-emphasized the importance of sustainable livelihoods and economic development in the fight against the illegal trade, recognizing that:

- Sustainable livelihoods resulted from local people being allowed to benefit.
- *'The active engagement of local people is also key to effective monitoring and law enforcement'*.

Here, four actions were required:

- Helping rural communities to live and coexist with wildlife
- Building conservation constituencies
- Promoting sustainable development
- Promoting *'the retention of benefits from wildlife resources by local people where they have traditional and/or legal rights over these resources'*.

All of this means more working together.

## 30.4   The Buckingham Palace Declaration

We have already seen that wildlife traffickers need to get their commodities from the producer countries to their final destination and that some traffickers have set up their own companies to do this. However, most carriers are probably quite ignorant of what is happening, that they are being used by the smugglers. So in 2014 the United for Wildlife Transport Taskforce was set up, *'to work with private sector businesses from the transport industry who may be unwittingly drawn into the illegal wildlife trade'*. And on 15 March 2016, 40 companies signed up to this, the Buckingham Palace Declaration.[3]

Committed to take real steps to shut down the smugglers' routes, it has two aims:

---

[3] United for Wildlife Newsletter, April 2016; United for Wildlife info=unitedforwildlife.org@mail78.suw17.mcsv.net on behalf of United for Wildlife info@unitedforwildlife.org.

- To identify any role the transport sector plays in the trade.
- To examine ways to break the chain between suppliers and consumers.

Its members are from:

- The global transportation industry, including many airlines and shipping groups.
- The World Customs Organization and national customs authorities.
- Conservation groups, including TRAFFIC, UNEP – Global Environment Facility, WWF-UK and the Wildlife Conservation Society.

They have agreed to 11 commitments, which focus on:

- Securing information-sharing systems for the transport industry to receive credible information about high-risk routes and methods of transportation.
- Developing a secure system for passing information about suspected illegal wildlife trade from the transport sector to relevant customs and law enforcement authorities.
- Notifying relevant law enforcement authorities of cargoes suspected of containing illegal wildlife and their products and, where able, refusing to accept or ship such cargoes.

The importance of working together is obvious. There are sections such as *'Information sharing and detection'*, which includes training, *'Practical measures to stop the transportation of illegal wildlife products'* and *'New mechanisms tackling illegal wildlife trade'*.

China was one of the 40 countries that signed the pledge, and in April 2016, TRAFFIC held a workshop with a leading Chinese courier company, Shunfeng Express (SF-Express), teaching staff how to detect suspected illegal wildlife products.[4] That same month, India's transport sector, in the form of Air India, Jet Airways and Apeejay Shipping, three of the biggest shipping and airline companies in the country, also signed the Declaration, each of them making clear how keen they were to join this important global initiative.[5]

---

[4] TRAFFIC – Wildlife Trade News *TRAFFIC helps train leading Chinese courier com...* See http://www.traffic.org/home/2016/4/7/traffic-helps-train-leading-chinese-courier-com Accessed 12/04/2016.

[5] IBC World News *Indian transport sector joins fight to shut down illegal wildlife trafficking* 12 April 2016. See https://www.ibcworldnews.com/2016/04/12/indian-transport-sector-joins-fight-to-shut... Accessed 19/04/2016.

## 30.5   The Hanoi Conference on 'The Illegal Wildlife Trade' and the Hanoi Statement

Held in November 2016, in Hanoi, and hosted by the Ministry of Agriculture and Rural Development of Viet Nam, this Conference was attended by Viet Nam's Vice President, the Deputy Prime Minister of Lao PDR, Prince William and the Under-Secretary General of the UN, together with 7 UN organizations, delegates from 47 countries and many others. Once again, all were concerned to take practical actions to protect wildlife from extinction from illegal trading.

The format was similar to the earlier conferences. The main threats were identified, and the Hanoi Statement on Illegal Wildlife Trade set out a number of actions the participants had agreed were crucial to undertake, if endangered wildlife was to be protected. Vice President Dang Thi Ngoc Thinh went straight to the point:

*Along with the commitments and statements, it is critical to turn the commitments into practical actions and create a mechanism to oversee these actions. In so, we can ensure the sustainability of our commitments.*

Reduction of demand and supply for illegal products was again recognized as the key to success, and this required not only law enforcement to be strengthened but also governments to be supported in their attempts to eliminate associated financial crime and corruption. And once again, the high importance of developing sustainable livelihoods and benefits for the local communities was stressed.

This time however, one of the **actions** specifically stated the need '*To clarify the link between wildlife crime and other criminal organized activities, even terrorism*'.[6]

The UK would host the next conference in October 2018.

## 30.6   A Historic United Nations Resolution

In July 2015, all 193 Member States signed the ground-breaking UN General Assembly Resolution *Tackling the Illicit Trafficking in Wildlife*, thus acknowledging the fact that only by working together could the world bring an end to

---

[6] Hanoi Conference on Illegal Wildlife Trade. See http://www.iwthanoi-conference-illegal-wildlife-trade/. Accessed 22/03/2018.

the devastating consequences resulting from the seemingly never-ending rampage of wildlife trafficking. Initiated by Germany and Gabon and sponsored by 84 other countries, it had taken 3 years of non-stop diplomacy to achieve.

Not just about ending poaching, the Resolution also:

- Encompasses good governance and the rule of law
- Takes into account the well-being of local communities
- Recognizes the fact that as well as damaging conservation, the illegal trade was also financing criminal networks and funding armed conflict
- Highlights both the impact transnational and organized crime has on the environment and *'the need for countries to counter corruption and address money laundering linked to wildlife crime'*.

Because the signatories agreed *'to enhance regional and international co-operation along the entire illegal wildlife trade chain, including measures to stop the poaching, trafficking and buying'*, the Resolution encourages countries to strengthen their law enforcement and judicial processes. And because it is also concerned with sustainable use, local communities are encouraged to become actively involved in the fight against the illicit trade, by enhancing their rights and capacity to manage and benefit from wildlife resources.

Its progress is to be monitored. Every year the Secretary General of the United Nations will present a report on:

- Global wildlife crime
- How countries are implementing the Resolution
- Recommendations for further action

And a special envoy is to be appointed.

# 31

## CoP16

## 31.1 Preparations

Luckily for us, we are only concerned with three of the multitude of documents and reports submitted to the Standing Committee and thence the Secretariat for consideration before CoP16. Two reports tell us what was happening to elephants and rhinos in the run-up period to the Conference, and we have already included everything that is necessary from the third, a literature review of the possible medicinal properties of rhino horn.[1]

### 31.1.1 Elephant Populations: 'Elephants in the Dust'

If the report *No Peace for Elephants* had painted a very sombre and disturbing picture of the dangers of unregulated ivory markets to elephant populations, *Elephants in the Dust - The African Elephant Crisis* made even grimmer reading. It was very bad news indeed. The UN Under-Secretary General and UNEP Director, Achim Steiner, recorded his fears in the report's Preface:

> *In Central and West Africa, the elephant may soon disappear from whole areas unless urgent action is taken.*[2]

---

[1] Species trade and conservation Rhinoceros *Assessment of Rhino Horn as a Traditional Medicine* A report prepared for the CITES Secretariat by Kristin Nowell on behalf of TRAFFIC, April 2012, SC62 Doc.47.2.

[2] UNEP, CITES, IUCN, TRAFFIC (2013). *Elephants in the Dust – The African Elephant Crisis. A Rapid Response Assessment.* United Nations Environment Programme, GRID – Arendal, p.4.

© Springer Nature Switzerland AG 2019
B. Martin, *Survival or Extinction?*, https://doi.org/10.1007/978-3-030-13293-4_31

By 2011 and 2012, MIKE was recording mass killings of the animals, to the extent that in some areas, the level of poaching was finally beginning to threaten the continued existence of one species, the forest elephants. ETIS was coming up with similarly alarming statistics. A rapid response assessment was urgently required. The results, published in 2013 in *Elephants in the Dust*, provided:

- A graphic overview of the current status of elephants
- How poaching was affecting them
- How ivory was being trafficked along the entire length of the ivory trade supply chain

Then it became possible to respond.

It was another product of working together, UNEP, CITES and TRAFFIC, the IUCN/SSC African Elephant Specialist Group (AfESG), the International Consortium on Combating Wildlife Crime (ICCWC) and others. The AfESG provided estimates of the different elephant populations, while MIKE and ETIS provided the African and Asian Elephant Database.

## 31.1.2   The Results

**Range Areas and Populations:** The starting point was to determine the elephants' range areas, no easy task because the animals move with the seasons, sometimes tramping vast distances in their search for food and water. Only when this was done could estimates be made of the size of a particular elephant population. Range areas have always played an important part in keeping elephants safe. For example, if they enter war zones, they will be at even greater risk.

The range of African Elephants continued to grow smaller, reduced to some 35–38 countries in sub-Saharan Africa, and it was uncertain whether there were any left in Somalia, Senegal or Sudan.[3] Similarly, elephant populations continued to decline, particularly in central Africa, and the populations in west Africa have been fragmented for many years. Fortunately, populations in much of southern Africa remained high and stable, some had even increased.

---

[3](CITES 2011) in ibid., p.15.

### 31.1.2.1   Poaching and MIKE

The evidence from MIKE was equally grim.

*The last seven years have seen a clear increase in the level of elephant poaching across all African sub-regions. The year 2011, and probably also 2012 saw an all-time high in poaching since systematic monitoring began more than a decade ago. It is estimated that in 2011, approximately 7.4 per cent of the total elephant populations in elephant sites across Africa were killed illegally. These sites represent 40 per cent of the total African elephant population, which means that 17,000 elephants were killed at these sites alone.[4]*

For 40%, read more than 230,000 elephants.[5]

Central Africa presented the most disturbing picture, despite the fact that some of the sites were also UNESCO World Heritage Sites, although that is no guarantee of safety. In 2011, some *14% of the entire elephant population in MIKE sites in the central African sub-region were killed*, and two sites, both located in DRC, showed that catastrophic events had taken place. All the elephant bodies found in Virunga National Park indicated they had been unlawfully killed, while in the Kahuzi-Biega National Park, because of persisting armed conflicts, the total elephant population had been reduced to no more than 20 animals. For the first time, the extinction of these beautiful animals was mentioned as a possibility.[6]

Poaching had also increased in eastern Africa where as many as 7% of the elephants living in MIKE sites had been unlawfully killed. In other words, poaching had trebled. Once again, World Heritage Sites had not been spared the onslaught. In 2011, over 65% of the 224 carcasses found on patrols in the Selous Game Reserve in Tanzania had been killed by poachers, and poaching levels had gone up in the Ruaha Rungwa National Park.

There was a similar pattern in Kenya, with elephant populations in both the Tsavo National Park and the Samburu Laikipia ecosystem suffering heavy losses to poachers (464 bodies, two-thirds of which had been illegally killed). Even the Murchison Falls National Park and the Queen Elizabeth National Park in Uganda, which had far fewer elephants, had been targeted.

Although four countries in southern Africa, Namibia, Botswana, Zimbabwe and South Africa, had been permitted by CITES to sell some of their lawfully held ivory stocks because their elephant numbers had increased (the two

---

[4] Milliken in ibid.
[5] CITES 2012a; see Blanc et al. 2007 for further definition of population categories, in ibid., pp. 32–33.
[6] Ibid CITES 2012a.

one-off sales), this had never been the case in Mozambique. We have already seen its seeming inability to control illegal sales of ivory within its ivory markets, so perhaps it should not come as a surprise that its Niassa National Reserve reported a very high PIKE level. And the data suggested an increasing trend in levels of poaching, with an estimated 4% of the total elephant population living in MIKE sites in Southern Africa being illegally killed in 2011.

Although it was much more difficult to work out PIKE values and hence poaching trends in west Africa with its small and fragmented elephant populations, the data and poaching levels were high enough to generate concern.[7]

### 31.1.3   Rhino Populations and 'the Deadly Nexus'[8]

About 95% of the total population of white rhinos in Africa now live in South Africa; their numbers gradually increasing from the desperately low threshold of between 20 and 50 animals to about 1800 in 1968 and an estimated 18,800 by 2010.

For black rhinos, however, the story is very different. Decimated in the first catastrophic rhino poaching crisis, their numbers fell from approximately 100,000 in 1960 to only 2410 in 1995. Fortunately, by 2010, the numbers had increased to 4880 animals, but, for a total population, this number was still frighteningly small, and black rhinos continued to be classed as *Critically Endangered* on the IUCN's Red List. Once again, South Africa played a key role in their conservation. Their population of black rhinos has been steadily increasing since the 1980s, reaching an estimated 1915 animals in 2012 or nearly 40% of all black rhinos alive today. The private sector has played a major role in the recovery of both species. In 2010, they owned about 25% of all South Africa's white rhinos and about 22% of the black rhinos.

In the early 2000s, everything started to go wrong. Dramatically wrong. White rhino trophy hunting had been carried out without problems for some 35 years, but, quite unexpectedly, in 2003, it became a favourite past-time of the Vietnamese. Not traditional hunters, they began, quite deliberately, *'to exploit loopholes in South Africa's legislation to obtain hunting trophies for a revived rhino horn trade in Asia'.* Rhino poaching started to spiral out of control. The numbers become bleak.

---

[7] Ibid, pp.33–39.

[8] Milliken T. and Shaw J. (2012) *The South Africa – Viet Nam Rhino Trade Nexus: A deadly combination of institutional lapses, corrupt wildlife industry professionals and Asian crime syndicates* TRAFFIC, Johannesburg, South Africa.

Between 1990 and 2005, an average 14 rhinos were poached every year, and then the numbers began to rise.

- In 2008: 83 rhinos were poached.
- In 2009: 122 rhinos were poached.
- In 2010: 333 rhinos were poached.
- In 2011: 448 rhinos were poached.
- In 2012: 251 rhinos were poached up to June.[9]

Not only that, organized crime syndicates were managing to get hold of rhino horns from stocks officially held by the private sector. And there were also the thefts, similar to those we have already seen taking place in some European countries. Public displays in South Africa lost at least 46 horns in this way. Even more disturbing was finding out that some corrupt personnel in the wildlife industry were prepared to illegally cut off the horns of live rhinos without getting the correct permits. Yet South Africa had such a proud reputation in rhino conservation and was such an important country for these animals' survival. So what had gone wrong?

On behalf of TRAFFIC, Tom Milliken and Jo Shaw undertook a '*comprehensive of overview of events and dynamics currently driving and escalating illicit trade in rhino horns from South Africa to Viet Nam*', documenting '*the scale and characteristics of the new illicit trade linking Africa with Asia*'. Their findings were set out in '*The South Africa-Viet Nam Rhino Horn Trade Nexus: A deadly combination of institutional lapses, corrupt wildlife industry professionals and Asian crime syndicates*',[10] a key document that helped to influence the outcome of CoP16, greatly strengthening the provisions regarding the conservation of rhinos.

As the Conference approached, it had become obvious to people all over the world that elephant and rhino poaching had reached such an intensity that unless something was done soon, very soon, there might be no more elephants or rhinos. Even governments were taking notice. Hillary Clinton (of the Clinton Foundation), who was then the US Secretary of State, called for action to be taken against wildlife crime. Importantly, she urged that enforcement should be expanded and strengthened. The UN General Assembly decided to act. So did the Asia-Pacific Economic Cooperation. Wildlife crime could fall within the jurisdiction of the Convention against Transnational Organized Crime, finally enabling it to be recognized as serious, transnational, organized crime.

---

[9] Ibid.
[10] Ibid.

## 31.2   CoP16

Held in Bangkok, Thailand, in March 2013, this CoP marked the 40th birthday of CITES, and perhaps because of this, there was a record attendance. There were of course the representatives of the Parties but also many observers from intergovernmental, international and national organizations, making a grand total of some 2000 people. The opening speeches were given by:

- The Secretary General of CITES, John E. Scanlon.
- Prince William, who gave a video address
- The Prime Minister of Thailand and, finally, by tradition
- Speeches from the Chairman of the important CITES Standing Committee and the Executive Director of UNEP, Achim Steiner

This was a very high-profile lineup.

In his opening remarks, the Secretary General observed that although there was renewed interest in CITES, this was *regrettably fuelled in part by an increase in wildlife poaching, particularly of elephants and rhinoceroses*, and Prince William spoke of the need to combat *shocking levels* of poaching and illegal trade. The Prime Minister of Thailand pledged to take forward the CITES agenda. She also announced that her own country, which we have already seen has a bad track record regarding illegal ivory trading, would *work towards amending the national legislation with the goal of putting an end to ivory trade and to be in line with international norms*.

As is usual at CoPs, in addition to giving specific consideration to a number of species, there are always wider key themes. We saw earlier, the critical role secret voting played in influencing the first of the one-off sales of elephant ivory. Since then, attempts have been made to strictly limit its use and move towards a fully transparent Convention. The EU raised this issue again at Cop16, without success.

Another key theme was strengthening the links with other multilateral environmental agreements, such as the one between CITES and the CBD. So it will come as no surprise that references to the CBD Strategic Plan on Biodiversity 2011–2020 and the relevant Aichi Biodiversity Targets in the Strategic Vision were incorporated into Resolution Conf. 16.3, as was the need to extend the Strategic Vision and Action Plan to 2020. This might provide CITES with access to certain funding, always a problem as we have already seen.

A 'roundtable' on combating transnational organized wildlife and forest crime was organized by the ICCWC working with the Government of

Thailand, so that countries and regions could share their experiences of how they tackled such crime and reaffirm a commitment to counter it at top political levels.

A number of Resolutions and Decisions were passed to strengthen enforcement and compliance. Some, like Resolution 11.3, contained new and innovative measures to help combat money laundering and to facilitate the forfeiture of the assets of criminals. It also highlighted the need to provide real-time support to park rangers confronted by heavily armed groups.

Decision 16.40 agreed to:

- Assess follow-up actions after large seizures of CITES specimens.
- Initiate a process to assess implementation and enforcement of CITES Appendix I species.
- Establish Wildlife Incident Support Teams (WISTs) to assist countries affected by significant poaching or making large-scale seizures of CITES specimens.

TRAFFIC and the Ministry of the Environment for Japan co-organized an event promoting the enforcement of trade regulations for CITES-listed species in southeast Asia.

The first-ever global meeting of the WENS (the Wildlife Enforcement Networks like ASEAN WEN) was held. Organized by the ICCWC, its aim was to enable them to share experiences and discuss the need for their enhanced cooperation and coordination. More than 10 WENs participated, agreeing that a network of WENS should be set up with the help of the ICCWC. The NWEN will:

- Fortify the role of WENs.
- Foster attention from civil society organizations.
- Draw more notice to the magnitude of wildlife crime.
- Work with them to reduce demand for endangered species.

CITES must have 'teeth'!

This was a sentiment reflected in many of the discussions. The criteria and measures for compliance with CITES were set out in Resolution Conf. 14.3 CITES Compliance Procedures. The 63rd meeting of the Standing Committee, held just before CoP16, drew attention to the importance of using these Procedures, so, applying them, the Conference agreed new texts regarding illegal trading in elephants {Resolution Conf. 10.10 (Rev. CoP16)} and rhinos {Decision 16.91}.

The USA raised the ongoing problem of national legislation. Unfortunately, not all Parties pass legislation that fully implements CITES, and this tends to annoy those Parties that are fully compliant. The USA proposed that Parties who had been members of CITES for more than 20 years could not justifiably claim that exceptional circumstances were preventing them from adopting appropriate measures for its effective implementation. The proposal was accepted.

## 31.2.1   Elephants

Because:

- The ETIS analysis showed that illegal trade in ivory in 2011 was three times greater than that in 1998.
- The MIKE analysis showed *the highest levels of poaching since the system began analysing site-specific data more than a decade ago*.[11]

The text of Resolution Conf. 10.10 Trade in Elephant Specimens was significantly strengthened. New measures introduced included:

- Compulsory annual reporting of all ivory stockpiles held by governments anywhere in the world
- Mandatory forensic examination of all large-scale ivory seizures (seizures of 500 kg or more)
- Inclusion of *demand reduction* as a necessary course of action in end-use markets for ivory
- The tracking of trade in live elephants
- A compliance mechanism in accordance with Resolution Conf. 14.3 CITES compliance procedures, including the threat of sanctions when Parties fail to implement the Resolution's requirements

Decisions were also passed to help support elephant conservation.

Decision 16.78, for example, mandates, subject to funding, the Secretariat to convene a CITES Enforcement Task Force. It will comprise the 9 countries we have already seen that form the principal illegal trade chains, namely, China, Kenya, Malaysia, the Philippines, South Africa, Tanzania, Thailand, Vietnam and Uganda. The first eight have also been identified by ETIS as

---

[11] Julie Gray, Tom Milliken, David Newton, Thomasina Oldfield, Glenn Sant and Sabri Zain *Report of the 16th meeting of the Conference of the Parties to CITES* TRAFFIC Bulletin, Vol. 25 No. 2, October 2013, p.60.

*countries of primary concern* and Uganda was identified by ETIS as being *of concern*, in a report to SC62. Working together with the ICCWC partner organizations as a CITES Enforcement Task Force, they will:

* Review current law enforcement strategies and practices.
* Promote law enforcement collaboration all along the trade chain.
* Examine DNA testing and other forensic identification techniques for ivory.
* Consider the broader use of controlled deliveries, anti-money laundering and asset recovery mechanisms as tools for combating wildlife crime.

A further 15 countries will be subject to a CITES oversight process that will examine their ivory trade policies and actions, the findings to be reported back to the next meeting of the Standing Committee (Decisions 16.79 and 16.80). These countries, again based on ETIS analysis, include Cameroon, Congo, DRC, Egypt, Gabon, Mozambique, Nigeria and Uganda (all identified as countries of *secondary concern*) and Angola, Cambodia, Japan, Laos PDR, Qatar and the United Arab Emirates (all *important to watch*). Where the findings reveal problems, the Standing Committee can take additional action/s against that particular country.

Decision, 16.81, was concerned with CITES and the UN Office on Drugs and Crime (UNODC) working together.

Decision 16.83 was directed at all Parties involved in any ivory seizure of 500 kg or more. They are required to collect and submit samples to appropriate forensic laboratories for analysis within 90 days of the seizure. Milliken comments that, provided it is implemented effectively, this *'should reveal sources of large quantities of ivory that would otherwise be likely to remain unknown'*.

ETIS analysis was also responsible for other important developments. These were discussed at the 62nd, 63rd and 64th meetings of the Standing Committee, which were held just before and just after CoP16. The destination countries China and Thailand, the transit countries/territories Malaysia, Hong Kong, the Philippines and Viet Nam, and the source countries Kenya, Tanzania and Uganda were all required to prepare detailed national ivory trade action plans (NIAPS) and submit them to the CITES Secretariat by May 2013. They all complied.

CoP16 had been good for elephants. The NIAPs and other developments held *'great potential for putting "teeth" into the CITES framework for dealing with illegal trade in ivory, and holding governments accountable for their ivory trade policies and practices'.*[12]

---

[12] Milliken 2013.

## 31.2.2   Rhinos

Resolution Conf. 9.14[13] highlighted:

* Record poaching losses in South Africa
* The highest number of rhino horns in trade in two decades
* The advent of Europe and North America as sources of rhino horn through thefts from museums and other institutions and the illegal acquisition of sport-hunted trophies
* The use of European sport hunters in South Africa as conduits for illegal horn trade to Vietnam
* The emerging roles of Mozambique and China in the illicit trade

At the Conference, it became obvious from the measures agreed upon that the Parties were going to take action on the report's conclusion that *rhinos are facing a crisis and there is no room for complacency.*
Let us look at some of these measures.
Decision 16.84 directs all Parties:

* To report rhinoceros horn seizures to the secretariat and all countries along the trade chain to enable follow-up investigations
* To submit seized horn samples to accredited forensic laboratories for DNA analysis
* To pass national legislation (if necessary) to underpin the use of specialized investigative techniques (e.g. undercover investigation) and other law enforcement tools (e.g. anti-money laundering and asset forfeiture legislation) to combat rhinoceros crime
* To employ a combination of relevant legislation in prosecutions so that penalties serve as effective deterrents

The Decision also:

* Strengthens CITES procedures for issuing permits for rhinoceros horn specimens
* Calls for measures to regulate internal trade in rhinoceros specimens such as sport-hunted trophies or antique specimens, including all parts or derivatives (which of course, includes horn and powdered horn)
* Stipulates that parties should consider stricter domestic measures to regulate the re-export of rhinoceros horn products

---

[13] This mandated the report compiled by the IUCN and TRAFFIC.

Decision 16.85 deals with demand and applies to all Parties identified as range or consumer States. It calls for these Parties:

> *to develop and implement long-term demand reduction strategies or programmes and immediate actions aimed at reducing the illegal movement and consumption of rhino horn products, taking into consideration the draft demand reduction principles (found in another document) to achieve measurable change in consumer behaviour.*

In range States *'strategies or programmes to enhance community awareness with regard to the economic, social and environmental impacts of illicit trafficking in wildlife crime'* should not only be developed, but actions taken need to be reported to the CITES Rhino Working Group by 31 January 2015 so that *best practices* and challenges can be elucidated for a report to SC66.

Decision 16.86 is addressed solely to Viet Nam, which was identified at CoP15 as the principal destination for rhino horn. Once again, several measures were included.

It exhorts Viet Nam to:

- Implement those elements of the South Africa-Viet Nam action plan that will serve to strengthen national management of rhinoceros horn trophies (including issues of tracking, possession, alteration and transfer of ownership)
- Improve investigations and prosecutions of Vietnamese nationals or others implicated in rhinoceros crime
- Conduct consumer behaviour research (in order) to develop and implement demand reduction strategies or programmes aimed at reducing the consumption of rhino horn products
- Report on this and other activities, including arrests, seizures, prosecutions and penalties in rhinoceros crime cases, to the CITES secretariat by 31 January 2014

Mozambique was the target of the next Decision, because most poaching incursions into South Africa's Kruger National Park come from Mozambique. And at the time of the Conference, it was also the main country in Africa exporting illegal rhino horn.

Decision 16.87 therefore calls on Mozambique to:

1. Implement Resolution Conf. 9.14 (Rev. CoP15) and Decision 16.84 effectively, with special attention to enacting legislation to establish penalties for rhinoceros crime that serve as a deterrent

2. Produce a comprehensive report on its efforts to the Secretariat by 31 January 2014 for consideration by the CITES Rhino Working Group

Decision 16.88, directed at both Mozambique and South Africa, calls for:

- Enhanced bilateral cooperation to combat rhinoceros poaching and illegal horn trade
- A comprehensive report to the Secretariat by 31 January 2014 on actions taken

As with elephants, subject to external funding, a CITES Rhinoceros Enforcement Task Force was proposed. Convened by the Secretariat and drawn from Parties affected by rhinoceros poaching and illegal trade, the ICCWC partner organizations, Parties and experts, it must:

- Develop strategies to improve international cooperation.
- Develop guidelines on best practices, protocols and operational procedures that will promote the use of wildlife forensic technology.
- Examine the implementation of Resolution Conf. 9.14 (Rev. CoP15) in … South Africa and Zimbabwe ….
- Examine progress with curtailing illegal trade in rhinoceros parts and derivatives by citizens of implicated States, particularly Viet Nam …[14] and submit a revised version of Resolution Conf. 9.14 (Rev. CoP15) for consideration at CoP17.[15]

The Decision also makes a request for external funding:

- *'to undertake a technical mission to the Lao People's Democratic Republic to assess current enforcement activities relevant to illegal trade in wildlife, in particular in rhinoceros parts and derivatives, and the implementation of Resolution Conf. 9.14 (Rev. CoP15)'*

And finally, to Decision 16.92, the Standing Committee had to review the definition of 'hunting trophy' (in Resolution Conf. 12.3 (Rev. CoP16)) as applied to rhinos and determine whether it needed revising so as to eliminate any possible misuse that might be making it easier to trade illegally in rhino horn.

---

[14] Revise Resolution Conf. 9.14 (Rev. CoP15), taking into consideration the contents of Decisions 16.84 and 16.85.
[15] Decision 16.89.

Once again, according to Tom Milliken, we should be encouraged:

*Collectively, these measures establish a far-reaching plan of engagement under CITES for international action to tackle the escalating illicit rhinoceros horn trade. The degree of specificity and the breadth of content in these Decisions hold considerable hope for improving the conservation status of the world's five beleaguered rhinoceros species.*[16]

## 31.3    Standing Committee 66 and the NIAPS

CoP17 took place in September 2016. Before that a number of materials were submitted to the Standing Committee, for consideration. This was the 66th meeting of the Standing Committee.

One of the outcomes of CoP16 was that certain countries were required to prepare and submit NIAPs, their own National Ivory Action Plan, which would then be assessed to see what progress a country had been making.

Some countries have done well. Uganda's report was upbeat, and the country was *'happy to report that the level of awareness on wildlife conservation and illegal ivory trade has significantly improved to the extent that it is no longer a concern of enforcement agencies alone but a general concern among citizens'*. To raise the level of awareness so that there is now *a general concern among citizens* is indeed an achievement to be proud of.

---

[16] See Milliken (n.12).

# 32

# Updates on Some Countries

## 32.1 The USA

2014 was a key year. Recognizing *the serious and urgent conservation and global security threat posed by illegal trade in wildlife*, President Obama's new national strategy to combat illegal wildlife trafficking and the organized crime that was associated with it was published in February 2014 and unveiled at the London Conference.[1]

It consisted of three components:

* Enforcement: domestic and global enforcement to be made tougher
* Demand reduction: for illegally traded wildlife, both at home and abroad
* International partnerships: strengthened to combat wildlife poaching

Then the states of New York and New Jersey strengthened their legislation, passing bills restricting sales of ivory. In New York State, the legislation also applied to sales of mammoth ivory and rhino horn, while the legislation in New Jersey applied to any ivory, including, but not limited to, elephant, hippopotamus, mammoth, narwhal, walrus and whale ivory, as well as to rhino horn products.

---

[1] Laura Parker *Fighting Wildlife Crime: New US Strategy Broadens Scope* National Geographic, 11 February 2014. See http://news.nationalgeographic.com/news/2014/02/140211-united-st Accessed 13/03/2014.

The new legislation increased penalties. From August 2014, all breaches of the law in New York State involving more than USD25,000 ivory became a class D felony, thus increasing fines. A defendant found guilty could:

- Be fined up to USD5,000
- Double the amount she/he gained from the crime, whichever is the higher
- Sent to prison for between 1 and 7 years

While in New Jersey, anyone committing a first offence under the state's wildlife trafficking legislation could:

- Be fined a minimum of USD1,000
- An amount equal to twice the total value of the products involved, whichever is the greater

For second and subsequent offences, the fine rises:

- To a minimum USD5,000
- Twice the total value of the products involved, whichever is the greater

As well as this, the ivory or rhino horn can be seized, and the New Jersey Department of Environmental Protection has discretion to:

- Destroy
- Donate the ivory or rhino horn to an educational or scientific institution or organization, such as a museum, university or research group[2]

There is, however, no power to impose a prison sentence.

The restrictions have resulted in a dramatic drop in both international and domestic ivory trade.[3]

## 32.2   China and Hong Kong SAR

In 2015, President Xi Jinping and President Barack Obama, two key players in controlling the ivory trade, agreed to ban almost all ivory sales, in the USA from June 2016 onwards, while China decreed that, apart from the few per-

---

[2] Daniel Stiles *Ivory Trafficking in California, USA* 2014.
[3] *UK is the largest supplier to the world's ivory markets* EIA, 10 August 2017. See https://eia-international. org/uk-largest-supplier-worlds-ivory-markets Accessed 17/08/2017.

mitted exceptions, their ban would be complete by 31 December 2017. And to ensure its success, the State Forestry Administration, together with various partners,[4] would run a campaign saying *no to ivory*, using posters, articles and videos and making use of social media. Despite this though, and although some factories and retailers had closed in March that year, a survey carried out by WWF and TRAFFIC found that only 19% of the people surveyed knew about the ban. More encouraging was the fact that 86% of those who didn't, said they would support it now they knew about it. Sufficient education and enforcement will be key to its success.

And Chinese residents in Kenya began to change their attitudes, partly because prices for ivory skyrocketed but also because of Kenya's new Wildlife Conservation Management Act, which, among others things, greatly strengthened enforcement. The importance of conserving wildlife and the consequences of breaking the law were promoted by both the Chinese Embassy and by many Chinese businesses. To good effect. After 2015, ivory seizures from local Chinese nationals gradually diminished. Whereas in April 2014, 4 Chinese nationals were fined RMB520,000 for possessing ivory, and in May a Chinese national was sent to prison for 6 months for possessing 0.75 kg ivory, by 2016, a Chinese businessman offered a raw ivory tusk by one of his employees and refused it.[5]

And, as usual, education is playing an important role. One Chinese state-owned enterprise instructs its employees about Kenya's ivory legislation both before they go there and after they arrive, particularly regarding the consequences of breaking the law.[6]

Meanwhile, Hong Kong SAR was not only rather slower to respond, but when it did, its response was disappointing. In June 2017, the Environment Minister announced that the Protection of Endangered Species of Animals and Plants was being amended to phase out the ivory trade, unfortunately though, the trade would not fully cease until 2021.

However, the good news was that the Minister made it quite clear that the government was not prepared to pay compensation to the ivory traders, who were demanding it. The government would not buy up their ivory, and the market had to close. Erik Mararv, the manager of Garamba National Park

---

[4] The partners included: the Natural Resources Defence Council, Wildaid, China Wildlife Conservation and the SEE Foundation, see Rachel Bale *China Shuts Down Its Legal Ivory Trade* National Geographic, 30 December 2017. See https://news.nationalgeographic.com/2017/12/wildlife-watch-china-ivory-ban-goes-into-effect.html Accessed 09/01/2018.

[5] Jing Wang *Law enforcement changes Chinese involvement in ivory trade in Kenya* Posted by and Special to The China Africa Project. See http://www.chinaafricaproject.com/china-kenya-kws-wildlife-service-ivory-enforcem Accessed 17/08/2017.

[6] Ibid.

(more of both later), was so concerned that compensation might be paid, that the previous week he had gone in person to speak to the legislators, to describe to them the horrendous suffering of both elephants and people caused by poaching and the illegal ivory trade. Perhaps his intervention had helped (more of this later).[7]

## 32.3   The European Union

In 2016, the EU published its *Action Plan to tackle wildlife crime*. Concerned with enforcement, cooperation and prevention, it will run for 4 years and be overseen at the highest level, implemented jointly by the EU Institutions, including Europol, and the Member States.[8]

One proposal was that wildlife trafficking should be considered a serious crime falling within the United Nations Convention against Transnational Organized Crime, so making it punishable by prison sentences of a minimum of 4 years. Other measures include improved collaboration on enforcement, strengthening links with the ICCWC and examining demand reduction. With help from the private sector.

Although the EU has been regulating wildlife trade under CITES since 1982, it only became a full Party to CITES in 2015, and CoP17 was the first time it attended as a full member. Its position on elephant ivory was not free from controversy. While it fully supported the existing ban on international trade in the ivory and opposed proposals for a resumption of the trade, it was not prepared to go further and support a total ban on ivory sales.[9]

However, in 2016, some 2972 kg of ivory were seized in the EU, the result of just four seizures. Which made it a very bad year indeed! Records were broken yet again. They had been broken in both 2014 and 2015 as well *'with re-export volumes surpassing the previous 8 years combined'*. And a smuggling ring was broken up.

Then Germany had an unpleasant surprise. A seizure close to Berlin's Schoenefeld airport and from a workshop in Koblenz netted 1200 kg of ivory from a Vietnamese gang. It was worth over EU1m. The workshop was quite

---

[7] *Hong Kong launches ivory ban bill* Phys.org, 14 June 2017. See https://phys.org/news/2017-06-hong-kong-ivory-bill.html Accessed 06/03/2018.

[8] European Commission *The EU Approach to Combat Wildlife Trafficking* See http://ec.europa.eu/environment/cites/trafficking_en.htm Accessed 15/02/2017.

[9] *EU position for the 17th Conference of the Parties (CoP) to the Convention on International Trade in Endangered Species of Wild Fauna and Flora (CITES)* European Commission – Fact Sheet Brussels, 23 September 2016. See http://europa.eu/rapid/press-release_MEMO-16-3145_en.htm Accessed 16/02/2017.

unexpected, the equipment suggesting it was being used to cut the ivory into pieces suitable for transport. There were no machines for fine artwork.

By the start of 2017, the Commission's position appeared to have shifted. *The Guardian* (a UK newspaper) had seen a draft EU guidance document indicating a ban on raw ivory exports from 1 July to *'make sure that tusks of legal origin are not mixed with illegal ivory'*,[10] and towards the end of the year, the Commission was consulting on the possible adoption of additional measures regarding the ivory trade.[11]

## 32.4   Viet Nam

In January 2016, the city and seaport of Da Nang, *Southeast Asia's gateway to the Pacific Ocean*, which lies roughly halfway between the cities of Hanoi and Ho Chi Minh and *on the north-south transport axis of Viet Nam's land, sea and air routes*, issued a Directive on fighting wildlife crime, which:

- Urged *'government officials strictly to comply with all wildlife management and protection laws'*
- Called on them *'to refrain from advertising, giving or receiving wildlife products as gifts'*

Its intention was *'to strengthen the city's enforcement capabilities and ensure unified action against illegal wildlife trade through all levels of government'*, so it provided specific guidance for the relevant authorities on *'monitoring all wildlife-related activities in the city carefully and systematically'*. Among other matters:

- Tourist markets and restaurants under the control of the Department of Culture, Sport and Tourism must stop selling any threatened wildlife products, which, of course, include elephant ivory and rhino horn
- Enforcement agencies, some of which had been selling *confiscated wildlife, its products and derivatives* (e.g. confiscated rhino horn powder), can no longer do this.

---

[10] Arthur Neslen *EU set to ban raw ivory exports from July* The Guardian, 22 February 2017. See https://conservationaction.co.za/recent-news/eu-set-ban-raw-ivory-exports-july/ Accessed 28/02/2017.

[11] *Survey by the Directorate General for Environment of the European Commission on Ivory Trade in the European Union.* See https://ec.europa.eu/eusurvey/runner/SurveyOnIvoryTradeInTheEU2017 Accessed 26/10/2017.

Although the Directive's message is tough and its terms are strong, so that, in effect, the People's Committee of Da Nang is promoting zero tolerance, the business community and government agencies had already been taking action. A year earlier, the first ever Pangolin Range States meeting was held in the city, a most welcome development in the fight to save another severely endangered animal. And in the same week, there was a bicycle ride for business leaders. Hosted by TRAFFIC and the Da Nang branch of the Viet Nam Chamber of Commerce, and as part of the Chi Campaign, its aim was for the leaders to show their commitment to wildlife protection. It was so enthusiastically successful that a second ride was then arranged. Indeed, working together with TRAFFIC, Da Nang *seeks to* become *an environment-friendly city by 2020.*[12]

And in November 2016, there was more good news for this country which is now struggling so hard to reduce its illegal wildlife trade. After the 2014 London Conference and the 2015 Kasane Conference came the 2016 Hanoi Conference on illegal wildlife trade (as we have seen), where representatives from 54 countries made further commitments to protect wildlife, ultimately by eradicating the illegal trade and ensuring its effective protection from imminent threats of extinction.[13]

## 32.5   Swaziland

White rhinos live here, although they were once hunted to extinction. That was in Colonial times. Now there are two rhino parks, a rhino sanctuary and a total of 73 rhinos.

They were reintroduced into Mlilwane Wildlife Sanctuary in 1965 as part of South Africa's translocation programme *to spread the risk against extinction* and into Hlane Royal National Park later that decade. It only became safe in the 1980s to place some in Mkhaya Game reserve.

The numbers increased, and by the time of the first major episode of poaching, 1988–1992, there were some 120 animals. The rhino war reduced that to 24. Then Swaziland passed new legislation, *'preventative rather than remedial'* and *'equipped the Game Rangers to deal with a new kind of poaching'*. For the next twenty years it did not lose a single rhino to poachers.

---

[12]TRAFFIC *Da Nang City issues wildlife crime-fighting Directive* TRAFFIC Wildlife Trade News, 25 January 2016.

[13]IUCN *Further commitments made to protect wildlife at the Hanoi Conference on Illegal Wildlife Trade in Viet Nam* 22 November 2016. See https://www.iucn.org/news/further-commitments-made-protect-wildli Accessed 15/02/2017.

In 2008 it began again. There was a new poaching blizzard, and even though Swaziland is the closest target to the criminal syndicates operating out of Mozambique, their magnificent Game Rangers kept all their rhinos safe. Except 3, and those crimes were solved. A proud record.

But by 2016, Swaziland was so concerned about its ability to keep protecting its rhinos that it submitted a proposal for CoP17, that it be allowed to legally sell some of its rhino horn (more of this later). The problem was money, or rather, lack of it. The parks rely on self-generated income to survive. Trophy hunting is forbidden.

Money was and still is needed to meet the increasing costs of protecting their rhinos. And a pay increase was long overdue for their park employees, particularly the Game Rangers *who serve far beyond the call of duty to protect the country's rhinos against ever-increasing dangers*. Infrastructure and equipment needed improvement. The country was enduring the worst drought in living memory, so supplementary food was required, and it was scarce and expensive. Rhino numbers had dropped to 73 animals. If they were allowed a sale, some of the proceeds would also be used *'to provide for sustainable and long-term developments, all of which would ultimately benefit neighbouring communities and Nature conservation initiatives'.*[14]

In other words, the government was putting forward the case for the sustainable use of some of their rhino horn. It was rejected.

But the poachers are still out there, threatening. In February 2017, when 2 Taiwanese nationals were arrested at King Mswati International Airport on their way to Taiwan, they were found to be in possession of 24 pieces of rhino horn which had come from 3 rhinos poached in South Africa and 1 in Swaziland.

In November 2017, the poachers/traffickers were sentenced by the High Court of Swaziland to:

- 11 years each, in prison for killing 4 rhinos
- 9 years and 11 years each, in prison for trafficking the horn
- 9 years each for exporting the trophies

As the sentences were to run concurrently (at the same time), each of them would be in prison for a total of 11 years. They were also ordered to pay compensation to the rhinos' owners, USD13,000, and if they did not do this, they would each serve 4 more years in custody.[15]

---

[14] Swaziland's Proposal for CoP17.
[15] *Seizures and Prosecutions* TRAFFIC Bulletin, Vol. 30 No. 1 (2018).

## 32.6   Mozambique and Angola

We return to these two countries that are so important for elephants and start with a case study.

**Case Study 9: The Shuidong Connection**

A combination of elephant numbers dramatically falling in the Selous-Niassa ecosystem and a huge improvement of enforcement in Tanzania had forced the ivory traffickers into neighbouring Mozambique. And that was where EIA investigators following leads, discovered a Chinese organized crime gang that had been operating for over two decades, sending ivory back to Shuidong in southern China.

By posing as potential purchasers, gradually over several months, the investigators won the smugglers' trust, eventually being invited to China to inspect the Mozambique tusks. While there, the traffickers explained how their trade was carried out.

It had all begun with sea cucumbers, Chinese traders importing sea cucumbers, a legitimate business, operating out of Zanzibar (Tanzania), except that what had started out as lawful trade ended up as cover for illegal ivory trafficking. However, since 2011, a number of ivory seizures had exposed Zanzibar as *a significant ivory trafficking hub*, forcing the smugglers to take every precaution to avoid contact with the ivory, building up a network of Tanzanian accomplices who sourced the ivory and stored it in safe locations on the mainland until about three tonnes had been amassed. This would then be taken on small boats to Zanzibar, to be stored in warehouses until it could safely be hidden in a legitimate cargo of low-value goods, such as plastic pellets, the Chinese traders observing from *a vantage point overlooking the port*, ready to leave the country immediately should anything go wrong. So only the Tanzanians names appeared on the shipping documents as the freight agents. The operation was all about a rapid turnaround, a quick sale even if it meant a smaller profit, so that the traffickers could then get on with their next consignment/sale.

After a series of seizures and arrests, the switch was made to Pemba, Mozambique, and it was there, in April 2016, while travelling on a fact-finding mission to check out the rumours that the traffickers were now operating from that country, that the EIA investigators encountered the three men they were *to do business with*, two of the men turned out to be second-generation ivory smugglers, and their families were very wealthy.

In October 2016, the investigators carried out their sting. Only when they had been vetted as genuine were they permitted to see the ivory, and this involved a late night drive as the tusks had been removed to a secure and remote location. There they discussed price and payment and how to convey the tusks to the buyers' warehouse. The traffickers would get back to them the next day. Except that by then, the investigators were far away. They had obtained their evidence.

However, the smugglers did not give up so easily, and, a few months later, one of the smugglers again contacted the EIA with an offer of a consignment of tusks from Nigeria. Not only did forest elephant ivory fetch higher prices, but there was *lax enforcement and corruption in Nigeria*, whereas, according to Ou, '*In East Africa, only in*[16] *Mozambique can these things still be moved out; it's not possible in any other place. West Africa is easy*'. The EIA did not respond.

Mozambique, we know, has been struggling to do its best for wildlife, but poverty has a nasty habit of getting in the way. Elephant poaching takes place in Mozambique itself, but the poachers also find it easy to cross into South Africa for in 1994; the fences came down, and a corridor was made linking Kruger National Park with the Greater Limpopo Transfrontier Park. There they poach both elephants and rhinos.[17]

In 2014, Mozambique was identified as a hotspot for poaching, and it continues to be so. For example, in 2016, four groups of elephant poachers, intent on killing the animals for their ivory, were identified by officials from Kruger National Park.[18] We have also seen that its capital Maputo is frequently used by the traffickers to smuggle out both ivory and rhino horn going to Viet Nam.

In May 2015, the police in Maputo seized 340 elephant tusks and 65 rhino horns, together weighing about 1.3 tonnes, the largest such seizure in Mozambique's history. 2 Chinese nationals were arrested, one after he offered a bribe of USD34,000 to the police to drop the case. Days later, 12 of the rhino horns were stolen from the police's provincial command headquarters and replaced with crude replicas. Another six people were arrested including four policemen, one of whom was the Head of the Criminal Investigation Department.

The case went nowhere, and the two Chinese nationals, who were given bail, vanished.[19] But in July, Mozambique burned a stockpile of 2434.6 kg of

---

[16] *The Shuidong Connection: Exposing the global hub of the illegal ivory trade* EIA, 2017, p.16.

[17] S Sautner *Mozambique makes a stand against wildlife crime* Wildlife Conservation Society, Newsroom, 6 July 2015.

[18] Hanti Schrader *Four groups identified in elephant poaching* News24, 14 June 2016 See http://www.news24.com/Green/News/four-groups-identified-in-elephant-poaching-20 Accessed 11/08/2016.

[19] Julian Rademeyer *Beyond Borders Crime, conservation and criminal networks on the illicit rhino horn trade (Part 2 of a 2-part investigation into rhino horn trafficking in Southern Africa)* Global Initiative against Transnational Organized Crime, July 2016.

ivory and 86 pieces of rhino horn including the horn from the earlier seizure,[20] a signal that the country *will not tolerate poachers, traffickers and the organized criminals which employ and pay them to kill our wildlife and threaten our communities*.[21]

But Mozambique is doing its best to fight back. It has passed new legislation to make poaching and wildlife trafficking criminal offences and has set up an Environmental Police Unit that will work with the ANAC (National Agency for Conservation Areas) scouts to enforce the law. The protected-area scouts will benefit from improved training which is to include the development of an intelligence-led enforcement capability.

And of course Mozambique will work together with partners and international organizations. One such collaboration, with Wildlife Conservation Society and Stop Ivory, should produce an Ivory Inventory, which will go to the CITES Secretariat. Mozambique has also signed Memorandums of Understanding with both South Africa and Tanzania, to strengthen cross-border collaboration to tackle both poaching and trafficking.

As we have seen throughout our story, corruption continues to be a major stumbling point. In Mozambique, it weaves its sticky tentacles throughout the country, aided by the poverty still endured by so many of its inhabitants. Diplomatic immunity still seems to be regarded as sacrosanct. There are ugly signs of new tensions between the old civil war enemies of Frelimo and Renamo. And there is still no formal extradition treaty with South Africa, although there is cross-border cooperation.[22]

But its problems are being recognized. The new Conservation Law (Act 16/2014)[23] does at least plug some loopholes, and anti-corruption legislation is being strengthened. As the result of another amendment, passed in 2017 and closing yet another gap, in early 2018, 4 offenders caught selling ivory tusks in the Central Market of Maputo were successfully prosecuted and sent to prison.[24]

However, enforcement still leaves a lot to be desired. Mozambique must demonstrate that laws are there to be obeyed, and anyone who breaks them will be punished accordingly. Now though, there is hope.

---

[20] See Sauter (n.17).

[21] See Rademeyer (n.19).

[22] Ibid.

[23] The Law on Protection, Conservation and Sustainable Use of Biological Diversity (Law 16/2014 of 16 June).

[24] *UK aid to crack down on criminal gangs driving the illegal wildlife trade* UK Government, See https://www.gov.uk/government/news/uk-aid-to-crack-down-on-criminal-gangs-driving-the-illegal-wildlife-trade Accessed 11/10/2018.

And so to Angola, to its results in the Great Elephant Census. These were unexpected and they were shocking.

Mike Chase, who led the survey/investigation, and who is a long-time expert on elephant populations in Angola, had not expected to find the rate of poaching there as *'among the highest in Africa'*, particularly since the country *'has the ability to provide elephants with the largest elephant range remaining in Africa …'*.

Although the population was decimated during the civil war, he had been there at the start of their recolonization, moving back from Botswana through Namibia into Angola. He had documented it. And he believed *'this home-coming of elephants in southeast Angola'* to be *'one of the greatest conservation success stories in the past 50 years'*.

He had also thought *'Angola would be one of those last sanctuaries that escaped elephant poaching'*. But he was wrong, for the Census found the country was losing 10% of its elephants each year, *'a higher mortality rate than any other country on the GEC'*, although populations of other animals were healthy. Angola though, is part of the Kavango-Zambesi Transfrontier Conservation Area (KAZA), which also straddles Botswana, Namibia, Zambia and Zimbabwe, and, at the moment, a weak link with *'limited capacity to manage its wildlife'*. Mike Chase, however, continues to be hopeful, because *'one of the major goals of the GEC is to empower governments to save what they have left. In the case of Angola, I hope they were shocked out of apathy'*.[25]

## 32.7   Lao PDR

The news only broke in September 2016, when *The Guardian* (a UK newspaper) exposed yet another criminal family, the Bach brothers (we met them earlier). Although some of their trading has been legal, they have also smuggled enormous quantities of endangered species, including thousands of tonnes of elephant ivory and rhino horn, sometimes aided and abetted by senior officials of the Laos government. They operated out of Nakhon Phanom, a town in northeast Thailand and next to the Mekong river, so it was a primary gateway to Laos and Viet Nam.

Freeland, the counter-trafficking NGO which exposed them, was working with a Thai government surveillance team, but because law enforcement agencies had failed to act, they also shared their evidence with *The Guardian*

---

[25] Christina Russo *Exclusive: This Is Africa's New Elephant Poaching Hot Spot* National Geographic, 5 July 2016. See http://news.nationalgeographic.com/2016/07/angola-elephants-great Accessed 11/04/2017.

hoping the press exposure would *finally jumpstart cross-border police action against this and other networks*. Journalists from the paper decided to investigate further.

The brothers, originally from Viet Nam, set up a global animal trafficking highway, dealing not only with wild but also farmed animals, such as tigers and lions, reared under the cruellest and most appalling conditions imaginable. One of their alleged associates sold fruit in a market in Bangkok but wanted to trade in wildlife, the same Chumlong Lemtongthai who played a pivotal role in the pseudo rhino hunts in South Africa. The younger brother was finally arrested in January 2018 (see earlier).

But our concern here is with Laos, which has been a Party to CITES since 2004 and has passed legislation accordingly. In fact, the government *'has publicly paraded its commitment to the convention'*.[26] Yet in 2014 alone, one company was issued with CITES permits allowing them to traffic USD16.9 m of animal products through the country, including 20 tonnes of elephant ivory (the EIA estimated this at 2985 dead elephants), valued at USD5 m. Permits issued by the government to carry out trade that is banned by CITES, the treaty they are a Party to. Another company did even better. Its permits allowed it to trade, among other things, 90 tonnes of ivory (or 13,432 dead elephants), estimated value USD22.5 m and 4 tonnes of rhino horn, which was valued at a mere USD240,000![27] And apparently this has been going on for years, although Laos was briefly suspended by the CITES Secretariat in 2015 for failing to comply with its NIAP.

---

[26] Nick Davies and Oliver Holmes *Revealed: how senior Laos officials cut deals with animal traffickers* The Guardian, 27 September 2016. See https://www.theguardian.com/environment/2016/sep/27/revealed-how-senior-laos-offi Accessed 27/09/2016.
[27] Ibid.

# 33

# Rhino Trophy Hunting in South Africa

By 2011, 75% of the wild rhinos in the world were living in South Africa, and their numbers were being decimated by the catastrophic poaching epidemic. White rhino trophy hunting resumed in South Africa in 1968, when there were only 1800 animals,[1] and by the beginning of 2011, their numbers had increased to 18,800. Indeed, hunting rhinos was a key component of the South African game industry.[2] But was there a link between this and the poaching epidemic?

We will start by re-examining some of the extraordinary goings-on that we glimpsed earlier.

South Africa comes top when a trophy hunter is choosing which African country to visit, one of the main reasons being that it provides an opportunity to hunt the prized 'big five', namely: buffalo, leopard, lion and our two animals, elephants and rhinos. The hunters come mainly from the USA and Europe, as well as 60 other countries. And with the hunt comes money. In 2000, the total value of the game industry was an estimated ZAR140 million (about USD20.2m).[3] This had risen to approximately ZAR730 million (about USD91.2m) by 2007[4] and to about ZAR1 billion (about USD124.8m) a year later, a peak.[5] By any standard these are enormous amounts.

---

[1] Adcock and Emslie 1994 in Milliken T. and Shaw J. (2012) *The South Africa – Viet Nam Rhino Trade Nexus: A deadly combination of institutional lapses, corrupt wildlife industry professionals and Asian crime syndicates* TRAFFIC, Johannesburg, South Africa.

[2] Ibid.

[3] Barnett and Patterson 1995 in ibid.

[4] PHASA 2009 in ibid.

[5] DEA 2010 in ibid.

© Springer Nature Switzerland AG 2019
B. Martin, *Survival or Extinction?*, https://doi.org/10.1007/978-3-030-13293-4_33

Although the data was incomplete, rhino trophy hunts showed major jumps in price in 1989, 2001, 2008 and 2010, effectively doubling after 2007. Furthermore, the numbers of rhinos hunted increased after 2004 with the figures showing demand exceeding supply. Milliken and Shaw observed that *from 2005 onwards, the sport hunting of rhinos has boomed in South Africa as never before, rapidly driving the number of hunts and their prices to unprecedented heights in an effort to service the appetite of a new, non-traditional trophy market: Viet Nam'*. And this is where it begins to get very interesting.

Viet Nam, unlike North America and Europe, had no sporting tradition of hunting, so it was a surprise when Asians, particularly Vietnamese, suddenly became the dominant force in trophy hunting for white rhinos. And the direct result of this was that hunt prices steadily rose from 2004 onwards, with over USD22m estimated to have been paid by Vietnamese hunters to trophy hunt rhinos between 2003 and 2010. It also became clear that these hunters were radically different. They were not doing it to show off their hunting skills and bring back a trophy. They were acquiring rhino horns for commercial trading.

The authorities gradually realized that there was something strange about these particular hunts. For a start, they were not booked by the normal methods but by word of mouth, a device described as having progressively expanded *into an insidious web of relationships binding key representatives of organized Asian syndicates, with a cadre of a few corrupt professional hunters and selected property owners.*[6] The key personnel in any hunt are the foreign client, the property owner and the professional hunter.

So what made these 'pseudo' rather than 'true' hunts?

There are repeated accounts of these 'trophy hunters' being taught to shoot during an actual hunt, because they were unable to handle a gun. Sometimes the professional hunters had to step in and kill the rhinos themselves. One of them, operating from the Loskop Dam Nature Game Reserve, was successfully prosecuted in 2006 *'for leading hunts feeding the horn trade'*, and *'he paid a token fine after his Vietnamese hunter casually told an official that he did not know how to shoot'*. When, two years later, he was prosecuted again, he was represented by a lawyer who succeeded in convincing the judge that the case should be dismissed. It was apparently *'a technicality that the client had not actually fired the fatal bullet, a privilege that may have gone to [his] teenage son'*. According to him, the Game Department was present on most of the hunts,

---

[6] Ibid, p.53.

but although he defended the practice, *'he stopped guiding the Vietnamese after the Game Dept. informed him they were involved in the horn trade'*.[7]

Another professional hunter, C. F. van Wyk, who also happened to be a taxidermist, was successfully prosecuted and fined ZAR30.000 (about USD4250) for illegally shooting a white rhino. The hunting trip, for a Vietnamese client, was organized by a different professional hunter who also owned a Safari company. Although he himself did not attend the hunt, his wife and father did, together with van Wyk and the client. It took four shots to gun down the poor rhino, four shots from a distance of between 50 and 100 m away. The Vietnamese did not participate. During the subsequent court proceedings, it was revealed that van Wyk did not have a permit to hunt the rhino, nor was he registered as a professional hunter in that area (Limpopo).[8]

Professional hunters have also been accused of other offences involving hunting permits, including:

* Allowing people to shoot rhinos when they were not named on the hunting permit
* Obtaining export permits under false pretences, for clients whose names were not on the hunting permit

Over time it became clear that the Vietnamese hunters, unlike other hunters, did not want their trophies either mounted or prepared in some other way by a taxidermist. What they wanted was the horns to be removed as quickly as possible from the dead rhino and to take them with them when they left the ranch. It also became clear that certain ranch owners *'were repeatedly hosting Vietnamese hunting parties on multiple occasions'*.[9]

By 2011, some of the hunts had become quite bizarre. Prostitutes and strippers from Thailand, who had been trafficked to South Africa and with whom Chumlong Lemtongthai (from the Bach Brothers gang) had made friends while they were working in the bars in Pretoria to pay off their debts, were hired by an international wildlife trafficking syndicate to pose as *hunters* in sham rhino trophy hunts. It was, apparently, better than bringing in phoney Vietnamese clients, cheaper![10]

---

[7] Borrell 2010 in ibid.
[8] Rademeyer 2011b in ibid.
[9] Ibid.
[10] Rademeyer 2011c in ibid.

Chumlong was one of the traffickers setting up the pseudo-hunts. Stooges were paid to stand by the professional hunters, while the rhinos were shot. Each stooge then had his/her photo taken with the dead rhino and their name put on the paperwork which allowed the horn to be 'legally' taken back to Asia, to the Bachs and another notorious trafficker Keosavang.

Chumlong started by bringing the stooges over from Thailand and paying them 5000 rand (about USD350). Using prostitutes saved him the air fare. The Guardian (the UK newspaper) has seen paperwork showing that at least six members of the Bach family also travelled to South Africa to take part in the hunts. Chumlong was paid USD 20,000 a time for their horns, and dozens of rhinos were killed.[11] There is also evidence suggesting he was trafficking in rhino horn from poachers.

Eventually however, South African revenue officers arrested him and went through the documents on his laptop. Packed full of incriminating evidence, one 6-month period showed a white South African hunter had been paid USD1,394,282.40 for killing rhinos in the pseudo-hunts and that *Chumlong was paying USD6,500 per kg for the horn'*. As a Chinese end-user would pay ten times as much, Chumlong's rhino horn trafficking was *'potentially worth USD13.9m'* to those involved.[12]

There were also hunters from other Southeast Asian countries. Between 2007 and 2012, Chinese hunters acquired and exported 20 rhino trophies, and at least one hunter came from Cambodia.[13] And then there was the curious case of the citizens from the Czech Republic. Suddenly there was a dramatic increase in the number of hunts they were taking part in. Could they also be working on behalf of Asian crime syndicates?

Unfortunately, the syndicates were assisted not only by some unscrupulous professional hunters but also by some of the people who owned white rhinos, people concerned mainly with making a quick killing financially rather than abiding by ethical standards and rhino conservation. Furthermore, export permits continued to be issued to some of the Vietnamese nationals who had previously been implicated in unlawful hunts and their aftermath. Suddenly they seemed to have acquired addresses, business and/or residential, in South Africa itself.

So what were the South African authorities doing while all this was happening? We are about to see.

---

[11] see Nick Davies and Oliver Holmes *Revealed: how senior Laos officials cut deals with animal traffickers* The Guardian, 27 September 2016. See https://www.theguardian.com/environment/2016/sep/27/revealedhow-senior-laos-offi Accessed 27/09/2016. Accessed 27/09/2016.

[12] Ibid.

[13] See Milliken and Shaw (n.1).

# 33.1    South Africa's Response

To begin with it was slow, because the legislation was inadequate. But then, what we might term *'the Vietnamese connection'* had taken everybody by surprise. One reason it was so unexpected was that before 2003 the Vietnamese had no great desire for rhino horn. The growing problem was recognized during the years of 2005–2007, when more and more operators from the private sector became involved in the *pseudo-hunts*. A surge in enforcement action across the provinces, recommendations from TRAFFIC and others that both trophy hunting and professional hunters should be more strictly monitored[14] and growing pressure from CITES finally enabled the South African government to recognize the threat this was posing to their lucrative hunting industry. They took action.

In August 2008, a major loophole was abolished by the TOPS regulations. Before this, *standing* permits had enabled white rhino hunts to take place on certain properties without the local conservation authorities being aware of what was happening. As they did not know whether or not a bona fide hunt, as opposed to a *pseudo-hunt*, had taken place, the provincial authorities could be issuing CITES export permits for animals killed in the phoney hunts. Then, in July 2009, new standards were imposed. These only allowed individual hunters to take part in one white rhino hunt a year and also required national approval be obtained before provincial licences could be issued.[15]

And a moratorium was imposed on domestic sales of rhino horn.

- In April 2012, the amended norms and standards for sport hunting of white rhinos came into effect, bringing about the final demise of *pseudo-hunting*. No more shape-shifting of hunting trophies into plain rhino horns. It was the result of a challenge to the courts, brought by a private hunting operator who was testing the new legislation to see how restrictive it really was. So what happened when the challenge arrived?
- Provincial authorities were advised by the DEA not to issue hunting permits to Vietnamese citizens because of concerns regarding illegal hunting practices. Mr. Slipper, who brought the challenge, had applied for hunting permits for five Vietnamese citizens. These were authorized at provincial level, a decision that was overturned. The court ordered that the permits could only be issued provided certain conditions were satisfied. The intending

---

[14] Milledge 2007a,b in ibid.
[15] Milliken et al. 2009b in ibid.

hunters had to submit themselves for interviews with relevant government officials in order to verify:

- Their personal details
- Their backgrounds
- Their financial capabilities
- Other related information.

This was required by the amended norms and standards.

Although all the arrangements were made for the hunters to attend the court hearings, they did not appear because they were not yet in the country. The government then discovered that there were legitimate grounds for refusing the permits so the court ordered that the permits should not be issued. Mr. Slipper had to pay the costs of the action.[16]

The objective was admirable. The measures were designed to reduce white rhino sport hunts to the numbers that had taken place before '*pseudo-hunting*' took over. Law enforcement officers were mandated to be present at every hunt to make sure the new legislation was strictly adhered to. All should have ended well, but unfortunately this did not happen. The problem was poaching or rather the spike in poaching. So were the new measures contributing to the increasing levels of poaching now taking place on both state and private land?

## 33.2   The Unintended Consequence

### 33.2.1   Disaster!

Earlier in our story, we learnt that some officials in China were stockpiling rhino horn, because, as a commodity, it was so very valuable and that had led to a catastrophic rise in rhino poaching. During that episode, Zimbabwe was the worst affected of the southern African states. Matters improved again after 1994. The major consumer countries, which at that time were China, Taiwan and South Korea, took drastic action to restrict rhino horn use by their traditional medicine industries, and Zimbabwe not only moved its rhinos into Intensive Protection Zones but also dehorned them.[17]

---

[16] Ibid.
[17] Milliken et al. 1993 in ibid.

Then came the current crisis. Suddenly, a vibrant new market had surfaced. The consumers were back in big numbers. And they had quantities of money. It was a very threatening situation for the rhinos.

A trickle of poaching started in the early 2000s. TRAFFIC noticed and published a warning for CITES Parties at CoP14. The trickle developed into a torrent and then a tsunami. By 2008, both Zimbabwe and South Africa were experiencing major losses to poachers. To fully comprehend the depth of the disaster, we need look at some numbers. These figures, published by the South African DEA in 2015, show the number of rhinos poached each year from 2006 to 2015:[18]

- 2006: 36 (at that time a record, the highest number for decades)[19]
- 2007: 13
- 2008: 83
- 2009: 122
- 2010: 333
- 2011: 448
- 2012: 668
- 2013: 1004
- 2014: 1215 (one rhino killed every 8 h)
- 2015: 749 (as at 27 August 2015)

The TRAFFIC/IUCN report (tabled in late 2009) to CoP15 was blunt, pointing out that *'since 2006, 95% of all detected or presumed rhino deaths in Africa from illegal killing, have occurred in Zimbabwe and South Africa'*, countries that are *'the epicentre of an unrelenting poaching crisis in Southern Africa'*. And unfortunately, the killing was having an adverse impact not just on Kruger National Park (its border with Mozambique) but on other protected areas (in KwaZulu-Natal) and, for the first time, on a range of private sector game ranches spread across the provinces (Limpopo, Gauteng, North West and Eastern Cape).[20]

The numbers showed only too clearly, just how dramatically the situation was deteriorating.

---

[18] Save the Rhino International *Poaching: The Statistics 2016* See https://www.savetherhino.org/rhino_info/poaching_statistics Accessed 21/01/2016.

[19] See Milliken and Shaw (n.1).

[20] M. Knight in litt 2009 in Milliken et al. 2009b, ibid.

# Part XI

Research and Education

# 34

# Research

## 34.1 Data

Our story is compiled from data. Data is evidence. Research is the key to data production. The CITES Standing Committees and the Secretariat have recourse to research data when taking their decisions. It provides evidence of what is happening. TRAFFIC *'applies a research-driven approach to its planning strategy, and execution initiatives against the illegal wildlife trade'.*[1]

Without evidence, without facts we could not take informed decisions on how to better protect our animals. We would not even know how many elephants and rhinos there were. Let us look at an example from the UK.

MIKE and ETIS were designed to produce data. They are a continuous collection service, providing a steady stream of information about elephants.

One of the ETIS analyses undertaken for the CITES CoPs is *'cluster analysis'*, which *'identifies those countries or territories most prominently implicated in illegal trade in ivory'.*[2] ETIS cluster analysis works like this:

- It prescreens the data, separating out countries only marginally implicated in ivory.
- It then works out what is happening in the countries more heavily implicated, including factors such as underlying trade dynamics and law enforcement.

[1] Compton 2014, in *Symposium on Combating Wildlife Crime: Securing Enforcement, Ensuring Justice, and Upholding the Rule of Law. The Proceedings* Editors Kala K. Mulqueeny and Francesse Joy J. Cordon, 2014, p.40.

[2] Lau, W., Crook, V., Musing, L., Guan, J., and Xi, L. (2016) *A rapid survey of UK Ivory Markets* TRAFFIC, Cambridge, UK.

© Springer Nature Switzerland AG 2019

B. Martin, *Survival or Extinction?*, https://doi.org/10.1007/978-3-030-13293-4_34

By using this analysis, it has been possible to show that the UK, for example, has a problem, because the data suggests that it has been featured in six successive cluster analyses since 2002 (when the first cluster analysis was carried out). This, in turn, suggests that it has *consistently played a role in illegal ivory trade globally*.[3]

Other revealing sets of data are the CITES exporter and importer data, though these are not wholly reliable.[4] An example of how this data can be used is shown in TRAFFIC's *Rapid Survey of the UK's Ivory Markets*. Trade records from importer and exporter countries/territories are recorded on the CITES trade database, and these can highlight discrepancies in the number of transactions. Such discrepancies showed up in the UK, thus indicating there might be a problem.

The relevant time: 2005–2014: the UK reported the total number of ivory re-export records for commercial purposes. The exporters' and importers' records did not coincide.

The exporters: submitted 567 records composed of about 990 kg and 54,000 specimens.

The importers: submitted 274 records composed of about 20 kg and 26,500 specimens.

Although this difference could be explained away without concern, being attributed to the records of *'worked ivory'*, there was another potentially more serious problem. This concerned records of *'raw ivory'*.

The exporters: submitted 10 records, composed of 49 specimens.

The importers: submitted 58 records, composed of 585 specimens.

Ivory was coming into the country, and most of it was staying there rather than being re-exported. But was this really happening?

While there were virtually no records of trade in *'ivory pieces'*, so virtually no re-exports, the importer data showed that 39 of the 58 submitted records were of ivory pieces, imported into the UK. Similarly with tusks, importer data was greater than re-exporter data, 109 tusks compared with 17 tusks. The importer data suggested that:

In 2012, 38 tusks were re-exported from the UK to Macau; and 11 tusks were re-exported from the UK to Hong Kong SAR.

In 2013, 28 tusks were re-exported from the UK to Hong Kong SAR.

None of these were reported by the UK. Why not, when the differences were so large? And there were other transactions that were not recorded by the exporters as well, including particularly large discrepancies between UK

---

[3] CITES Secretariat 2013 in ibid.
[4] Ibid.

records and those of certain trading countries/territories such as Hong Kong SAR, which does not have a good track record on ivory trading.

Data such as this provides a firm basis for further investigation. Other data is collected by rangers. GPS tracking data can help protect collared animals from many threats. Data comes in many shapes and forms, and all of it is valuable.

## 34.2   Research

Some of this is ground-breaking, as exemplified by the audacious plan to save the northern white rhinos from extinction.

### 34.2.1   Saving the Northern White Rhino

With only three of these rhinos left on the planet, a male, Sudan, and two females, Najin and Fatu, there were problems. They were closely related, Najin is Sudan's daughter and Fatu is Najin's daughter, and Sudan had a low sperm count. However, in 2014 there were plans to mate a male southern white rhino with the females in the hope of breeding hybrid offspring *that would at least conserve some of the northern white genes*.[5] But Najin had injured her leg so could no longer bear the weight of a mounting male or of pregnancy, and Fatu had a reproductive problem which would prevent an embryo from implanting. So they were both unable to breed.[6]

It was an ambitious project to save them, conceived as a desperate measure to prevent yet another species of rhino from becoming extinct. Now there is another one, using IVF. So Ol Pejeta is collaborating with Dvur Kralove Zoo; the Leibniz Institute for Zoo and Wildlife Research in Berlin, Germany; Avantea Cremona, Italy; and KWS. There are specialists in stem cell and reproductive biology as well. The intention is to blend *high-tech-assisted repro-duction methods with the latest research in cell biology*[7] so that more rhinos come into existence.

The rescue plan was published in May 2016, although the first, tentative steps were taken early in 2015 when conservationists meeting in Nairobi,

---

[5] Press Release *The Last Male Northern White Rhino Dies* Ol Pejeta, 20 March 2018.

[6] Ewen Callaway *Stem-cell plan aims to bring rhino back from brink of extinction* Nature, 3 May 2016. See http://www.nature.com/news/stem-cell-plan-aims-to-bring-rhino-back-from-brink-of Accessed 10/05/2016.

[7] Helen Pilcher *The End of Extinction?* BBC Wildlife, September 2016, Volume 34 Number 10.

Kenya, decided to harvest eggs from the two remaining females. These were stored in the Institute, which already held frozen sperm taken from male northern white rhinos.[8] A priceless hoard. Initially though, these eggs would not be used. Fresh egg cells would be collected from Najin and Satu and then fertilized with some of the frozen sperm, with resulting embryos implanted into a surrogate mother, a southern white rhino.[9]

However, because eggs from just two rhinos severely limits the gene pool, to the extent that any offspring would be unlikely to thrive in the wild, the second stage of the rescue plan would be an attempt to convert frozen rhino cells into stem cells, as these 'have the capacity to develop into any type of tissue, including eggs and sperm'. Once again, the technique is highly experimental, and there is no guarantee it can be made to work. But the Leibniz Institute is working on it. Thomas Hildebrandt, the veterinarian in charge of the project, and his team are already trying to conjure these stem cells into eggs.[10]

The wildly ambitious ultimate aim of the plan, therefore, was not just to bring these rhinos back from the brink but to create (this verb is used deliberately) a viable population. Research like this takes a long time to come to fruition and costs a great deal of money. It is also controversial. For example, the Chair of the IUCN's African Rhino Specialist Group described the science as 'Star Trek-type' and was concerned the plan would divert money from conserving other rhinos. 'If you want to save a [rhino] species, put your money into southern white conservation'.[11]

But the scientists were determined, collecting eggs from southern white rhinos living in zoos in Europe. In Longleat Safari Park, a UK zoo, in July 2017,[12] they harvested nine eggs from three rhinos considered to be suitable egg donors because they had not mated naturally with the zoo's male. The eggs were taken to Italy, to a clinic specializing in assisted reproduction in animals, where IVF techniques and embryo transfer would be used to try eventually to produce a northern white rhino or, if that could not be achieved, at least a hybrid between southern white and northern white rhinos, thus meaning that 'at least some of its genetic material survives'. Whether or not such

---

[8] Paul Rincon *First step in plan to save northern white rhino* BBC News, website Science and Environment, 28 January 2015. See http://www.bbc.co.uk/news/science-environment-31001941 Accessed 12/01/2016.

[9] See Callaway (n. 6).

[10] See Pilcher (7).

[11] See Callaway (n.6).

[12] *Longleat in bid to save the world's rarest rhino* 19 July 2017. See https://www.longleat.co.uk/news/Longleat-in-bid-to-save-the-world Accessed 20/07/2017.

hybrids could breed to produce further offspring is still an unknown quantity.[13]

And then in May 2018, the results of a study of the genetic material taken from northern white rhinos and stored in the San Diego Zoo Institute for Conservation Research's Frozen Zoo showed that it would be possible to use it *'to successfully seed a recovered population in the future'*. Many methods would be available, including *'turning northern white rhino cell lines into stem cells and then into eggs and sperm or by cloning (placing northern cell line DNA into a southern rhino egg cell)'*.[14]

However, by then the inevitable had happened. In March, Sudan had died of old age, so very sad. A new challenge but one the scientists are already meeting head on (as we shall see later).

## 34.2.2   Rapid Assessment of Populations of Forest Elephants

Although forest elephants are threatened with extinction, there is still an important population in Cameroon, where the forests in the south-east of the country *'represent a stronghold for the species'* and *'have been identified as a priority for conservation efforts'*. But forest elephants are difficult to count, and that is what makes this research project so important. It examined the potential of a new method, using an interview-based technique *'as a tool for the rapid assessment of the distribution and threats to the forest elephant in the eastern region of Cameroon'*.[15]

Like savannah elephants, forest elephants are a *'keystone species'*, megagardeners of the forest.[16] Cameroon is not a wealthy country, and as it relies almost entirely on its timber concessions, maintaining a healthy elephant population not only plays a key role in good forest management but also benefits its human population.

---

[13] Rebecca Morelle *UK rhino eggs 'could save last northern whites'* BBC, 19 July 2017. See http://www.bbc.co.uk/news/science-environment-40655273 Accessed 20/07/2017.

[14] Jake Buehler *The Practically-Extinct Northern White Rhino Just Got Some Good News* Earther. See https://earther.com/the-practically-extinct-northern-white-rhino-just-got-s-182629365 Accessed 29/05/2018.

[15] Stephanie Brittain *A rapid assessment of the status and distribution of Loxodonta cyclotis in South East Cameroon September 2013* Supervised by Dr. Marcus Rowcliffe & Paul DeOrnellas. See https://www.iccs.org.uk/wp-content/thesis/consci/2013/Brittain.pdf Accessed again 22/11/2018.

[16] Beaune et al. 2013 in ibid.

Although most forest elephants qualify for full protection under Cameroonian legislation, those with larger tusks (weighing more than 5 kg) can be killed if the correct licence is granted.[17]

If an elephant population is to be effectively managed certain basic information is needed, such as, how many elephants there are and where they are to be found.[18] In 2007, the IUCN estimated a possible 1600 elephants left in Cameroon, although only 179 animals were definitely recorded. In 2011, the FAO noted there were still elephants living in eastern Cameroon despite the fact that the country had the highest deforestation rate in Central Africa. So it was essential to obtain more accurate numbers. The problem was how to get them. The research team, which consisted of two scientists and their driver, decided to take a novel approach, using a social survey based on interviews rather than direct (aerial surveys) and indirect (e.g. dung pile density) observations.

There were six objectives:

- To determine the distribution, occupancy and status of forest elephants in the eastern region of Cameroon using evidence-based occupancy analysis
- To assess the reliability and suitability of this method of rapid assessment in the context of forest elephants in Africa
- To gain a deeper understanding of the threats to elephants
- To gain a deeper understanding of the perceived level of population change
- To increase understanding of people's attitudes towards elephants
- To make recommendations for conservation action in the eastern region[19]

Although some informal interviews of suitable interviewees and a pilot study were carried out, the main research consisted of easily replicated, semi-structured interviews of the following personnel:

- Timber company workers: These fell into three groups, the 'prospect' team, the 'fauna' team and the 'pre/post evaluation' teams.
- Authorities: These were officials from the Ministry of Forests and Wildlife (responsible for protecting forested areas and their biodiversity).
- Managers of the Department of Fauna (those managing the eastern region of Cameroon) and the chef de postes (from the Ministry of Forests and Wildlife) whose job was *to monitor what is happening in the forest and report (any illegal actions) on a regular basis*.[20]

---

[17] Stiles 2011b in ibid.
[18] IUCN 2008 in ibid.
[19] Ibid, p.7.
[20] ForestMonitor.org-Cameroon 2013 in ibid.

Because of their differing functions, the timber company workers made different contributions. The prospection team spent as long as 3–4 weeks at a time within the forest, often being the first people to enter a particular area. They used machetes to clear paths, did not operate noisy machinery and were 'one of the teams ... most aware of animals, as they often disturb them and need to be aware for their own personal safety'.[21] One of the main tasks of the fauna team was to deter poachers within the Forest Management Unit, which made them particularly aware of the different animals, while the pre/post evaluation teams were required to report on signs of animals within the Unit. All teams operated on foot only. Data was collected from 196 participants over a 10-week period and then carefully analysed.

In addition to the interviews, detection/non-detection data was collected (from 342 sites within 34 Forest Management Units), together with data on perceived threats, abundance and change in population.

The results were not unexpected, showing that it was 'in sites adjacent or in close proximity to the borders of the CAR (Central African Republic) to the East, and the Congo to the South' where the largest decline in elephant population had taken place. Fortunately, in sites adjacent to two national parks, there was a perceived stable population, as there was in other sites.

Poaching was the severest perceived threat. The qualitative data suggested that despite being aware of the legislation, 'the financial reward of elephant poaching outweighs the risk for many in the eastern region of Cameroon' and that 'overall, the feeling was that elephants are of financial worth, and they can make you rich quick'.[22] And according to the authorities, the influence of international markets was 'a key force behind poaching across the region'.

Both authorities and timber concession workers recognized corruption was making conservation much harder to achieve. People were unlikely to get caught so were prepared to use poaching 'as a quick fix to poverty'. And it was the local people, the people on the ground, who took the blame. The report recognized that more resources were needed and recommended a transparent system of reporting and monitoring be set up if corruption was to be reduced.[23]

Civil unrest, in both CAR and the Congo, were also 'highlighted as additional key causes of the elevated risk of poaching in some areas'.[24]

What was encouraging however was the fact that the timber concession workers became much more enthusiastic about the need to conserve elephants

---

[21] Ibid, p.22.
[22] Ibid, pp.47–48.
[23] Ibid, p.59 and LAGA 2013.
[24] Ibid, p.50.

once they learnt how important the animals were for the survival of the forests, although *'simple awareness raising and education'* were not enough. A *'more nuanced approach'* was required if attitudes and behaviour were to be changed. People needed to be motivated and incentivised.

This valuable piece of research, once again the result of collaboration, had shown that social surveys could be a reliable way to obtain essential information quickly and were another effective tool for counting elephants. Stephanie Brittain's thesis was the result of enthusiastic collaboration. It was supervised by both Imperial College London and ZSL who also helped with the fieldwork in Cameroon. MIKE provided support, as did personnel from the government of Cameroon. Funding came from ZSL, Chester Zoo and Imperial College London.

### 34.2.3   The UNWTO Briefing Paper

Published in 2015, *Towards Measuring the Economic Value of Wildlife Watching Tourism in Africa* was a collaboration between the United Nations, the World Tourism Organization and the Convention on Migratory Species of Wild Animals (UNEP/CMS). It was *'a first step towards a more systematic measurement of the economic value'* of wildlife-watching tourism in Africa and its role in the fight against poaching. In other words, it was specifically concerned with nonconsumptive forms of wildlife tourism, with looking at wildlife in their natural habitats. This is particularly important because the animals tourists most want to see include elephants and rhinos, both at high risk from poachers.

Although data came from a wide variety of sources, we will concentrate on one, a survey that was carried out among *'African tourism ministries and authorities, protected area and wildlife conservation institutions and international and African-based tour operators'.*

The approach was multilevel, namely:

- International
- National
- Local

63% of UNWTO African Member States participated, including *'48 governmental institutions (tourism authorities and protected area and wildlife conservation agencies)'* from 31 African countries and 145 tour operators from 31 different countries selling trips to Africa.

Among its many findings, we need only record here that wildlife watching:

- Already generated considerable finance, and *'revenues are expected to grow further'*.
- That it was *'a very important segment of tourism for most African countries… with potential to benefit the local community'*.
- The poaching crisis was clearly *'viewed as a threat to the long-term sustainability of tourism'* and the development opportunities linked to it. In fact, the first chapter is headed *'Wildlife crime challenges nature conservation'*.

The survey found that most of the protected area authorities were involved in anti-poaching measures, whereas tourism authorities were only slightly involved; and most did not distribute information on poaching to their customers.

Of those tour operators participating in the survey, the findings showed that about 50% were *'funding anti-poaching initiatives and/or engaging in nature conservation projects'*; but only a few were proactively telling their customers about the issues, thus suggesting that the sector could be mobilized in antipoaching campaigns.

The survey found *'significant'* potential for education/information to *'play a key role in raising awareness'*, possibly also in *'financing (or cofinancing) anti-poaching initiatives'*. It recommended further research which included emerging markets such as those in Asia.

# 35

## Education

We have already seen that unless demand is significantly reduced, the future for elephants and rhinos is bleak. There is also no disguising the fact that this will be an enormous task. China and Viet Nam are key consumer countries, but they are tackling this seemingly intractable problem.

## 35.1 Curbing Demand in China

Earlier in our story, we saw how TRAFFIC had worked with 15 of China's leading e-commerce companies, how, as a result, they had adopted a zero-tolerance approach to the sale of illegal wildlife products on their websites, and how this had led to a decline in demand. By 2014, 24 companies were involved.[1]

TRAFFIC is playing a pivotal role in curbing demand. At the Symposium on Combating Wildlife Crime, James Compton, TRAFFIC International's senior director for Asia-Pacific, explained how this was being achieved. It was by making use of the key fact that *'61% of consumers in the PRC are more likely to follow the recommendations of family members and friends than recommendations in other forms or from other sources of recommendations'* and by working together.

---

[1] WWF/TRAFFIC 2014 See http://d2ouvy590dg6k.cloudfront.net/downloads/wildlife_crime_big_wins_Ir.pdf p.13.

© Springer Nature Switzerland AG 2019
B. Martin, *Survival or Extinction?*, https://doi.org/10.1007/978-3-030-13293-4_35

TRAFFIC's partners include:

- The China Wildlife Conservation Association (NGO), which is closely connected with the State Forestry Administration (Government)
- China Radio International, as this reaches Chinese communities in Africa
- The Traditional Chinese Medicine sector
- The National Tourism Administration of China
- The private sector (or rather, some of it)[2]

IFAW is also playing a key role. After learning that 70% of people had no idea that elephants had to be killed if they were to have ivory to buy, one of IFAW's campaigns sets out to explain where ivory comes from, combining it with a 'lesson' on the illegal wildlife trade and inviting participants to go to their website and download a report on the ivory trade. And, like TRAFFIC, they have worked successfully with Chinese Internet companies to reduce demand not only in ivory but also rhino horn and other illegal wildlife commodities.[3]

Chinese people are very civilized and keen to learn the facts, but problems remain, and time is running out. Reaching everyone with the message is still a challenge.

## 35.2 Curbing Demand in Viet Nam

The task of explaining to the people of Viet Nam why it is essential that they transform their desire for rhino horn into a desire to prevent living rhinos from becoming extinct is possibly even greater, except that the current elevation of the commodity into this wonderful must-have substance only happened very recently. Fortunately there are some very dedicated Vietnamese who are determined to explain what is happening, that the final, dreadful result of these actions might be extinction.

### 35.2.1 CHANGE's Campaigns

CHANGE, one of the key NGOs involved in this battle to change minds, aims 'to change the behaviour of the community and environmental problems through creative media activities and youth capacity building'. To do this, it 'acts

---

[2] Compton in *Symposium on Combating Wildlife Crime: Securing Enforcement, Ensuring Justice, and Upholding the Rule of Law. The Proceedings* Editors Kala K. Mulqueeny and Francesse Joy J. Cordon 2014.

[3] Grace Ge Gabriel in ibid.

*as a bridge between business, community and government in the formulation of policies and appropriate solutions to pressing environmental problems'.*

In March 2015, in Ha Na, CHANGE, AWF and WildAid started their campaign *'No More Buyers = No More Killers'*, using the media to raise awareness of the use of animal products. Various media partners pledged support, including the radio, news writers and magazines, and, together with other sectors and the goodwill ambassadors, they succeeded in raising over 60 billion VND through publishing their messages on television and in advertisements reading *'To change hearts, minds and behaviour and to save the endangered rhino'.*

In April 2015, they held a seminar/workshop in Ho Chi Minh City, a meeting of Vietnamese journalists and artists there to discuss *'The role of the media – Raising public awareness and improving the "image" of Viet Nam internationally about rhino horn'.* Over 60 journalists and editors from 40 news sectors attended, as well as a number of goodwill ambassadors drawn from the world of celebrities, including the singer/artist Thanh Bui; Thu Huong, the runner-up for Miss Universe; the actress Hong Anh; and the musician Duong Thu.

The aim of the meeting was to exchange information and points of view on the killing of rhinos *'as it is becoming more serious'* and to call for support from the media for the international campaign *'Stop Using Rhino Horns'.* There was also an exchange of plans and ideas *'on how to stop collecting and selling rhino horn illegally, hoping to change the image of Viet Nam in the eyes of society'.* Professor Nguyen Chan Hung, the President of the Viet Nam Cancer Group, was present. He was interviewed about the real effect rhino horn had in curing cancer, *'as well the superstitious hope of the Vietnamese people on using the horn as medication'.*

The goodwill ambassador Thanh Bui, who had been on a trip to Africa the previous year to see for himself what was really happening there, shared his experiences with the delegates, and Thu Huong and Hong Anh spoke about their plans to support the campaign.

In August, the *'Stop Using Rhino Horn'* campaigners, supported by about 20 famous Vietnamese artists and celebrities, put on a special art performance dedicated to rhinos, *'The Call of the Wild'.*

Then in March 2017, 17 pieces of graffiti appeared in Ho Chi Minh City, all of them rhinos in a variety of shapes and sizes, all very colourful, very eye-catching and all bearing the same message *'Curu te giac'* (Vietnamese for 'Save the rhino'). CHANGE's most recent campaign was designed to educate local people.

Working with Suby One, the French graffiti artist, 11 local and international artists were each invited to contribute a design. The hope was that while the pictures were being painted, people would be curious to find out what it was all about, thus presenting the perfect opportunity to tell them what was happening to these exotic creatures and why. And it worked. Although initially they were suspicious, once the people saw their neighbourhood was being improved, they brought cake and water for the painters, and the children started to play and talk about rhinos.

Nhi Thoi, CHANGE's Programme Manager, spoke passionately:

> *I want to create love for rhinos. In some images we had differences, like the rhino's horn exploding or the shadows of other rhinos that weren't actually there. We wanted to convey the message that the rhino is close to extinction and needs protection.*[4]

## 35.2.2   The 'Strength of Chi' Campaign

This is also proving very successful. It harnesses the Vietnamese concept of 'Will' or 'Chi', the power of what lies within, and can loosely be described as an 'educate-re-educate-nudge' campaign. It was launched on World Rhino Day 2014.

The Chi campaign, a collaboration between TRAFFIC and the Central Committee of Propaganda and Education (CCPE) in Viet Nam, *'promotes the concept that respect and success come from a person's inner strength or 'chi' and not from a piece of rhino horn'.*[5] The key to success is the importance of the individual character, their determination and their will. *'Education'* for those who need to start from the beginning, *'re-education'* for those who knew but who had ignored the problem and *'nudge'*, to gently nudge people in the right direction.

*'The most charismatic and successful men create their own good fortune'* is the essence of the campaign,[6] which is targeted at urban men between 35 and 50 years old, the key consumer groups, in Viet Nam's two main cities. The aim is to reshape public views about the use of rhino horn, and it is working

---

[4] Michael Tatarski *A Graffiti Campaign in Ho Chi Minh City Aims to Educate Locals About Rhino Conservation* Pacific Standard, 13 April 2017 (This story originally appeared at the website of global conservation news service Mongabay.com). See https://psmag.com/a-graffiti-campaign-in-ho-chi-mingh-city-aims-to-educate-locals-ab Accessed 20/04/2017.

[5] TRAFFIC *Chi on display at Noi Bai Airport* TRAFFIC Wildlife Trade News, 4 February 2016. See http://www.traffic.org/home/2016/2/4/chi-on-display-at-noi-bai-airp Accessed 03/03/2016.

[6] TRAFFIC *Innovative campaign promotes success from within* TRAFFIC Wildlife Trade News, 22 September 2014. See http://www.traffic.org/home/2014/9/22/innovative-campaign-promot Accessed 18/10/2016.

well. Focus groups indicated interest in the campaign from the start, one business man explaining *'I love the 'Chi' logo and tagline. It is extremely powerful. I get it and it's very Vietnamese'.*

Activities and events during the first year included *'Chi-themed networking and life-style events, as well as* suctaichi.com, *a forum for discussing and learning about Chi'.* And, like the *'No more buyers = No more killers'* campaign, prominent Vietnamese citizens were involved, including the designer Khai Silk, the composer and producer Huy Tuan and LUALA's CEO in a short video. Outdoor billboards helped spread the message.

A memorandum of understanding with the Viet Nam Chamber of Commerce and Industry (VCCI), due to last 3 years, and TRAFFIC holding a capacity-building workshop for high-level VCCI staff, has enabled the Chi Campaign to reach *'businesses and their wealthy urban male employees, across the country through specially designed materials and workshops that promote the integration of wildlife protection into corporate social responsibility practices'.*[7]

The outcome has been even more education, more training and more people keen to spread the word. By the end of 2015, there were 29 VCCI master trainers, and in December alone, 100 workshops were held, the aim, to create a ripple effect. And there was a bike ride. A great success. So now there has been a second one.

World Rhino Day 2015 was celebrated with *'Nail it for Rhinos'*, at a special reception for the Vietnamese government, embassy officials, conservation NGOs and Chi campaigners. *'By clipping their nails and signing TRAFFIC's life-size rhino model, those attending helped spread the message that rhino horn is made of the same material as human nails, and that they refuse to use, gift or accept rhino horn'.*[8]

In February 2016 the campaign was extended to the transport sector, after the Vice Chairman of the CCPE, Mr. Bui The Duc, asked the Ministry of Transport to involve the air transport sector, to commit them to fighting the illegal trade. So posters went up in Noi Bai Airport, a bustling airport with *'the capacity to serve 10–15 million passengers'* a year and identified as *'a significant throughway'* for wildlife trafficking. During their flights, passengers could flick through the pages of *'Heritage Fashion'*, Vietnam Airlines inflight magazine, and read the story of a man who pursues success by relying on his inner strength, his 'chi'.

---

[7] Jill Capotosto *One year's progress of the Chi campaign* Save the Rhino, 2015. This article was originally published in The Horn, Autumn 2015. See http://www.savetherhino.org/our_work/demand_reduction/consume Accessed 18/10/2016.
[8] Ibid.

Mr. Bui The Duc made it quite clear that *'by committing to display the Chi Campaign posters'*, the airport authorities *'are taking important steps'* not only to protect wildlife but also *'to counter the threats to national security posed by criminals poaching and trafficking wildlife in Vietnam'*. The Central Committee fully understood the ramifications of the trade. There would even be a 'chi'-themed calendar to provide a 1000 government officials with additional information about 'chi' and wildlife protection.

Once again we see the benefit of working together. TRAFFIC working with PSI (Population Services International), a global social marketing organization which devised the campaign, and using grants coming from Save the Rhino and the UK Government's 'Illegal Wildlife Trade Challenge Fund', a Fund which resulted from the London Declaration.

## 35.3   Japan's Changing Consumer Habits in Elephant Ivory and Rhino Horn

Elephant ivory and rhino horn were intimately connected with Japan's culture for centuries, but their attitudes have now changed, and the market in these two commodities is greatly reduced. So how have they have achieved this and can other countries learn from their experience? Another report, *'Setting Suns: the Historical Decline of Ivory and Rhino Horn Markets in Japan'*, can provide some answers.[9]

Once again the trade must be put into its historical context. We have already seen that the medicinal use of rhino horn dates back some 300 years to the Edo period. It is the same with the traditional carving of elephant tusks. But it was not until the middle of the 1970s and 1980s that consumption became a major cause for concern. And the main reason was an economic boom. People had lots of money to spend, and some of the things they spent it on were luxury items such as ivory jewellery. As for rhino horn, at that stage it was still an important component of many medicines. Japan was *'recognized as one of the world's largest end-use markets for wildlife products'*.[10] But all this changed because of CITES. In 1980, there was an international ban on trading in rhino horn, and 1989, as we know, a similar ban was introduced for elephant ivory. Japan observed both bans because it was a Party to the treaty. After that, the market for the products began to decline.

---

[9] Tomomi Kitade and Ayako Toko (2016) *Setting Suns: The Historical Decline of Ivory and Rhino Horn Markets in Japan* TRAFFIC, Tokyo, Japan.
[10] Ibid.

But not before two events had taken place. In 1984, at the first CITES regional seminar for countries in Asia and Oceania, Japan was non-compliant and therefore condemned by the others, not a good international image to have. Then the President of WWF, HRH Prince Philip the Duke of Edinburgh, visited Japan and asked the country to cooperate in wildlife conservation.

Although this was the tipping point, other drivers had also piled on the pressure. For example, TRAFFIC's monitoring/investigating activities 'constantly uncovered a series of problematic wildlife trade issues',[11] while lobbying and the mass media all helped to raise awareness so that by 1987, a government public opinion poll indicated that 73% of the population was aware of issues around illegal wildlife trading. And education was a key factor. People of influence advocated for nature conservation. A consumer survey carried out in 2014 showed that older people, those around at the time all this was going on, were still well aware of the issues, although younger people appeared to be more indifferent.

Just how important the import ban was as a driver for the reduction in use of rhino horn in medicinal products can be seen in the figures. Whereas in 1978 there were nearly 100 medicines listed in a National Formulary, after 1980 the number dropped dramatically, and currently there are very few listed. If rhino horn is used, it comes from old stocks. In 1973 Japan imported nearly 1.8 tonnes and continued buying until 1980 and the ban.

Kitade and Toko suggest they could have been stockpiling in anticipation of the ban, but whatever the reason, it did mean the industry had legal stocks available to them, while the products were phased out and alternatives offered, in this case, Saiga antelope horn (*Saiga tatarica*), which, they believed, had a similar effect. The unfortunate consequence has been that now these antelopes have themselves become *critically endangered* and are listed in CITES Appendix II. In fact, the consumer survey showed that people bought the medicine *'for the overall trust in the brand-named medicine itself'*, rather than because its contents contained a rhino horn equivalent.[12]

Other drivers included:

* The modernization of the medical system;
* Changes in the distribution and sales pattern of medicines
* Changes in the family structure in Japan, resulting in a break in tradition

People became aware of the problems generated by the medicinal use of rhino horn.

---

[11] Ibid, p.v.

[12] Ibid, p.vi.

An important distinction between Japan and current Viet Nam is that unlike Viet Nam, in Japan, social pressure has not been found to be a driver, which is good news. But the bad news is that as few as 29% of people in their 20s were aware of the implications of using wild biological resources in traditional medicine. Fortunately only 2% of all consumers (in the survey) showed any interest in buying rhino horn-related medicine in the future.

There was a similar decline in the market for elephant ivory, although the industry in ivory carving also dates back to the Edo period. To this day, hankos (personal signature seals) and bachi (a plectrum for a musical instrument, the shamisen) remain popular, mainly because ivory is still the best-quality material available for their manufacture.

During the economic boom, ivory, including polished whole elephant tusks, became highly desirable, to the extent that Japan imported as much as 950 tonnes of raw ivory in 1983–1984. And once it became obvious that ivory sales would be banned by CITES, a degree of speculation and stockpiling even took place.

Although the use of ivory declined because of the tightening of international trade through CITES regulations, followed by the complete ban in 1989, social pressure also played a role. The results included phasing out the use of ivory in piano keys and forcing department stores to stop selling ivory jewellery. Some companies and public offices even stopped using ivory hankos.

Japan was, in effect, forced to observe the CITES ban by the socio-political climate around the illegal wildlife trade. There was evidence suggesting many Japanese people were only too aware of elephant poaching and its consequences. Then came the recession from which the country has never really recovered. It too has played an important role in helping to damp down the demand.

With the reduction in demand, the one-off sales of ivory and the whole polished tusks already in the country, Japan has managed to balance supply and demand. But even with education and this success, there is the inevitable dark side. Loopholes in the law, regulations and management have seen some illegal export of and trading in ivory. The evidence comes from recent seizures. In 2015, the EIA reported on problems with the *'whole tusk registration scheme'.*[13] And it almost goes without saying that there are also problems with the Internet.

We can see therefore, that with a few reservations, in Japan, education has worked to reduce demand, although it has been only one of the factors.

---

[13] EIA *Japan's Illegal Ivory Trade and Fraudulent Registration of Ivory Tusks* EIA-GLOBAL.ORG, 2015.

Furthermore, the report documented *'considerable ignorance and indifference … amongst current consumers, especially the younger generation'*:

- 9%: showed interest in buying ivory products
- 26%: of the people buying ivory hankos were aware of the legal factors regarding the purchase of ivory products
- 15%: actually over that figure, *'expressed indifference toward illegal ivory'*[14]

These slightly disturbing findings were balanced by the 2014 survey which showed that over 80% of people who used hankos would be happy to consider a substitute, possibly titanium, and the traditional music sector was moving the same way, but there is still a need for education.

## 35.3.1   But There Are Still Problems

Because of the disturbing results of a market survey they carried out in 2017, TRAFFIC proposed some specific policy changes and new legislation to make sure that Japan's domestic market did not contribute to elephant poaching and/or illegal ivory trafficking. At the end of December 2017, China's ivory trade ban took effect, and Japan introduced new domestic regulations in June 2018, so TRAFFIC repeated the exercise later that year.

The results were again disappointing, for although there had been a slight improvement, the illegal exporting of ivory was still widespread. New artefacts calculated to appeal to Chinese buyers were openly on sale, and a majority of traders were still willing to sell to foreign buyers or Japanese buyers intending to export. The new regulations were, in many cases, not being complied with, and enforcement was poor.

TRAFFIC recommended that the government should implement its 2017 recommendations regarding policy and legislative changes in full and increase enforcement so that the new regulations were implemented effectively and illegal export deterred. It also recommended that the CITES Standing Committee should consider whether Japan should be included in the NIAPS process.[15]

---

[14] See Kitade and Toko (n.9), p.viii.
[15] Tomomi Kitade and Ryoko Nishino *Slow Progress: A Reassessment of Japan's Ivory Market in 2018* TRAFFIC Briefing, September 2018.

## 35.4   Education and China's Tourism Industry

In 2016, the China Tourism Administration celebrated World Wildlife Day by pledging zero tolerance towards illegal trade in wildlife at a special event convened by China's CITES Management Authority. Out of the 30 travel companies who organize trips to Africa, 10 key ones signed the pledge, which was particularly important because in the past 10 years, more and more Chinese nationals have been visiting other countries.

Let us look at some numbers:

* In 2012, there were over 80 million visitors (arrivals and departures).
* In 2015, there were 120 million visitors (arrivals and departures).

Chinese travellers going to popular African destinations:

* In 2011, there were one million.
* In 2014, there were over 1.52 million.

And to southeast Asia:

* In 2012, there were 7.3 million.
* In 2015, there were 17 million.[16]

There was an immediate increase in reports of smuggling, not just ivory but other endangered wildlife products. As *'up to one-third of suspects still claim(ed) to be completely ignorant of international trade controls on ivory and other wildlife products'*, it demonstrated the urgent need for education, to raise awareness of the existing legislation, which was just what the tourism industry was planning to do.

The Director of the Supervision Division, China Tourism Administration, Tang Bing had no doubts:

> *We will further strengthen governance of the tourism sector's operations, improve training materials for tour guides, strengthen advocacy for those servicing the industry and for overseas travellers and include wildlife conservation as one of the tourism sector's assessment criteria for selection as an industry 2016 Civilized Tourism Champion.*[17]

---

[16] TRAFFIC *Tourism industry in China says 'no' to illegal wildlife trade* Wildlife Trade News, 3 March 2016. See http://www.traffic.org/home/2016/3/3/tourism-industry-in-china-say Accessed 03/03/2016.
[17] Ibid.

The Head of TRAFFIC's China Office, Zhou Fei, warmly welcomed the development, describing it as *a fine example of how the tourism sector can play an active role in combating* the illegal trade, continuing, *'the tourism sector is now well-placed to help develop effective ways to reduce demand for products such as elephant ivory and rhino horns: 'Don't buy illegal wildlife products' is a pertinent message aimed at every Chinese overseas traveller'.*[18]

## 35.5    A Different Kind of Education

And then there are celebrities, people like Yao Ming, the former US National Basketball Association star, who are passionately concerned about wildlife and are using their status and influence to tell their fans, and anyone else who cares to listen, about the terrible cruelties being inflicted on elephants and rhinos, to explain they are now in danger of dying out. Yao Ming has collaborated with WildAid, working in China PDR on an ambitious attempt to end elephant poaching, once again using education. So has the actor Li Bing Bing, working with Save the Elephants.

Education is an important part of the work of the Black Mambas, the first all-female antipoaching unit who patrol the Balule Reserve. Although they come from different backgrounds, some from very poor families and villages, being a Black Mamba means they have a well-paid job, and they can exploit this when they go visiting the local schools as part of the *'Bush Baby'* programme.

They understand the poverty. Only enough food for one meal a day. Children with poor job prospects who frequently come from 'poaching' families, families who poach to pay the bills and to buy food to eat. They explain to the children that they must not be tempted by the prospect of poaching because they think it is easy money. They, the Black Mambas, do understand, only too well. They understand, but they don't agree. And they make perfect role models themselves.[19]

Local communities are also active in this are area. For example, in Laikipia, Kenya, the Loisaba Community Conservation Foundation held an event, 'Education for Conservation', for students. Its aim was to find *'better solutions for conservation and education within communities neighbouring Laikipia's conservancies'.*

---

[18] Ibid.

[19] Jay Akbar *The REAL lionesses of Africa: Stunning 'Black Mambas' are first all-female anti-poaching unit risking their lives to protect big cats, rhinos and elephants from men with guns* MailOnline, 22 February 2016. See https://www.dailymail.co.uk/news/article-3458085/The-REAL-lionesses-Africa-Stunning-Black-Mambas-female-anti-poaching-unit-risking-lives-protect-big-cats-rhinos-elephants-men-guns.html Accessed again 22/11/2018.

## 35.6   Training: Workshops, Seminars and Meetings

Successful enforcement requires training.

We have already seen that INTERPOL and the ICCWC hold regular seminars and workshops, and a selection of the events recently attended by Malaysian enforcers as recorded in their CITES SC66 NIAP Progress Report provides a good example of what is on offer.

Training Malaysian enforcers:

- The Deputy Director of Enforcement Division of Malaysia's DWNP attended the *Preoperational Training and Planning* meeting for Operation COBRA II in Kenya, in October 2013.
- The Director of DWNP attended a Workshop on *Combating Wildlife Trafficking in Indonesia*, in May 2014.
- Two enforcement officers attended a workshop on *Wildlife Trafficking-Related Customs Practices* in Cebu, in August 2015.
- Two enforcement officers, one from the RMCD and the other from DWNP, attended ICCWC workshops related to collaboration with the World Bank, INTERPOL, UNODC, World Customs Organization and CITES Secretariat.
- Three enforcement officers attended a *Workshop on the Application to Illegal Consignment of Wildlife Products* in the Czech Republic, in 2015.

## 35.7   Rangers and Dogs

Rangers benefit greatly from training. They carry out dangerous work and often use sophisticated equipment. In addition to protecting their animals, many of them also collect data. Conservancies such as Lewa and Ol Pejeta ensure their rangers are well-trained, but others are not so lucky.

Garamba National Park in the DRC was the final home of the last wild northern white rhinos. It continues to be a dangerous place, but the rangers have to protect the last remaining elephants that live there. Animals and humans bear the full brunt of foreign armed bands, the dreaded men on horseback coming from South Sudan and the frightening Lord's Resistance Army. Until recently, the rangers were ill-prepared for these onslaughts. Now however, African Parks has taken over Garamba, and they are working hard to train the rangers so that they are better equipped to fight the battle on the ground.

**The Southern African Wildlife College** a private higher-education and training institution, was set up to train rangers and conservation personnel and teach all the skills needed to sustain and rehabilitate wildlife areas. Its advanced certificate *Nature Conservation-Transfrontier Conservation Management* provides essential management skills for those working in locations where many protected areas cross frontiers and animals such as elephants migrate. And ground-to-air training using helicopters as well as fixed wing aircraft forms part of the course undertaken by patrol leaders training for work in the Kruger National Park.[20] It also works with the local communities.

In 2014, the College, which runs a Rhino Protection Programme, was runner-up for the Rhino Conservation Awards in the category *Best Awareness, Education and Fundraising for Rhino Protection and Conservation.*

Already supported by WWF South Africa and Peace Parks Foundation, when Prince Harry visited the College in December 2015, he announced that in the future, United for Wildlife too would work with it, another partner, providing funding *'so that its graduates are equipped with the best techniques and technologies available to protect some of the world's most endangered species'.*[21]

**Sniffer and Tracker Dogs** These all receive intensive training before they can become members of their enforcement teams. The dogs are so effective because they form close bonds with their handlers. There must be trust and understanding, but this takes time and careful training to develop, which is why the Ol Pejeta Conservancy has included the development of *'a K9 school of excellence to provide dogs in a deterrent and offensive role against poaching'* in its 2020 MAP.[22]

---

[20] Peace Parks. See http://www.peaceparks.org/college.php?pid=28&mid=29 Accessed 27/03/2017.

[21] A speech by Prince Harry to the Southern African Wildlife College, Kruger National Park, South Africa. See https://www.royal.uk/speech-prince-harry-southern-african-wildlife-col Accessed 12/01/2017.

[22] Ol Pejeta, 2016.

# Part XII

Sustainable Use

# 36

# The CBD Re-visited

At the beginning of our story, we learnt that the Convention on Biological Diversity, CBD, would be playing a crucial role. We are about to see what it is.

We also learnt that sustainable use, ecology and genetics are the three guiding principles of the Convention and that the greatest of these is sustainable use. Indeed, it could be argued that this principle is the main reason for the Convention's popularity. South Africa, a key player in the story of elephants and rhinos, has even written sustainable use into its constitution. There is, however, a big 'but', and that is that sustainable use must not operate at the expense of either of the other two principles, namely, ecology/ecosystems and genetics. So we need to see how they relate to each other.

## 36.1 Elephants, Rhinos and Ecosystems

Ecology is the relationship and interactions between organisms and their environments. It is the study of ecosystems. All of us live in ecosystems.

**Ecosystems, Ecology and Keystone Species**
Ecosystems come in many forms. They can be areas of woodland, arable pasture, ponds or even deserts. An ecosystem is a functioning unit of the physical environment of an area and the organisms that live in it. All its component parts, such as animals, plants, soil and even the atmosphere, interact with each other.

Ecosystem ecology is holistic in that it attempts to understand how the system functions as a whole unit by looking at how different populations

© Springer Nature Switzerland AG 2019
B. Martin, *Survival or Extinction?*, https://doi.org/10.1007/978-3-030-13293-4_36

react/interact with each other. An example we are about to consider is the relationship between elephants and frogs, an unexpected combination.

Then there are the 'keystone' species. A keystone species can be defined as one whose loss would have a disproportionately large effect on the other species in the community.[1] Take an arch, for example. Stones are built up on either side and then curved gently round until there is just room at the top for the final stone, the keystone, the stone that completes the structure. Without it, the whole arch (in our case, ecosystem) would begin to crumble and gradually disintegrate. The same thing happens within an ecosystem. If the keystone species is removed, the entire ecosystem gradually becomes degraded, and secondary extinctions can occur.

Trophic cascades are intimately linked with keystone species. A trophic cascade is an ecological phenomenon that is triggered by the introduction or removal of a keystone species. If a keystone species is introduced, it can balance an unbalanced ecosystem, whereas removing one can unbalance the system. The term 'trophic' refers to different levels within the food chain. Grass would be at the bottom, with perhaps insects next.[2]

Elephants and rhinos are herbivores, because of their size ($\geq$1000 kg), megaherbivores. They are apex consumers and have a particularly large impact on the ecosystems they inhabit.[3] Both fulfil the role of ecosystem engineers; in other words they are species that create or modify habitats. Both are key drivers of ecosystems. Both are keystone species.

## 36.2   Elephants

So how do we know elephants are ecosystem engineers? We know, because elephants create and maintain ecosystems through physically changing the habitat.[4] They do this by:

- Browsing on woody trees
- Stripping their bark
- Breaking major and minor branches, sometimes
- Uprooting them

---

[1] www.ecologydictionary.org/Keystone_species Accessed 26/11/2015.

[2] Mission Wolf *A Wolf's Role in the Ecosystem – The Trophic Cascade.* See www.missionwolf.org/page/trophic-cascade/ Accessed 03/12/2015. See now https://missionwolf.org/trophic-cascade/ Accessed 22/11/2018.

[3] Owen-Smith 1988.

[4] Jones, Lawton & Shachak 1997.

Even breaking minor branches and stripping off bark can create habitats that would not otherwise exist.[5] The trees are replaced by quick-growing vegetation, transforming dense woodlands into open grasslands.[6] The changed habitat makes it possible for other different animals and plants to move in. The transformation can provide an important opportunity for increasing biodiversity.

But could it also degrade the habitat? One research project set out to determine what was going on.[7] They took samples of the herpetofauna, that is, amphibians (frogs and toads) and reptiles (skinks/lizards and non-skinks) from an acacia tree habitat that varied in the amount of *'damage'* caused to it by elephants. This is what they were looking at to help them determine whether elephant foraging degraded or enriched a habitat.

They were measuring three aspects of the herpetofauna in the different habitats:

- Abundance: how many were living there?
- Diversity: how many different species?
- Richness.

If elephant activities enriched the habitat, there should be more of possibly all three of the above criteria, whereas if they degraded the area, the numbers would be lower.

There was also a control site from which elephants were excluded.

The results were encouraging for the elephants and thus for the herpetofauna as well. The highly damaged area contained *'significantly greater species richness than the control region'.*

- Highly damage areas: 18 species
- Medium damage areas: 12 species
- Low damage areas: 11 species
- Control area: 8 species

If herpetofauna are to live in a habitat, there has to be suitable *'accommodation'* for them. It is essential that species such as frogs, toads and lizards have sufficient suitable, viable refuges. The way elephants damage trees, breaking

---

[5] Pringle 2008.

[6] Laws 1970; Shannon et al. 2006.

[7] Nasseri N. A., McBrayer L. D. and Schulte B. A. (2011) *The impact of tree modification by African elephant (Loxodonta Africana) on herpetofaunal species richness in northern Tanzania.* African Journal of Ecology, 49: 133-140.

off branches, uprooting them, produces a lot of coarse, woody debris which provides excellent habitat for the herpetofauna. Places where they can live and breed, and from where they can hunt for food. This is because the mounds and craters resulting from the uprooted trees not only *'house'* more organisms the herpetofauna can eat but may also give rise to temporary breeding pools during the rainy season.[8]

So elephant engineering produces complex habitats that, in their turn, support a richness and diversity of species. The research showed that frogs in particular thrived in highly damaged areas; there were many more species there than in the control area which had excluded the elephants. Other research has shown that elephant engineering is a key driver in ant colonies living in acacia trees[9] and the many species of spiders living in the sand forests of South Africa.[10] In other words, it plays a fundamental role in developing ecosystems.

### Elephants as Megagardeners
Why Megagardeners?

Plants are an essential component of many, if not most ecosystems. They help shape their composition and structure as well as the functions they carry out. Plants need to reproduce. One of the ways they do this is by dispersing their seeds to new locations. This can be achieved by various methods. Some seeds, like the winged seeds of the sycamore tree, are dispersed by wind, while others are dispersed by animals, including elephants. In fact *'animal body size, ecological niche, diet and ranging patterns determine the dispersal potential, competitive balance, and survival fate of hundreds of billions of seeds every year'.*[11]

So how do African forest elephants, savannah elephants and Asian elephants disperse seeds?

All elephants are large animals, megafauna. They exist in many kinds of habitat, and this dictates what they eat, where they must go to find their food and their social behaviour. Because they are so large, they need to eat a lot, from a diversity of plants, and what they eat depends on what is available. For example, in Namibia which is mainly desert, they eat just 33 species, in Uganda they eat over 200 species, while in Congo, in the Ndoki forest, the elephants there consume at least 500 plant species (a record-breaker, more

---

[8] See Greenbeg 2001; Guo 1996; Olff & Ritchie 1998.

[9] Palmer et al. 2008.

[10] Haddad et al. 2009.

[11] Campos-Arceiz Ahimsa, Blake Steve *Megagardeners of the forest – the role of elephants in seed dispersal* Acta Oecologica 37 (2011), pp.542-553.

than any other animal).[12] Asian elephants feed from about 100 different plants.[13] This all adds up to a very large number of species and highlights the importance of elephants in seed dispersal and hence in enabling ecosystems to function.

Elephants are particularly efficient at foraging fruit. Indeed, it has been shown that forest elephants seem to use their spatiotemporal memory to go to the right trees at the time they are fruiting, even to the point of creating permanent elephant trails in some places.[14] But because it is difficult to actually see elephants devouring fruit, most of the research is based on the analysis of their dung, the seeds and fragments of fruit providing evidence of which species had been eaten. When analysed, this showed that forest elephants, those critically endangered animals, were eating more seeds and fruit from more species and from a larger number of species of woody plants than either the savannah or Asian elephants. They were also dispersing more seeds. Furthermore, the closer to the equator they lived, the higher the diversity of plants and seeds.[15] Asian elephants did not appear to be eating much fruit, but there were a number of reasons for this, including the fact that trees using wind dispersal dominate in those tropical forests.

Some species may have evolved a special relationship with elephants, the *'megafaunal syndrome'*, which only uses them to disperse their seeds, a very rare occurrence as most plant species are generalists.[16] These plants bear very large fruits and seeds. Elephants have a particularly large mouth and gape, which enables them to eat the fruit other fruit-eating mammals could not possibly manage. The size of the seeds adds to their attraction, and the gastric acids in the elephant guts work on their hard shells, breaking them down, breaking dormancy and stimulating germination.

Another benefit to the plants, and hence to the ecosystems, is that elephants tend to travel very considerable distances, and although 88% of the seeds are moved about 1 km from the parent plant, the furthest recorded distance was 57 km (data collected from the Ndoki forests, Congo).[17] This helps with the colonization of vacant areas and ecological succession and can enable fragmented landscapes to reconnect.[18] An ecosystem whose elephant population is either declining or has disappeared would be more

---

[12] Blake 2002 in ibid.

[13] See Campos-Arceiz et al. 2008a in ibid.

[14] See Blake and Inkamba-Nkulu 2004 in ibid.

[15] Ibid.

[16] Janzen and Martin 1982; Guimaraes et al. 2008 in ibid.

[17] See Blake et al. 2009 in ibid.

[18] Fragoso et al. 2003; Trakhtenbrot et al. 2005 in ibid.

vulnerable, and changes would occur in the demography and distribution of many species of plant.[19]

There can also be a benefit to animals. Elephant dung is full of beetles and other invertebrates, and this, together with the seeds in the dung, entices in other larger animals, such as red river hogs, to forage for their food. It has even been suggested that these may act as secondary seed dispensers.[20]

But they can also damage their habitat. For example, Ol Pejeta has been forced to exclude elephants from some of their acacia woodlands to give them a chance to be rehabilitated.

## 36.3   Rhinos

Rhinos also play an important part in maintaining healthy ecosystems. Like elephants, greater one-horned rhinos eat large tree seeds from the forest canopies and disperse them to the grasslands, where they normally germinate successfully.[21] Unlike elephants however, which are megagardeners, white rhinos are megagrazers, which makes them eco-engineers.

Elephants are really browsers. When they eat grass, they pluck tufts with their trunk. White rhinos crop the grass, so should thought of as lawnmowers, a very different function.[22] Until recently, little was known of how this impacted on the ecosystem they lived in. However, recent research carried out in Kruger National Park is changing this, and in the light of the current poaching epidemic which is now threatening the very existence of wild rhinos, the researchers intended their results to *'contribute to understanding the effects of their possible elimination'*. In fact, this was *'some of the first empirical data on potential ecosystem-scale impacts of returning a megagrazer into the landscape'*.[23]

Kruger National Park with its savannah grasslands presented an excellent location, because in 1896, there were no white rhinos there. They had been hunted to extinction. However, they were reintroduced during the 1960s[24] and, since then, have been steadily recolonizing the Park, so that by 2010, there were about 10–11,000 individuals.[25] What made this natural experiment

---

[19] Ibid.

[20] Ibid.

[21] Dinerstein 1992.

[22] Owen-Smith 1988.

[23] Cromsigt J. P. G. M., te Beest M. (2014), *Restoration of a megaherbivore: landscape-level impacts of white rhinoceros in Kruger National Park, South Africa.* Journal of Ecology, **102**, 566-575. doi:https://doi.org/10.1111/1365-2745.12218, p.567.

[24] Pienaar 1970 in ibid.

[25] Ferreira, Botha & Emmett 2012 in ibid.

of rhino impact possible was the interesting fact that not all parts of the Park had been recolonized to the same extent, so their effects varied.

The Park's geology influenced the types of vegetation growing there. Essentially there were two types of soil. In the west, they were derived from granite and supported denser, bushland savannah, and in the east, from basalt, leading to more open, productive grasslands,[26] the climate (rainfall) keeping the savannah arid to semiarid. The central area of the Park is roughly contained between two rivers. In other words, there are three distinct areas.

A total of 336 rhinos were released between 1961 and 1972,[27] 330 into the southern part and 6 into the northern part. None were released into the central area.[28] The animals appeared reluctant to cross the rivers, so even in 1991 there were very few rhinos there. Only 15 of the 330 rhinos were released onto the basalt-derived soils in the east of the Kruger, which still has a much lower density of rhinos than the open grasslands in the western borders of the Park. That was where the rest of the rhinos were released, an area that is now completely recolonized.

The researchers measured the numbers of *grazing lawns* and the proportion of short grass (both strictly defined), in the landscape, and they looked at the impact rhinos were having on the landscape, both high and low rhino impact areas, and whether/how this affected other animals.

They found that:

- There were more impala and warthogs in areas of high rhino impact.
- Termites and termite mounds might be playing an important role.
- The changes in short grass cover might lead to changes in the fire regime and to other cascading ecological effects, such as more short grass cover creating different habitats for more and different species to live in, small mammals, grasshoppers and birds, in other words, effects that can have surprisingly far-reaching results on other species.

Grazing was the key. Different grazers produce different effects. If grasslands can have a number of different species growing in them, they will support more species, and this can be critical to maintaining biodiversity. Grazing white rhinos could be playing a key role in this. Hence the research paper's final sentence, *'not only is rhino poaching threatening the species conservation status, but also the potentially key role of this apex consumer for savanna ecosystem dynamics and functioning'.*[29]

---

[26] Venter, Scholes & Eckhardt 2003 in ibid.
[27] Pienaar, du Bothma & Theron 1992 in ibid.
[28] Pienaar 1970 in ibid.
[29] Ibid, p.574.

## 36.4   Genetic Diversity

With the continuing decimation of both elephant and rhino populations comes a more hidden threat, how to maintain sufficient genetic diversity in a steadily shrinking gene pool, so we turn now from ecosystems to genetics, to genetic diversity in particular, because this is the second key to whether or not sustainable use is permitted. Why is genetic diversity so important? Put at its most extreme, because extinction of a species is almost always preceded by loss of genetic diversity within that species.

Every living thing on the planet carries genes. Genes are a vital part of all organisms. They are responsible for carrying out vital functions, functions such as the regulation of body shapes and sizes, as well as *'physiological processes, behavioural traits, reproductive characteristics, tolerance of environmental extremes, disease resistance'*.[30]

All organisms have genetic codes. The genotype is the genetic code found in the nucleus of each cell. Genotypes are partly responsible for determining the physical forms and functions of the organism. Other genetic codes are found in other parts of the cell. A good example of one of these codes is the mitochondria that are found in the cells of animals.[31]

Genes are passed on during mating. In other words, they are inherited. Inbreeding, or breeding within a circle of close relatives, has long been recognized as potentially undesirable. It can lead to a wide range of problems, perhaps causing the individual to be susceptible to a number of illnesses, like sickle cell anaemia in humans, or even causing physical disability. It happens when the resulting offspring inherits the same 'nasty' gene from each of its parents, rather than one 'nasty' and one 'good' gene, where the 'good' gene is dominant and can, in effect, 'over-ride' the bad gene.[32] In fact, an especially important factor in disease resistance is a diverse array of genotypes.[33] In other words, genetic diversity is good for fighting disease.

Another human disease is cystic fibrosis. While this is not caused by inbreeding, it is the result of the same 'nasty' gene being carried by both parents and being passed on by both of them to their offspring during a

---

[30] Raven, Peter H., Ray F. Evert, and Susan E. Eichorn. 1986. *Biology of Plants* Fourth ed. New York: Worth Publishers.

[31] US National Parks Service *Why is Genetic Diversity Important?* See https://www.nps.gov/plants/restore/pubs/restgene/1.htm Accessed 03/08/2016.

[32] Information given to the author in 2016, by Professor Sir Patrick Bateson.

[33] Schoen, D. J., and A. H. D. Brown. 1993 *Conservation of allelic richness in wild crop relatives is aided by assessment of genetic markers* Proc National Academy of Sciences (US) 90: 10623-10,627; McArdle B. H. 1996. *Levels of evidence in studies of competition, predation and disease.* New Zealand Journal of Ecology 20:7-15.

particular mating. If only one parent passes on the gene, the offspring will not have the disease.

We begin to see how very important it is to have genetic diversity. But it is not only a useful tool to fight disease; it also helps cope with adverse environmental conditions, another essential requirement in this world of climate change. Unfortunately, severe droughts are becoming more normal in parts of Africa where elephants and rhinos live. The results can be extreme. Animals, especially the young, elderly and weak, are gradually dying, their strength failing for lack of food and water. Human-elephant conflict is intensifying. Genetic diversity can help with survival. In other words, *genetic variation holds the key to the ability of populations and species to persist over evolutionary time through changing circumstances*.[34]

The implication is that a certain level of population is required for genetic diversity to occur. But we already know that at the beginning of the 1900s, the population of white rhinos had dwindled to as few as 25–50 individuals. Their fate would seem to have been sealed. Extinction loomed. It did not happen though. The question is why not? And the answer is vital, because we also know that populations of all the other species of rhino are small, some of the Asian ones, tiny, and there are now only two northern white rhinos left. Ignoring all the other factors, does the genetics mean that they are all doomed? The answer is 'No' provided they can escape what is known as an 'inbreeding depression'. We know that some genes are nasty, like those causing cystic fibrosis, for example. If the species is lucky, the small populations, remnants of the large populations, will successfully breed through the problem, breeding out the nasty genes and leaving the remaining population healthy and able to expand again. This process is known as genetic purging, and it *may account for why sea mammals, that went to the brink of extinction, have recovered remarkably well*.[35] It could well be what happened in the case of the white rhinos, and we must hope it will happen with the rest of them.

But can northern white rhinos be saved? Some of their genetic material is stored in San Diego Zoo Institute for Conservation Research's Frozen Zoo, and the results of a study to see whether the genes would be diverse enough to successfully establish a recovered population of the animals were published in May 2018. They were described as *very promising*, *the genetic diversity among the nine northern individuals was surprisingly high, nearly equivalent to that of*

---

[34] Freeman, Scott and Jon C. Herron 1998 *Evolutionary analysis. Upper Saddle River, New-Jersey* Prentice-Hall.

[35] Professor Sir Patrick Bateson (n.32).

*the southern rhinos-which came from a much bigger wild pool', 'the results suggest that from a genetic standpoint at least, a recovery scenario is still very possible'.*[36]

With this encouraging result, and if a genetic rescue succeeds, it could be possible for the northern white rhinos to recover to the same extent as the southern white rhinos did so long ago.

## 36.5   Sustainable Use

All countries are different, different peoples, different groups of people within the same country, different cultures and so on, which means of course that each country will have its own form of sustainable use, of raising possibly badly needed income. Our story will concentrate on two key methods of raising income, eco- or photo-tourism and trophy hunting, although whether the latter can be considered to be sustainable use continues to provoke fierce debate.

South Africa's wildlife industry is *'one of the most developed in the world'*[37] and has been described as *'a conservation revolution'.*[38] There, most wildlife thrives, in the national parks, the formal game reserves and on private property. Where *'the business of commercial game ranching or wildlife management … has contributed immeasurably to improvements in techniques of translocation, capture and immobilization, resulted in a resilient market in the sale of live animals, provided a source of protein for local and export markets, and encouraged the growth of a new professional class of wildlife managers and consultants'.*[39]

Most ranches compete to attract customers and maximize profits. They use the revenues to fund conservation activities. And they make a lot of money, which is partly why numbers of game animals on private land are steadily increasing at the expense of cattle, sheep and goats. The wildlife industry in South Africa is important; it is a viable economic enterprise.[40]

It was inevitable, therefore, that the industry would spawn a number of associations representing various facets:

---

[36] Jake Buehler *The Practically-Extinct Northern White Rhino Just Got Some Good News* Earther. See https://earther.com/the-practically-extinct-northern-white-rhin-just-got-s-182629365   Accessed   29/05/2018. The results were published in *Genomics Research*.

[37] Milliken T. and Shaw J. (2012) *The South Africa – Viet Nam Rhino Trade Nexus: A deadly combination of institutional lapses, corrupt wildlife industry professionals and Asian crime syndicates* TRAFFIC, Johannesburg, South Africa.

[38] Bothma et al. 2004 in ibid.

[39] Carruthers 2008 in ibid.

[40] Du Toit 2007 in ibid.

- Wildlife Ranching South Africa (WRSA) (which also includes the Private Rhino Owners Association): This liaises closely with game ranchers, NGOs and government and aims to ensure that both the ranchers and government agree on policies that will be implemented by the ranchers.[41]
- The Private Rhino Owners Association: Its primary object is *'to help preserve and secure rhinos under private ownership within South Africa through close collaboration with relevant law enforcement and government regulatory agencies'*. These owners had different visions of what to do with their rhinos, ranging from ecotourism and conservation to farming for live sales and trophy hunting, much more profitable.

The threat of increased rhino poaching brought them together, and they both developed National Rhino Security and Coordination Plan and organized a National Rhino Summit.

Other associations include:

- The Wildlife Translocation Association (WTA)
- The Veterinary Council
- The Veterinary Association
- The Professional Hunters Association of South Africa (PHASA)
- Two main taxidermy associations[42]

## 36.5.1 Community-Based Natural Resource Management (CBNRM)

We know that South Africa has a flourishing wildlife industry. We also know that most African rhinos live in South Africa and Namibia, both of them countries in southern Africa. And that it is in this region that the majority of savannah elephants are to be found. So what policies are in place to ensure that they are protected from use that is not sustainable? This is where community-based natural resource management (CBNRM) comes in. In effect, this is an integrated approach to resource management, its key components being:

- *'Sustainable use of wildlife resources*
- *Distribution of revenues to local communities'*

---

[41] WRSA 2009.
[42] See Milliken and Shaw (n.37).

And its aim is '*to devolve common property rights over use and wildlife resources into the hands of rural communities*'.[43]

Namibia, which is home to some wonderful wildlife, was not only the first African country to write environmental protection into its constitution, but it also introduced the concept of CBNRM.

Animals thrive there now, but it was not always like that. Namibia suffered under military occupation, was held in the grip of a severe drought and was ravaged by poachers. Fortunately for the wildlife and the Namibians, a home-grown organization had the brilliant idea which was to evolve into CBNRM. In 1990 the country became independent, and the communities gradually took shape. The new government needed a bit of help, but working with WWF, they were able to look at each community's specific needs and make sure they could all participate in the benefits of the conservancies, the wildlife resources in their area.

The communities have rights, both to manage and distribute the benefits of their conservancies. They need them, because they depend on them for their livelihoods and for their differing cultures. They are the stewards of their land and the wildlife on it, and their voice must be heard.

WWF has no doubt about the importance of communities to both wildlife and people:

> *In conservancies, wildlife is now a valued livelihood asset, with the result that poaching is no longer socially acceptable and populations of lions, cheetahs, black rhinos, zebras and other native species have been restored. Human welfare is also improving, thanks to an annual income of $5.5 million in annual income and benefits the conservancies for communities.*[44]

In Namibia, the income from sustainable use mainly comes from phototourism and trophy hunting. The penalties for poaching and illegal hunting, contained in the Nature Conservation Amendment Act No. 3 of 2017, are severe, with fines of up to N$25 million for elephant and rhino poachers and up to N$10 million for illegally hunting protected species with terms of imprisonment of 5 to 10 years.[45] This enables both communities and conservancies to thrive.

---

[43] Rodgers Lubilo *Sustainable use, local communities and tackling illegal wildlife trade in Managalane, Mozambique* SULi News, Issue 10. See http://www.iucn.org/about/union/commissions/ceesp_ssc_sustainable_use_and_liveli Accessed 13/08/2015.

[44] WWF 2016.

[45] *UK aid to crack down on criminal gangs driving the illegal wildlife trade* UK Government 2018. See https://www.gov.uk/government/news/uk-aid-to-crack-down-on-criminal-gangs-driving-the-illegal-wildlife-trade Accessed 11/10/2018.

WWF offered similar assistance in Mozambique several years later, to the Managalane community.

## Case Study 10: The Managalane Community and Sabie Game Park

This case study shows how even in a troubled country like Mozambique, CBNRM can turn a very difficult situation around.

Sabie Game Park is a private reserve that was established in 2000. Before that, it was home to the rural community of Managalane. No room for both people and Park. The people had to go, either displaced or relocated. However, they believed those who told them that their lives would improve. When this did not happen, they took to poaching instead.

Poaching did improve their incomes, particularly because the location proved to be an excellent transit point into Kruger National Park. External gangs of poachers took full advantage of this. All those rhinos and some elephants just waiting to be 'harvested'! And they 'bought' the local community with cell phones and money. The youngsters were particularly pleased and were quite prepared to join in.

It was inevitable that deaths and disappearances would follow. Some of the 'recruited poachers' were shot by rangers; others simply did not come back. Their fate remained unknown. Widows and orphans were left. Something had to be done if both people and animals were to be kept safe. Once again working together provided the answer. This time the partnership was between WWF-South Africa, Southern Wildlife College (a CBNRM unit) and Sabie Game Park.

The first step was to organize the communities into five village conservation areas, each with an elected committee to spearhead the programme that started with CBNRM governance training. As a baseline of information was required, situational analysis and livelihood and community governance surveys were conducted. The community awareness programme explained to the local people that wildlife was potentially valuable. Sabie Game Park was generating a considerable amount of money.

The people listened and decided they too wanted to be part of CBNRM. The correct community structures were in place; all that was needed was an injection of money. This came in 2014, from the Mozambique government, 20% of the revenues generated by trophy hunting. It went to 382 households, and it was the share that they had waited to get for 7 long years.

That changed everything. Before the initiative, the community had nothing but complaints about the Park, about human-wildlife conflict; indeed, over 70% of them did not like wildlife and did not want the Park. A year later, by 2015, attitudes had changed completely, and over 90% had decided that wildlife was, in fact, valuable. Furthermore, poaching had *drastically reduced*.[46]

And things got even better. The community set up a voluntary community scouts programme, whereby 21 young men (some reformed poachers) and women would *closely monitor any illegal wildlife and domestic crimes such as theft of cattle*. They graduated in summer 2015 and now carry out neighbourhood watch duties in their community to make sure there is no illegal access to wildlife and that no illegal hunting occurs.

And now Sabie Game Park is proving to be a good neighbour, with over 35 members of the community employed there full time, directly benefitting over 150 local people. It has also embarked on an ambitious programme working with the Southern African Wildlife College:

- To improve community governance
- To devolve revenues
- To expand the wildlife economy through investment in wildlife areas

This requires land to be set aside for community conservancy, land which the people can access and use for trophy hunting and things that bring in tourists.[47]

## 36.5.2   Kenya's Community Conservancies and the Northern Rangelands Trust (NRT)

The Kenyan community conservancies manage their wildlife resources in a rather different manner. Northern Kenya presented a challenge. There were a number of pastoral communities, ethnic rivalries were rife, and the land was difficult to farm because of frequent droughts and its poor quality. There was little help from the government, and dangers came from elephant poachers and cattle rustlers. Yet out of these inauspicious beginnings has grown a story of amazing success. A new movement based on community conservation was established, whereby:

- Areas of land are collectively owned and managed by the resident communities.
- They incorporate conservation and sustainable livestock practices alongside traditional land use.

---

[46] See Rodgers (n.43).
[47] Ibid.

- The revenues generated benefit the communities through education bursaries, healthcare and infrastructure development.
- They are governed by autonomous, legally registered institutions with democratically elected boards.

Recognized by the Kenyan government, these community conservancies were given legal status in 2013, in the Wildlife Conservation and Management Act.[48]

As well as rights, all members have responsibilities, which include managing the conservancy to good standards and to agreeing and implementing conservation and development programmes to these standards. As for benefits, which are essential if the conservancy is to thrive, they must be adequate, social as well as economic, and they must be distributed fairly to the members of the community, going hand in hand with changed attitudes and perceptions.

The first conservancy was established in 1995, and by 2014 there were 27, but it was obvious that if the emerging conservancies were to thrive, they would need some form of support. This has been provided by the Northern Rangelands Trust, an umbrella organization whose mission is:

*To develop resilient community conservancies that transform people's lives, secure peace and conserve natural resources.*[49]

Among other matters this includes:

- Setting standards
- Fund raising
- Training
- Satisfying the requirements of donors, investors and government funders

It has also set up a trading company that it owns in its entirety.

Every conservancy must enter into a memorandum of understanding with the Trust.

As many of the conservancies are made up of more than one ethnic group, democratic representation, which includes women, is essential on each board of directors. Their functions are:

- To employ a manager, administrators and rangers

---

[48] NRT *State of Conservancies* 2014.
[49] Ibid, p.8.

- To ensure that tourism, finance and grazing committees are elected
- To ensure the accounts are independently audited at the end of each financial year

By 2014, the conservancies were *'helping to improve the lives of over 250,000 people ... across nearly 32,000 sq. km of Northern Kenya'*. Samburu, home to Save the Elephants, is also an NRT conservancy, while both Lewa Wildlife Conservancy and Ol Pejeta are partners.

### 36.5.3   The Social Assessment of Protected Areas (SAPA) Initiative

Ol Pejeta Conservancy has its own programme in this area, SAPA, whose objective is to benefit the poorer members of the Conservancy. An international programme, it aims to improve their lives, and it is currently working with 18 of its local communities.

The Community Development Programme supports a wide range of projects, which include:

- Support for schools
- Bursaries
- Support for health centres
- Provision of water
- Sustainable agriculture[50]

The Dynamic Conservation Education Programme enables more than 20,000 school children to experience wildlife, often for the first time.[51] The hope is that many of them will want to become wildlife defenders.

In Kenya therefore, apart from agriculture, a major source of income is wildlife-based tourism, which provides every member of the conservancies with an incentive to protect their precious wildlife, which includes savannah elephants, black rhinos and white rhinos. A sustainable use of the land and its resources is not controversial.

We can see quite clearly now the important role local communities are playing in saving the elephants and rhinos they coexist with. But to want to protect them, they need to live in harmony with the animals. Yet they are the

---

[50] Foster
[51] Foster

people most likely to suffer any adverse consequences where this does not happen. This is where sustainable use comes in.

Essentially, it is all down to economics. If local communities can benefit financially from the presence of elephants and rhinos living alongside them and the benefits that flow are considerable and are a direct result of having these animals alive rather than dead, then the people have a major incentive for keeping these animals alive. Wildlife, and its management, must generate an income for the local people. So sustainable use if it is going to work must:

- Sustain: Better still, increase wildlife populations
- Generate: High revenues compared to costs
- Provide: Benefits to the local communities[52]

---

[52] Pack, S.P., Golden, R.E., Walker, A. 2013. *Comparison of national wildlife game management strategies: What works where and why.* Consultant Report from the University of Maryland Sustainable Development and Conservation Biology Program to the Heirig Centre for Science, Economics and the Environment.

# 37

# Sustainable Use: Contentious Issues

Practical dilemmas arise from the principle of sustainable use, dilemmas that have a direct bearing on the future of elephants and rhinos. They arise because wildlife, or rather its sustainable use, needs to generate income, money, which can be spent not only on conserving the animals but also on improving the lives of the local communities who live alongside them. Money gives local communities an incentive to protect their wildlife.

## 37.1 A Rapid Review of the Background to Trophy Hunting

The tragic death of Cecil the Lion and the appalling manner in which it was brought about provoked widespread revulsion throughout the world and stimulated fresh discussion on the issues raised. For many people, the very idea of trophy hunting is abhorrent, and we have already seen the ambivalent role it has played in the conservation of rhinos, how even a well-regulated industry can be exploited and corrupted. But even a single trophy hunt can bring in much needed income, money that can pay rangers or provide fencing to keep the poachers out and the animals in.

Many countries in Africa permit trophy hunting. Kenya abandoned it years ago, but it continues to be popular in South Africa, Namibia and Zimbabwe, and the South African model of wildlife management provides the reason why.

Our starting point is the land. It is all too easy to have a romantic view of the land in Africa, especially if we watch films and television documentaries. The vast plains, trees and shrubs, with animals and occasional people wandering through.

© Springer Nature Switzerland AG 2019
B. Martin, *Survival or Extinction?*, https://doi.org/10.1007/978-3-030-13293-4_37

All very photogenic. But this is not the reality, especially in South Africa. There, the towns and cities are increasing in size, encroaching into the countryside. Outside them is the communal land, much of it used for agriculture, much of it grazed, even overgrazed. And then there are the game farms and the parks.

Back in 1960 the situation was not looking good. Not much wildlife, too much grazing. Something had to be done. So there was a change in land use, a retreat from agriculture back to wildlife. To be accurate, to game farms and game parks. Back to enclosing the land. So has it worked? Can people make a living from wildlife? The simple answer is 'yes'.

- Game farms are managed for their wildlife. They have to be if they are to be successful. There must be sufficient water holes, good grazing and fences.
- Fences are a key component. They keep in the wildlife and help protect it from poachers. And in the case of lions, they separate the lions from the pastoralists.
- The governments of South Africa, Namibia and Zimbabwe have all given private landowners full control over the animals on their land. This enables them to take advantage of the most profitable uses, which turn out to be connected with tourism and wildlife trade (a different form of agriculture if you like).
- Wildlife and its management generate significantly higher income than livestock ranching, which also produces more negative environmental impacts.[1]
- Because this form of management requires the natural and semi-natural habitat to be properly maintained if their animals are to thrive, it enables wildlife generally to thrive and has led to the over-all recovery of some populations.[2]

So in South Africa, '*wildlife management centers on the privatization and commercialization of wildlife resources, with devolution of rights over wildlife to private landowners and local communities*'.[3]

But to what extent do local communities benefit? In Zimbabwe, Operation Campfire certainly brought benefits to local communities. Monies from hunting permits contributed to health and education projects benefitting the villagers involved, as well as to conservation. It gave a sense of ownership to them. There was a quid pro quo.

---

[1] Pack, S.P., Golden, R.E., Walker, A. 2013. *Comparison of national wildlife game management strategies: What works where and why.* Consultant Report from the University of Maryland Sustainable Development and Conservation Biology Program to the Heirig Centre for Science, Economics and the Environment.
[2] Muir-Leresche and Nelson 2000 in ibid.
[3] Ibid.

However, research published in 2013 by the pro-hunting International Council for Game and Wildlife Conservation and the neutral UN Food and Agriculture Organisation suggested there was little benefit, concluding that '*authors from all sides of hunting and conservation debates agree that local communities… generally receive minimal benefits from trophy hunting*'.[4]

In 2014, Botswana banned elephant trophy hunting. Elephants then flooded into that country, so many of them that they became a nuisance to some of the population, and now, in 2018, the president is consulting on whether trophy hunting should be reintroduced.[5] The question of just how beneficial trophy hunting is to local communities is likely to rage on.

## 37.2   Lifting the Moratorium on Domestic Sales of Rhino Horn

We start with the following points:

- There is a critical difference between rhino horn and elephant ivory.
- Rhino horns can be harvested without injuring the animals, and they grow back.
- In theory, rhinos could be farmed for their horns; no animals need to be killed.
- Elephant tusks can be harvested naturally as well, but only from dead animals. That is why poachers have to kill them.
- Rhino poaching has again skyrocketed to record levels. But
- The large stockpiles of legally held rhino horn could, even if only a small portion was sold, provide a lifeline by way of funding to help protect the living rhinos from poachers.

In 2009, after the abusive use of rhino trophy hunting had been exposed, the South African government imposed a moratorium on domestic trade in rhino horn. Needless to say, this was not popular with everyone, so in 2015, the South African Department of Environmental Affairs (DEA) appointed a committee of experts, including conservationists, scientists and immigration authorities:

---

[4] Economists at Large 2013 *The $200 million question: How much does trophy hunting really contribute to African communities?* A report for the African Lion Coalition prepared by Economists at Large, Melbourne, Australia. Lead author – Roderick Campbell.

[5] Jane Flanagan *87 elephants killed in poaching frenzy* The Times, 5 September 2018.

- To see whether legal trade in the horn was viable
- To identify '*additional measures to curb the illegal killings*' which would include '*enhanced intelligence to break up syndicates*'[6]

Matters finally came to a head later that year, when two South African game breeders, John Hume and Johan Kruger, major players, took the government to court in an attempt to overturn the moratorium. Hume, who owned more than 1500 rhinos which he regularly dehorned in an attempt to keep poaching to a minimum, argued that:

- Allowing legal sales of rhino horns could help end the illegal trade
- It was their constitutional right to sell the horn, which was '*a renewable resource*' (sustainable use)

However, the court documents clearly show that the government intended to keep the ban:

- '*It (the government) is of the view that, at present, the moratorium constitutes a positive step toward the conservation of rhino*'.[7]

The Pretoria High Court found for the breeders because the government hadn't followed the correct procedure on public consultation before imposing the moratorium. The ban would have been lifted, but the government appealed.

In January 2016, the South African High Court dismissed the appeal, so did the Supreme Court of Appeal in May.[8] Although the ban was constitutional, there had been a procedural fault.[9]

This did not affect the fact that all international trade continued to be banned under CITES.

That same year, CITES CoP17 was held, and although South Africa did not make any attempt to change the status quo, Swaziland submitted a proposal that would:

---

[6] *South Africa weighs legal rhino horn trade* Phys.org, 10 February 2015. See https://phys.org/news/2015-02-south-africa-legal-rhino-horn.html Accessed 28/02/2017.

[7] *S. African breeders ask court to end rhino horn trade ban* Phys.org, 22 September 2015. See https://phys.org/news/2015-09-african-breeders-court-rhino-horn.html Accessed 28/02/2017.

[8] *S. African appeal court backs domestic rhino horn trade* Phys.org, 23 May 2016. See https://phys.org/news/2016-05-safrican-appeal-court-domsetic-rhino.html Accessed 28/02/2017.

[9] Bryan Christy *South Africa Just Lifted Its Ban on the Rhino Horn Trade* National Geographic, 23 May 2016. See http://news.nationalgeographic.com/2016/05/160523-rhino-horn-ban... Accessed 31/05/2016.

- *'Permit a limited and regulated trade in white rhino horn which has been collected in the past from natural deaths, or recovered from poached Swazi rhino, as well as horn to be harvested in a non-lethal way from a limited number of white rhino in the future in Swaziland'.*

The Conference overwhelmingly rejected the proposal, but the arguments rumbled on, and in February 2017 matters took an unexpected turn when South Africa's DEA dropped a bombshell by publishing draft legislation:

*To regulate the domestic selling or otherwise trading in, giving, donating, buying, receiving, accepting as a gift or donation, or in any way disposing or acquiring, rhinoceros horn within the borders of the republic, and the export of rhinoceros horn for personal purposes, from the republic.*

The regulations would cover the horn and any part, product or derivative of it.

Although these activities would still be restricted under the Biodiversity Act, people could apply for a permit to obtain permission. However, strict conditions would have to be met. Applicants would have to supply:

- Proof that the horn was acquired and possessed legally
- Photographs of the horn
- Details of the horn itself in the form of a microchip or serial number

Applicants other than South Africans or people not living permanently in the country would have to provide documentary evidence that the horn would not be used in ways that contravene CITES. And breaches of the regulations would be punishable thus:

- A first offence: a maximum fine of R5 million or a maximum prison sentence of 5 years.
- A second offence: the maximum penalties would be doubled.

The public were given 30 days to comment on these draft regulations.[10] The NGO Conservation Action, describing the developments to be *'of serious concern'*, immediately raised a number of important issues, such as:

---

[10] Matthew Savides *Minister paves way to legalise rhino horn trade* HeraldLIVE, 9 February 2017. See http://www.heralslive.co.za/news/2017/02/09/minister-paves-way-legalise-rhino-horn...     Accessed 14/02/2017.

- Whom did the Minister consult when drawing up the draft regulations?
- How were they to be enforced and where would the funding come from?
- Had '*the pro-trade breeding and hunting lobby*' been involved?
- What guarantee was there that the horns will not enter the illegal markets?
- Why were the penalties not more severe?

Most important of all:

- Why had no distinction been made between the more numerous white rhinos and the critically endangered black rhinos?[11]

If this was a deliberate omission rather than a mere oversight, what effect might it have on the fragile population of black rhinos in South Africa?

Anticipating the need for clarification of international sales should the proposed regulations be passed, in March the CITES Secretariat issued a Q & A, making clear they were not '*offering any comments on the proposed measures themselves*'.[12]

In April 2017, the Constitutional Court dismissed the government's appeal, so that once again, domestic trade in rhino horn was legal.

In July, two things happened.

At a press briefing, the government, in the form of the Honourable Minister of Environmental Affairs, confirmed that the draft regulations were still under review by the DEA and clarified some of the provisions, stating that she, the Minister, would personally be the issuing authority for permit applications relating to the sale and purchase of rhino horn, with the proviso, of course, that the nine provinces agreed. South Africa would still comply with CITES.

And John Hume announced that he would be holding an online auction of part of his huge stockpile of rhino horn (264 horns weighing about 500 kg) in August, followed by a live auction in September.[13] Once again the DEA tried to delay the inevitable, failing to grant him the requisite permit that would allow the auction to be held, so, at very last minute, Hume again went to the court, again winning so that he was finally issued with his permit.

---

[11] Ian Michler *Is the SA government preparing for a legal trade proposal in 2019?* Conservation Action, February 2017. See https://conservationaction.co.za/resources/reports/sa-government-preparing-legal-trad... Accessed 28/02/2017.

[12] See        https://cites.org/eng/news/Background_issuance_CITES_permits_export_of_rhinoceros_horn_15032017 Accessed 28/02/2017.

[13] Tony Carnie *Rhino baron shifts blame for 'disappointing' first horn auction* timeslive 26 August 2017 See https://www.timeslive.co.za/news/sci-tech/2017-08-26-rhino-baron- Accessed 28/02/2017.

However, in a brief statement, even his lawyer admitted the auction had been a disappointment, with *'fewer bidders and fewer sales than anticipated'*.[14]

In the meantime, a disturbing new development had occurred. Instead of attempting to smuggle out whole or pieces of rhino horn, which at least had a chance of being detected, the Asian trafficking networks based in South Africa had started to turn the horn into large beads, often disguised in necklaces and bracelets, and *'disks'*, as well as *'packaging offcuts, shavings and rhino horn powder in bags'*, so much easier to hide.[15]

While it is too soon for sufficient data to have been collected, it is easy to see that this inevitably presents another opportunity to launder in poached rhino horn piggy-backing on the legal trade.

In March 2018, Michael Eustace, Director of the Majete, Liwonde and Nkhotakota National Parks in Malawi and Bangweulu National Park in Zambia wrote to the Times, calling for regulated trade in rhino horn. In his opinion, if South Africa could sell part of its large stock of legally obtained rhino horn, enough to satisfy the *'Far Eastern annual demand for about 1,200 horn sets'*, this would, in effect, stop the killing of both rhinos and poachers; there would be no need. And it would be sustainable.[16]

Further auction sales were delayed because it was taking so long for the requisite paperwork to be completed, and, in June 2018, John Hume started looking for *'a buyer to take a stake of up to 50 per cent in his farm and assets'*. A wealthy buyer, as he valued his assets at £45 million. They include 1600 rhinos including more than 300 pregnant females which are costing him £302,000 a month including measures to protect them from poachers.[17] He was running out money, having spent all his savings on protecting his precious herd. But rhinos, especially almost 2,000 of them, take a lot of protecting when rhino horn is trading at £75,000 on the black market.

And in July, Bella the white rhino was killed. The previous week she had been dehorned to protect her from poachers, so there was only a tiny stump left. But the greedy poachers were determined to take even this from her, slicing her face open to do it. Viciously slaughtering a 20-year-old rhino with a young calf, Tank, 16 months old and still dependent on his mother for milk. A dead mother and a tiny orphan for just 1 cm of horn stump.

---

[14] Save the Rhino *John Hume's rhino horn auction*. See https://www.savetherhino.org/rhino_info/thorny_issues/johm_hume Accessed 28/02/2017.

[15] See Moneron, S., Okes, N. and Rademeyer, J. (2017) *Pendants, Powder and Pathways* TRAFFIC, East/Southern Africa Regional Office, Hatfield, Pretoria, South Africa, pp.12–13.

[16] Michael Eustace *Call for regulated trade in rhino horn* Letter to The Times, 26 April 2018.

[17] Harriet Salem *Rhino breeder seeks cash to save £45m herd* The Times, 22 June 2018.

So what will happen now? We can only wait and see, but we do know what seems to have happened as a result of the one-off sales of elephant ivory. It's not just John Hume and other game breeders who argue that the money raised can be spent protecting the rhinos, covering vets' fees and fencing costs; others with no financial incentive also see a possibility of using money from the stockpiles to help ensure the rhinos' survival. But there is no guarantee that illegal horn cannot be laundered into some of the sales, and surely it can only damage all the efforts at demand reduction that are making such successful progress in, for example, Viet Nam. No demand has to be the ultimate aim. No demand, no fortunes made from illegal trade; no trade, no need for poachers; no poachers, no poaching; no poaching, no need for very expensive measures to protect rhinos from poachers; and the rhinos would/should then be safe. Win-win all round. Enough money would then be raised from eco-tourism. Wouldn't it?

## 37.3   Banning Sales of Elephant Ivory

Here, sustainable use is about tourism. Will banning sales of ivory reduce or even eliminate poaching so that there will still be elephants to attract the tourists?

The 1989 ban on international ivory sales was a success. It brought the earlier frenzy of poaching to an end. The herds began to recover, and in the case of the savannah elephant, numbers began to grow to such an extent that, in effect, some countries could be said to have a surfeit of elephants. There were so many in Kruger National Park that they were even beginning to put a strain on the ecosystem.

Gradually, large stockpiles of legally obtained ivory were accumulated, ivory from elephants that had died of natural causes or perhaps had been killed under licencing provisions because they were causing damage to agricultural crops. These stockpiles were held by the governments of each country. They were worth considerable sums of money.

Eventually (as we know), the CITES CoPs voted to allow two one-off sales of these stockpiles. The producer beneficiaries were Botswana, Namibia, Zimbabwe and South Africa, and the resulting monies were used to help both conservation and the local communities.

It was sustainable use in practice, and it should have worked well. Except that it did not. Although some good did come of it, circumstantial evidence pointed to a darker outcome, the seemingly inexorable rise of poaching, with enormous numbers of elephants most cruelly slaughtered so that unless the situation improves, some elephant populations may become extinct.

Ivory has become such a desirable commodity that its monetary value has risen and skyrocketed, and there are humans who are more than happy to supply the insatiable demand while, at the same time, making fortunes for themselves. So demand must somehow be reduced. Education is the best way of achieving this, but it takes time and patience. The 1989 ban on international sales swiftly started improving matters, but it left domestic markets untouched.

These are only now becoming regulated but lack of enforcement, and inevitable loopholes continue to enable poached ivory to enter the system. Problems also arise in relation to antique ivory, which used to be exempt from controls because governments had not realized just how many fake antiques there were and how difficult it was, in many instances, to identify the fake from the real thing. And there are difficult issues like moving family heirlooms from one country to another or the sale and movement of valuable musical instruments containing small pieces of ivory. Museum exchanges. Unintended consequences.

The USA was the first country to announce a ban on commercial trade in ivory and to pass new legislation accordingly, but the legislation is complex because there are certain exemptions and the federal regulations are constantly under review and can be changed at any time. This has caused problems for antique dealers, especially as important auction houses such as Sotheby's and Christie's are based there.

Despite their best efforts, people still get caught out. For example:

- In January 2016, four British antique dealers were in for an unpleasant surprise. On their way to the original Miami Beach show, an antiques fair, they were stopped by US customs. They were carrying contraband. The problem? A single shipment of silver and *objects of vertu* which had been intercepted just days before the show opened. The offending artefacts, which included Georgian and Victorian silver teapots, figures and even gold boxes, all contained parts of endangered species, including elephant ivory. The dealers had not mentioned this in the import documents because they had mistakenly thought the amounts were so small as to be insignificant. They were wrong and were forced to destroy, among other things, teapot and coffee pot handles '*to remove ivory insulators*' and the carved ivory hands and face of a beefeater by Berthold Muller.[18]

---

[18] Roland Arkell *UK dealers told to destroy ivory antiques in US* Antiques Trade Gazette, 25 February 2016. See https://www.antiquestradegazette.com/news/2016/uk-dealers-told-to Accessed 08/09/2016.

These were respectable business men, and they were shocked by their experience.

The ban has also affected loans between museums. For example, in October 2015, the British Museum was intending to lend some ivory religious reliefs dating from the Byzantine era to the Museum of Russian Icons in Massachusetts. But as they were denied an import permit for entry into the America, the British Museum revoked its loan.[19]

People who sell ivory in the UK often rely on the Antiques Trade Gazette for guidance, in an attempt to make sure their sales are legal, while others rely on a very popular television programme, the Antiques Road Show. Because the show's experts were receiving so many queries from viewers about the legal sale of ivory, working with the USFWS, they decided to issue their own guidance, which sets out the current legislation governing the possession, sale, transfer, import and export of elephant ivory. It is continuously monitored and updated.[20]

After the UK tightened up its definition of '*worked*' ivory, the EU strengthened its legislation, and this led to the successful prosecution of two reputable firms of auctioneers:

- Christie's were offering for sale an ivory tusk mounted on silver which was illegal to sell, because it no longer fell within the definition of '*worked*' ivory even though it was over 100 years old. The auctioneers admitted it was '*an honest and genuine mistake*', '*an isolated incident*', and confirmed that their robust training programme to identify illegal ivory was regularly reviewed. They were fined £3250.[21]
- The second auction house, based in Scotland, had also offered elephant ivory for sale, three lots, on their website, all of which included elephant tusks, obvious '*unworked*' ivory. They were fined £1500.[22]

Then in March 2017, a British antique dealer who was confused by the definition of '*worked*' ivory suffered a different kind of shock.

- Russ Allen, a specialist in an antique clocks and barometers, was horrified when his shop, Steam Mill Clocks, was invaded by a posse of police officers carrying Tasers and wearing protective stab vests. He had advertised a

[19] Mia Tomijima *Towering Ban on Ivory Trade* Center for Art Law, 1 October 2015. See https://itsartlaw.com/2015/10/01/towering-ban-on-ivoty-trade/ Accessed 04/08/2016.
[20] Ben Phelan *An Overview of Current Ivory Law* Antiques Road Show. See http://www.pbs.org/wgbh/roadshow/stories/articles/2015/06/22/ivor Accessed 04/08/2016.
[21] RSPB *Elephant ivory: English auctioneer fined* Legal Eagle, November 2016, No. 80.
[22] Ibid.

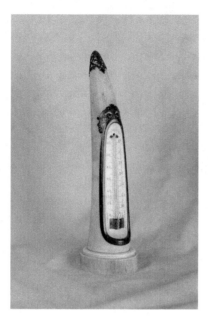

**Fig. 37.1** Russ Allen's Victorian ivory barometer and its base. Copyright Russ Allen

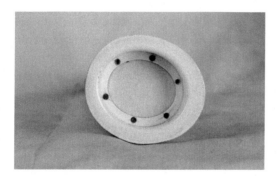

**Fig. 37.2** Russ Allen's Victorian ivory barometer and its base. Copyright Russ Allen

Victorian ivory barometer for sale on a website, unaware that this was illegal because he rarely came in contact with ivory. Arrested on suspicion of selling illegal ivory, he was given a community resolution order. Overkill perhaps on the part of the police, but they were acting on a tip-off and had no idea what they would find.[23]

---

[23] The Times *Antique experts demand clearer ivory law after raid* 4 March 2017. Mr. Allen also talked to the author.

In July 2016, delegates to the IUCN's World Conservation Congress (an international summit) eventually agreed a motion that called on countries to '*close the internal trade of ivory as a matter of urgency*'. The motion, WCC-2016-Res-011-EN: *Closure of domestic markets for elephant ivory*, stated that '*IUCN members are deeply concerned by the illegal killing of elephants that threatens the species' survival, harms national security and undermines sustainable economic development in countries where elephants are found in the wild*', and although it was not legally binding, most of '*the 217 state and national members of the IUCN, as well as 1,000 conservation groups that are part of the union*' backed it. The aim was to send a powerful message to those benefitting from the illegal trade.[24] A similar proposal was adopted at CoP17.

A complete ban? Not all countries agreed. Japan, Namibia and South Africa, for example, argued instead for better regulation. But Japan still has problems regulating its own domestic market, and both Namibia and South Africa have sizeable stockpiles so would benefit from any legal sustainable use of ivory.

A surprise contributor opposing the ban was the EU, which must have come as a shock to some of its Member States. At the beginning of July 2016, the European Commission issued a position paper which argued for '*a balanced position*' and stated that '*it would be better to encourage countries with growing elephant numbers to "sustainably manage" their populations rather than for a complete ban to be imposed*', although an official admitted it should be '*banned in those situations where it can facilitate illegal trade*'.[25]

The paper was not well received by many African States who were appalled by the decision, with Uganda, Cameroon, Kenya and the African Elephant Coalition (made up of 29 African States) warning of '*a mass extinction on the continent within 25 years*' unless the illegal trade was stopped.

A few days later, the UK, which we have already seen is one of the most important ivory markets, announced it too was imposing a ban on sales of elephant ivory. However, it was not a total ban, only applying to post—1947 items of ivory. But it did mean that anyone trading in antique ivory had to establish the ivory was older than 1947.

The measure met with fierce criticism, not just from conservationists but also scientists, politicians and even some celebrities. Prince William, patron of

[24] Oliver Milman *Ban on domestic ivory trade passes at international summit* The Guardian, 11 September 2016. See https://www.theguardian.com/environment/2016/sep/11/ivory-trade-international-agre Accessed 13/09/2016.

[25] Arthur Neslen *African wildlife officials appalled as EU opposes a total ban on ivory trade* The Guardian, 6 July 2016. See https://www.theguardian.com/environment/2016/jul/06/african-wildlife-officials-appa Accessed 27/09/2016.

TUSK, was particularly outspoken. Professor Jonathon Baillie explained why it was so important to close this loophole. '*Having this legal ivory trade [in pre-1947 ivory] masks the illegal trade … we want ivory to have no economic value so there is no incentive for people to poach elephants because there is nowhere to sell it*'. It would help China and other countries enact similar bans.[26] (Which, of course, they have now done, and in autumn 2017, the UK started to consult on whether there should now be an almost total ban on ivory sales in the UK.)

In October, CITES had encouraging news: ivory seizures were at their highest for 27 years, and poaching incidents had stabilized, even falling in some countries although they had risen in unstable countries such as the DRC and CAR. And there had been a 50% drop in the price of ivory, which, according to John Scanlon, CITES' Secretary General, was due to:

- '*The prospect of tougher enforcement*
- *Greater awareness about illegal wildlife products among the public*
- *A growing number of countries banning the trade in ivory*'[27]

But museums and historians in the UK explained that even something seemingly as clear-cut and popular as a total ban on ivory sales could have disastrous consequences of another kind, by damaging important collections. They feared that inability to trade in old objects made from or containing ivory would render them worthless and end up in their eventual destruction.

Among many irreplaceable collections in the British Museum are:

- The 3000-year-old Nimrud ivories from Iraq
- The Benin ivories, carved in Nigeria during the fifteenth and sixteenth centuries

A spokesperson for the Museum described ivory artefacts as '*integral parts of the collection*' and '*playing an indispensable part in the museum's presentation of the history of human cultural achievement*'. The Victoria and Albert (the V & A), another London Museum, voiced similar concerns. It too holds internationally important collections, including:

- Many miniature portraits painted on ivory from the eighteenth and nineteenth centuries, a common practice then.

---

[26] Ben Webster *Britain to crack down on illegal ivory trade, dealers must prove antiques are 70 years old* The Times, 21 September 2016.

[27] Aislinn Laing *Ivory seizures help to curb poaching* The Times, 25 October 2017.

The British Museum's argument against a total ban was that there was '*no public benefit in restricting the display or movement of ivory works of art made before 1947 and legislation should not extend to cover actions carried out before that date*'.[28]

In February 2017, the British House of Commons (Parliament) debated the UK's commercial ivory trade, concerned that the country should play its '*full part in increasing global efforts to halt poaching*'.[29] One of the speakers was the president of the British Antique Dealers Association. During the debate, it was pointed out that the British Museum:

- Has 13,000 ivory objects
- Loans ivory artefacts to countries throughout the world

One of the questions debated was:

- Should antiques continue to be exempt from any ban?

**Plan B?**
We will leave the last word to Dr. Daniel Stiles, Project Manager of Ol Pejeta, who has devoted so much of his life to saving endangered species. Once it became clear that a total ban on most legal ivory sales was fast becoming a reality and concerned that this might not be the answer the world was looking for, he asked:

- '*Is there a plan B for elephants? The next step in saving them is even harder*'.[30]

In 2015, he had led research into what was driving ivory demand in eastern Asia. The results showed, among other matters, that between 2009 and 2014:

- About 200 tonnes of illegal ivory was smuggled each year into China-Hong Kong SAR
- About 10 tonnes of it was processed each year, to satisfy consumer demand

---

[28] Gurpreet Narwan *Museums speak out against Duke's plea for total ban on ivory* The Times, 16 February 2017.

[29] Luke Hall (Conservative MP) in *Debate on Domestic Ivory Market* Hansard, 6 February 2017. See https://hansard.parliament.uk/commons/2017-02-06/debates/04A0F6F-C215-49EO-B     Accessed 09/02/2017.

[30] Daniel Stiles *Is there a plan B for elephants? The next step in saving them is even harder* The Guardian 19 November 2017. See https://www.theguardian.com/global-development-professionals-network/2016/nov/1 Accessed 12/01/2017.

- About 190 tonnes was probably being stockpiled by speculators, betting that prices would remain high
- About 3.5 tonnes of legal ivory, from the 2008 one-off sale was consumed annually in China. This means that China was consuming a total of
- About 13.5 tonnes of ivory each year
- About 3/4 of which was illegal.

When translated into items, more than 90% of these were therefore made from illegal ivory, because the legal ivory was mainly used to carve large and intricate artefacts rather than the small trinkets and pieces of jewellery that tourists tend to prefer.

Furthermore, as Dr. Stiles pointed out and as we have already seen, most illegal ivory sales now take place online, '*in members only chatrooms and through social media sites, which are difficult to monitor*'.

He concluded that:

> If it is speculation and stockpiling of tusks, rather than 'insatiable' consumer demand, that is driving poaching, plan B must take this inconvenient fact into consideration. Plan B had better be good.

## 37.3.1 What About Mammoth Ivory?

The Times carried a startling headline in August 2016, '*Mammoth hunters join new gold rush in Russia's arctic*'. Teams of men, '*tuskers*', were hunting for mammoth tusks which can fetch '*tens of thousands of dollars – and millions once they have been sold on and intricately carved*'.[31] While it is legal to collect and sell the tusks that are near the surface, the hunters were using banned methods to find their 'treasure', methods which were causing severe devastation to the ecosystem.

And in China, some traders have been passing off mammoth ivory as elephant ivory as well as using it as a means of laundering illegal elephant ivory into the market.

It would seem that mammoth ivory needs to be part of any solution.

---

[31] Tom Parfitt *Mammoth hunters join new gold rush in Russia's arctic* The Times, August 27 2016.

## 37.4   How Can We Make Sure that Local Communities Benefit from Protecting and Conserving Elephants and Rhinos?

It has become quite clear that local communities should be the guardians of their wildlife; that they have an essential part to play in the conservation of both animals and their ecosystems; that it is not always easy for people and animals to coexist, which is why so much is being done to try to minimize human-elephant conflict (this comes later in our story); that poverty might drive people into the arms of the organized crime gangs, into the dangerous world of 'easy-money' poaching appears to offer; and that not everybody would want to protect the animals they live alongside.

It is also clear just how deeply local communities do care about their animals. The obvious example is the rangers who, together with their families, are prepared to suffer hardship, even death, giving their all.

Admiration and praise, however, are not enough. If the world wants local communities to be the guardians of their and the world's wildlife, the world must help to cover the costs, particularly the financial ones. So where do monies come from?

Some communities benefit from wildlife-watching tourism, which generates a number of opportunities, including:

- Employment: In accommodation, restaurants and guiding.
- Supplying goods and services: Mostly food and drink.
- Indirect tourism benefits: They sometimes receive these through redistribution of revenues from protected area entrance fees and funds allocated to community development projects.

While this sector continues to grow, it is not enough to sustain many communities, and there are still countries where this tourism is very underdeveloped or non-existent or the countries themselves are too unstable.

Sometimes foreign governments help. For example, the UK government has a Department of International Development, one of whose functions is to administer overseas aid. During the Parliamentary debate (mentioned above), a number of speakers made it clear that they linked the sustainable development of local communities with conservation and that they would like aid money directed to further this end.

Elephants and rhinos are iconic species. They go back almost as far as the dinosaurs. They are part of the world's heritage, and it behoves us to make sure that, unlike the dinosaurs, they continue to exist and that they are here for future generations to enjoy. We cannot have it said of us that it was on our watch that the last elephant died, rhinos finally became extinct. What a legacy! So what can be done?

Perhaps we, the peoples of the world, should enter into a legally binding contract with the local communities.

A 'contract' is a legal transaction in which there is a 'quid pro quo'. In other words:

- Both sides get a benefit.
- Both sides suffer a loss.

This is how it could work:

- The benefit to us, the peoples of the world, would be that planet earth continues to have elephants and rhinos living on it for generations to come.
- The loss would be that we, collectively, would have to keep paying the local communities a realistic sum of money for them to achieve this and give them a decent standard of living, while the benefit to the local communities would be that the monies paid to them would enable them not only to protect the animals but also have a decent standard of living.
- The loss would be that they would have to do whatever was necessary to protect the animals and keep them safe.

A possible solution? Something to think about anyway.

# Part XIII

## Seeds of Hope

But there are many reasons to be encouraged. Many people, organizations and countries share a deep concern for and commitment to our beleaguered animals, and such passionate caring can produce game-changing results. Here, then, are some seeds of hope.

# 38

## People

Within such limited space, it is impossible to include everybody who is helping in the battle to save elephants and rhinos. But some of them are here, mentioned in this story, and we salute the many others out there, all of them unstinting with their time and efforts and all of them determined to save these precious animals.

And they do it because they care. Mike Chase, the lead survey investigator for the Great Elephant Census, explained that he was *a stubborn optimist* who did not want to spend his life *sharing depressing statistics and fighting a losing battle for elephants*.[1]

People have made and continue to make enormous contributions.

## 38.1 Trailblazers

We will start with two early pioneers:

**David and Delia Craig (1924–2009 and 1924–2014):** Delia, a Kenyan, and her husband David farmed in the lee of Mount Kenya, and some 30 years ago, they established the wildlife conservancy at Lewa, carving it from the enormous cattle ranch Delia had inherited from her father. The Craigs were early conservationists and David had remarkable foresight. Probably his finest achievement and lasting legacy was that he understood the link between

---

[1] Christina Russo *Exclusive: This Is Africa's New Elephant Poaching Hot Spot* National Geographic, 5 July 2016.

See http://www.nationalgeographic.com/2016/07/angola-elephants-great. Accessed 11/04/2017.

© Springer Nature Switzerland AG 2019
B. Martin, *Survival or Extinction?*, https://doi.org/10.1007/978-3-030-13293-4_38

conservation and local communities. If species and habitats were to be successfully protected, it would have to bring visible benefits to the local communities as well.

In fact, it is impossible to overestimate the conservation achievements of this intrepid pair and their son Ian, who is still actively involved. Anna Merz, another exceptional conservationist, persuaded them to fence off 5000 acres of former marshland for the black rhino sanctuary at Ngare Sergoi and established the Lewa Wildlife Conservancy.[2]

Ian Craig became Lewa's first Executive Director and, after he stepped down in 2009, its Strategic Adviser. He was also instrumental in forming the Northern Rangelands Trust and is now their Director of Conservation. In 2016, he was awarded an OBE by Queen Elizabeth II for services to conservation and security to communities in Kenya, a recognition of all his hard work and dedication. A family tradition, for his daughter Jessica, also a conservationist, has worked for the Panthera Africa Foundation and is married to Professor Jonathon Baillie, the Director of Conservation at ZSL.

Dr. Iain Douglas-Hamilton and his family are also conservationists. He founded Save the Elephants (STE) and has championed the cause of elephants in so many ways over the years, including long-term monitoring of the animals. His daughter Saba, also active in the field, is married to Frank Pope, Chief Operations Officer for Save the Elephants.

**Anna Merz (1931–2013):** Her name is synonymous with rhino conservation. Described in the literature of the Lewa Wildlife Conservancy as *'fearless, with a spine of steel and a wicked sense of humour'*, Anna Merz was devoted to her gentle, even-tempered animals, claiming *'you're far more likely to get kicked by a horse than charged by a rhino'*.[3]

This remarkable woman started out as a lawyer. A lawyer with a penchant for travelling, so she moved to Ghana to work. She was there in the 1970s when rhino poaching was rampant. And she was horrified by the plummeting numbers of black rhinos, by their terrible deaths, just so that their horns could be used for 'medication'. *'When we went to Kenya in '68, we saw rhinos everywhere … When we went back in '76 we saw corpses everywhere. The massacre was unbelievable … They were being killed at such a horrifying rate that I decided to do something'.*

---

[2] *Delia Craig: Obituary* The Times, July 2014.
[3] *Anna Merz: Obituary* The Telegraph, April 2013.

**Fig. 38.1** Two orphaned baby rhinos enjoying an exciting story with Jamie. Copyright Chris. Townend

She did. She convinced the Craigs of the need for the Ngare Sergoi Rhino Sanctuary, and once she had her land, she and the Craigs spent the next few years rounding up all the black rhinos in Northern Kenya, translocating them to the safe haven of the sanctuary. Obviously, they took time to settle in, so Anna used to read Shakespeare to them every day, for hours at a time. Fair enough, for she considered them to be at least as intelligent as chimpanzees. This, she claimed, *'completely tamed them'*. For once, we can turn away from the horrors and dwell on a most delightful picture.

Anna became very attached to Samia, a black rhino calf who had been abandoned by her mother. She reared the little rhino by hand. It took 3 years, and, like many young animals, Samia often required attention at night. So Anna, acting as a surrogate mother, eventually moved her into her own bed. Her husband retreated to his office, a baby rhino with a severe diarrhoea problem proving just too much for him! Samia later returned to the herd and to the David Sheldrick Wildlife Trust where she spent the rest of her life.

By the time of Anna's death, 66 rhinos were living in the Sanctuary, 36 of them black and 30 white. A remarkable achievement.

**Ian Parker (1927–2014):** *'The screams of agony from rhino that have had their horns chopped off while still alive should reach out into the hearts of all of us'.*[4]

---

[4] *Ian Parker: Obituary* The Times, August 2014.

This was Ian Parker's reaction to the dreadful deaths of over 1000 rhinos slaughtered by poachers in 2014. One of South Africa's leading conservationists, he was responsible for Operation Rhino, a game-changer when it came to protecting the future of white rhinos.

By the time Dr. Parker became senior warden in Hluhluwe-Imfolozi Park in the 1960s, white rhino numbers had clawed their way back to about 500 living there. Although the park was South Africa's oldest nature reserve, he realized that its close proximity to Durban potentially made the rhinos vulnerable, and he considered that if their survival was to be secured, some of the animals should be relocated to other reserves. He therefore set about convincing the authorities of this. Major battles ensued, some with bureaucrats, some with rhinos. Dr. Parker became expert in handling both.[5]

Rhinos are not the easiest of animals to move about. Translocation was, and continues to be even now, a very difficult and dangerous operation particularly for the rhinos. However, he worked out how to dart the animals with immobilizing drugs, a skill he gained as a child shooting airguns and catapults while exploring in the family car. Once darted he would follow the rhino, either on foot or on horseback, until it could be handled. Apparently *'he could often be found grappling with a rhino's tail, his knees wrapped with makeshift pads of cushions. 'You can imagine the difficulties … having to avoid those horns, which could easily disembowel a man'.* But it was a misfired dart, not a rhino, that caused most damage. He lost the sight in of one of his eyes.

Translocation proved to be a great success. Over 3000 white rhinos were moved, not only to other places in southern Africa, but even further afield to zoos like Whipsnade (ZSL). And it continues to play a vital role in rhino conservation.

Ian Player never stopped protecting rhinos. When he could no longer walk, he continued to campaign and raise awareness with his pen, writing of the wanton cruelty and terrible deaths meted out to these wonderful creatures just to rob them of their horns.

**Mark Shand (1951–2014):** This conservationist, who championed Asian elephants, died too young. He was only 62.

It was Tara the little female elephant, who changed the life of this born adventurer and swashbuckling hero. That was in 1988.

Tara was a captive elephant, whose owners used her to get money for them, by begging. He saw her and fell in love with her, love at first sight! He knew he had to take care of her, so he bought her. Then he rode her across India,

---

[5] Ibid.

from the Bay of Bengal to the Sonepur Mela, the ancient elephant trading fair at Patna on the Ganges, some 750 miles away. He wrote a book about his adventure, *Travels on My Elephant*. It turned out to be a bestseller and won him the prestigious award, Travel Writer of the Year. That was in 1992.

He then undertook another journey on the back of an elephant. A shorter journey, and this time he was accompanied by the only female elephant trainer in India, Parbato Barua. There was a BBC documentary and another prize-winning book followed. *Queen of the Elephants* won not only the Thomas Cook Travel Book Award but also the Prix Litteraire d'Amis. That was 1995/1996.

Driven by his passion, he founded Elephant Family, an NGO that is devoted to the long-term welfare and survival of the Asian elephant. That was in 2002.

It was Asian elephants that he found so attractive, and these are even more endangered than their African counterparts. Recognizing the harsh reality that *'every day an elephant kills a human being and a human being kills an elephant'*, he considered it to be *'our fault, because we humans have driven them away from their natural habitat. To cut the risk of human-elephant conflict and casualties, we are securing habitat all over Asia and purchasing corridors of land for elephants and helping local people relocate'*. Elephant corridors are yet another sign of hope.

He was able to harness his creative talents in an extraordinary demonstration of the frailty of existence. Some 250 elephants, 5 ft. tall, made of fibre-glass and decorated by celebrities were placed on plinths across London, only for them all to disappear mysteriously overnight later that year. A visual metaphor for extinction. The re-appeared elephants were later auctioned off to raise money. That was 2010.

Fortunately for Elephant Family, Mark Shand turned out to be rather good at fundraising. He had the right contacts. Royal connections, his elder sister Camilla is the Duchess of Cornwall, Prince Charles' wife as well as many good friends. By 2014, more than £6 million had been invested in a range of projects.

Only too well aware of the multitude of dangers facing wildlife, he accompanied the Prince for a time when he was visiting India, showing him the *'remains of Asia's incredible wildlife'*, highlighting *'the work of Elephant Family'*, making sure he understood the dangers of the wildlife trade. *'There has never been such an unprecedented threat to world wildlife. The illegal wildlife trade is now worth an estimated £6 billion a year, and it's growing'*. That was 2013.

It was a fall, a tragic accident at the end of a successful fundraising auction, raising money for both Elephant Family and underprivileged children, that led to his untimely death. He tripped and banged his head. He had a thin skull and the fall killed him. So very sad. That was 2014.

Mark Shand's legacy is Elephant Family.[6]

**The British Royal Family:** In addition to all their other duties, many of them play an active role in conservation, and they have made/continue to make a key contribution to the protection, not just of elephants and rhinos, but of wildlife in general. The Queen's current initiative is *The Queen's Commonwealth Canopy,* an ambitious project involving any commonwealth country wanting to participate. It will establish a unique network of established indigenous forests, precious habitat, raising awareness of their value, saving them for the future and encouraging more forests to be planted.[7] Prince Philip, the Duke of Edinburgh (the Queen's husband), was the first ever President of WWF-UK (from 1961 to 1981), then the first ever President of WWF-International (from 1981 to 1996) and, now, as the first ever President Emeritus WWF-International; he is still working hard for them. WWF has described his contribution as *'inestimable'*. Not only has he visited projects in many countries, because of his position, he has also been able to raise issues at the highest levels.[8]

His son Prince Charles, the Prince of Wales, is another respected environmentalist and conservationist. Way back in the 1970s he spoke, with prescience, on the potential dangers of plastic. He followed in his father's footsteps becoming President of WWF-UK in 2011, and, in 2014, he provided Tusk with vital funding that enable the NGO to set up a training programme for rangers, in antipoaching interception and information-gathering techniques. So far, 200 rangers across Africa have benefitted.[9]

Only too well aware of the need for sustainable development to support both people and countryside, he set up, for example, the Prince's Countryside Fund in 2010, a demonstration of this crucial linkage. He revealed the strength and depth of his understanding of the full implications of the illegal wildlife trade in a videoed speech for the London Conference, when he pleaded for greater efforts to be made to stop the trade that '*now poses a grave threat not only to the survival of some of the world's most treasured species but also to economic and political stability in many areas around the world'*.

---

[6] See The Telegraph, 24 April 2014 for Mark Shand's obituary.

[7] *The Queen's Green Planet* 12 April 2018. See https://queenscommonwealthcanopy.org/2018/04/12/queens-green-. Accessed 17/04/2018.

[8] WWF *Presidents Past and Present.*

[9] Charlie Mayhew MBE, CEO and co-founder of Tusk in *Sir, we salute you* The Daily Telegraph Magazine, 10 November 2018.

After Mark Shand's death, he and his wife became Joint Presidents of Elephant Family. Together they work tirelessly to promote its cause, and Elephant Family has described their patronage as having been *'crucial for Elephant Family's success so far'*, that *'they are deeply appreciative for their relentless support'*.[10]

Princes William and Harry, Charles' sons, are also passionately devoted to conservation, and *'supporting communities to protect and conserve their natural resources for future generations'* was an early priority for their Royal Foundation set up to channel *'their philanthropic activities, time and resources'*.

We have already seen how, in 2013, Prince Charles and the UK government hosted the Clarence House Conference, which took place in the Royal Residence of Clarence House and which led, a year later, to the London Conference on 'The Illegal Wildlife Trade'. All three Princes attended. The Kasane Conference in Botswana was followed by the Hanoi Conference, and Prince William went to both, thus ensuring they got good publicity. The Buckingham Palace Declaration, which set up a Transport Taskforce, came in March 2016. Buckingham Palace is the Queen's Royal Residence in London.

Prince William continues to seize his opportunities. For example, in March 2015 while visiting China, he delivered together with the Queen's invitation for President Xi Jinping to visit Britain, a speech condemning the illegal wildlife trade. And followed it up in October during President Xi's state visit, with another speech and a panel discussion held at Kings College, the University of London, on the protection of endangered species. Other panel members included Sir David Attenborough and Bear Grylls both of whom are well known in China, and everything was filmed by CCTV (the Chinese state broadcaster) as a special episode of *Let's Talk*, a weekly programme featuring a role model, and designed *'to give young people "realistic discussion and nourishment for the soul"'*.[11]

And in summer 2017, shock tactics were applied at the UK's famous Hampton Court Palace Flower Show. The Prince and Tusk (he is their patron) devised an exhibit to demonstrate to visitors why they should support the campaign to stop elephant poaching. *Not for Sale* consisted of an avenue of 200 artificial elephant tusks, with bones and bullet-riddled skulls strewn nearby.[12] In autumn 2018, it was the rhinos turn to benefit. The object of the Tusk Rhino Trail was to raise awareness about the threat of rhino poaching in

[10] Elephant Family *Who-we-are/royal-patronage*.
[11] Calum MacLeod *William uses state TV to save the tiger* The Times, 19 October 2015.
[12] The Times *Duke turns to shock tactics in battle to save elephants* 6 March 2017.

Africa, and to this end, 21 statues of the animals were placed in strategic locations across London.

While Prince William tends to deliver his message through diplomacy, Prince Harry, a former soldier and no stranger to war having fought on the frontline in Afghanistan, has now, among his many commitments, found time to fight on another frontline, the battle to protect elephants and rhinos; his experience enabling him to make suggestions about possible tactics.

As a wildlife conservation volunteer in summer 2015, shortly after leaving the army, the Prince spent 3 months working in Africa on a number of frontline conservation projects in Botswana, Namibia, South Africa and Tanzania. One of them included rhino dehorning in Namibia. Prince Harry worked his way up. Starting as a *mere* member of the helicopter team that tranquillized the rhinos, he progressed to monitoring heart rates and oxygen levels, taking blood and tissue samples and finally removing the horn with a chain saw.[13] He also managed to include a visit to a rhino orphanage.

He returned to South Africa in December that same year, where he revisited Kruger National Park and saw immediate evidence of the continuing poaching crisis, a mother white rhino and her calf, both dead, both with their horns hacked off. The Prince was visibly moved. *'Kruger was, he said, "one of the most beautiful places on Earth" but (it) has become "a major killing field"'.*[14]

And to coincide with his official visit to the Southern African Wildlife College, he also released a selection of his own photographs taken in the summer, each carefully captioned to explain why he took them. For example:

> *Trying to stop a three-tonne rhino with a rope and a blindfold isn't easy! Especially in this harsh terrain in Botswana. Mapp Ives and Kai Collins, with the help of Botswana Defence Force and the government, are doing everything they can to protect their newly reintroduced rhino population. This sometimes means having to sedate them to check on how they're doing.*[15]

Videos and photos all showing the depth of his concern and his appreciation of the situation. A revealing picture gallery.

Returning to Africa in summer 2016, Prince Harry took part in the *500 elephants* initiative, serving as *an integral part of our translocation team*. He also

---

[13] Caroline Davies *Prince Harry hugs elephant during Africa trip in plea to end poaching* The Guardian, 2 December 2015. See https://www.theguardian.com/uk-news/2015/dec/02/prince-harry-releases-images-from Accessed 30/03/2017.

[14] Ibid.

[15] Ibid.

helped translocate a male rhino. Peter Fearnhead, CEO of African Parks, spoke highly of him:

> He has had extensive field experience and was extremely comfortable with the animals, whether helping an anesthetized elephant to the ground and monitoring its breathing to affixing radio collars. He played a vital role in many aspects of this giant operation which requires not only all hands on deck, but a vigilance he exudes, and a commitment to the cause he embodies.[16]

In 2018, Prince Harry became the new President of African Parks.

## 38.1.1    A Sad Farewell

In 2017 and 2018, five deaths in particular stand out. Two of them were natural, three the result of murder.

**David Shepherd 1931–2017:**  A brilliant and very popular artist specializing in wildlife and other genres, his paintings sell for thousands of pounds. And his pictures of elephants, one of them '*The Ivory is Theirs*', are recognized over the world. Always an outspoken advocate for wildlife, in 1984 he founded the David Shepherd Wildlife Foundation which both raises awareness of wildlife issues, including the illegal trade, and funds conservation projects across Africa and Asia, including orphanages for elephants and rhinos. Concentrating on eight species, the Foundation makes a major contribution to wildlife conservation, carrying out a range of activities which include fighting to strengthen laws and their enforcement, working with local communities, education of various kinds and, of course, protecting its chosen species.[17]

In 1980 and 2008, David Shepherd was awarded first an OBE and then a CBE for his services to charity and wildlife conservation.

**Dr. Dame Daphne Sheldrick 1934–2018:**  Like some of our earlier trailblazers, she was born into a farming family in colonial Kenya. We have already seen some of her wonderful accomplishments, which were recognized in 2006 when she became the first person since Kenya's independence in 1963 to receive the title of Dame from the Queen. Married twice, after the premature

---

[16] African Parks *The Journey of Giants Prince Harry Takes Part in Historic '500 elephants'* 28 October 2016. See https://www.african-parks.org/newsroom/press-releases/the-journey. Accessed 30/03/2017.
[17] David Shepherd Wildlife Foundation *About DSWF – saving endangered species across Africa and Asia.*

death of her first husband, in 1977 she founded the David Sheldrick Wildlife Trust with the help of her eldest daughter Jill Woodley, and her Elephant Orphanage in Nairobi National Park grew from a request by park officials there. Would she look after two tiny elephant calves who had been orphaned by poachers?

Her approach to animals was unorthodox. She learnt at first-hand that elephants love, grieve and heal. That they are sentient beings. And she was a doughty fighter on their behalf. To quote Cynthia Moss, who founded and directs the Amboseli Elephant Research Project[18] and is another elephant champion, also wise in their ways as a result of all her years of research, *'she was a warrior, totally fearless. She stood up to government officials, scientists, anyone who threatened the elephants. Some said there's no point in saving individual animals, but of course there is. In order to curb the ivory trade, people had to first get to know and care about them'.*[19]

And she never relinquished her chairmanship of the Trust, even though she eventually handed over its running to her younger daughter Angela (Sheldrick) and her husband.

**Wayne Lotter 1965–2017:** Was tragically killed in his prime, deliberately gunned down while in a taxi returning from the airport in Dar es Salaam, Tanzania. Murdered, because he was so effective at his job, fighting international wildlife crime gangs, something he was well qualified to do.

Born in South Africa, after obtaining an MSc in Nature Conservation, Lotter became a ranger, gaining valuable experience and insights into what was happening in the field and rising to become invasive alien species programme coordinator for Ezemvelo KwaZulu-Natal Wildlife, a government wildlife organization. He also served as Vice President of the International Rangers Federation.

By 2006 he was in Tanzania, where, in 2009, he co-founded the PAMS Foundation (Practical Area Management System) which funded and supported Tanzania's National and Transnational Serious Crimes Investigations Unit (NTSCIU), realising that action was needed if the country's precious elephant population was to survive.[20] We have already seen how effective the poachers were in the Selous, and the Great Elephant Census showed elephant

---

[18] The world's longest running study of elephants.

[19] *Dr. Daphne Sheldrick: Obituary* The Times, 28 April 2018. *Redoubtable conservationist whose life in Kenya was dedicated to protecting her elephant 'friends'.*

[20] Fiona Harvey *Wayne Lotter: Obituary* The Guardian, 6 October 2017. See https://www.theguardian.com/environment/2017/oct/06/wayne-lotte. Accessed 05/07/2018.

numbers in Tanzania between 2007 and 2014 had plummeted a staggering 60%.[21]

Lotter believed that it was essential to involve the local community and have plenty of enforcers on the ground. In other words, robust policing. And it worked. Since 2012, the NTSCIU has made many arrests, over 2000 poachers and ivory traffickers, many of them notorious. They even caught the so-called Queen of Ivory, Yang Feng Glan. And the prosecutions have been successful, with an 80% conviction rate.

Sadly, his Lifetime Achievement Award had to be awarded posthumously.

In one interview, Lotter claimed the NTSCUI had helped reduce elephant poaching rates in Tanzania by 50%,[22] so it was inevitable he would become a target. The numerous death threats ending in a kill. Eight people were originally charged with murder and conspiracy to murder, and by October this had grown to 17, including the suspected hitman, but the organizers are still unknown.[23] The original suspects appeared in court in February 2018, not long after Esmond Martin, another irreplaceable conservationist, was himself murdered, although the two deaths were not connected.

**Esmond Bradley Martin 1941–2018:** In contrast to Wayne Lotter, Esmond Bradley Martin spent his life gathering data, painstaking research that has played a key role in strengthening protection for elephants and rhinos, making that protection more effective. In 1977, he was appointed to the IUCN's African Elephant Specialist Group, and he realized then that this was how he intended to spend his life, unravelling the complexities of the illegal trade in elephant ivory and rhino horn.[24]

Born into an extremely wealthy American family, after his education, his first degree in the USA, his master's in Kenya and his doctorate in the UK, he settled in Kenya travelling to the countries with unregulated domestic ivory markets, surveying them, wherever possible in minute detail, then publishing the results so that the world could see what was going on.[25] He worked mainly with three other distinguished people, his wife Chryssee, Lucy Vigne and Dr. Daniel Stiles. You will have read the results of a number of his surveys in this book.

---

[21] Elephant census data.
[22] Sophie Tremblay *Leading elephant conservationist shot dead in Tanzania* The Guardian, 17 August 2017. See https://www.theguardian.com/environment/2017/aug/17/leading-el. Accessed 05/07/2018.
[23] Roland Oliphant *Tanzania, once Africa's elephant graveyard, takes fight to poachers* The Times, 29 September 2018.
[24] Joe Shute *Who killed the elephant man?* The Telegraph Magazine, 31 March 2018.
[25] Ibid.

Esmond Martin was found stabbed to death by his wife on her return from a walk. Killed in his home on the outskirts of Nairobi, in the house they had built together so many years ago. His murder so far remains a mystery. Possibly carried out as a result of his work, because he was fearless in what he published and must have made enemies in the organized crime gangs, or possibly connected to an attempt at a land grab. Land there is short, at a premium. His safe had been broken open and cash and property deeds were missing. By all accounts he was a gentle man and an activist but using his fountain pen as his weapon. Such a cruel way to go and such a very sad loss.

But perhaps the most vivid way to remember him and his contribution to conservation is by a very few of the voices of his many friends in these selected snippets:

- *'The magnitude of Esmond's loss to elephant and rhino conservation cannot be underestimated':* Steven Broad, Executive Director, TRAFFIC[26]
- *'A massive blow'*, he provided *'irrefutable evidence and effectively alerted the world'* to the devastating crisis happening to elephants and rhinos: Charlie Mayhew, Chief Executive of Tusk[27]
- *'His surveys shone a powerful spotlight on the wildlife markets around the world that are sucking ivory, rhino horn and countless other African species into their maw'*, *'elephants have lost a great champion'*, and *'the shock of Esmond's death will be felt around the world':* Save the Elephants.[28]

His long-time friend and professional colleague, Dr. Dan Stiles, wrote a deeply moving tribute, painting a vivid picture of this most remarkable man, a polymath, whom he described as *'my mentor and friend, very much like a big brother'*. They met while he was attending Dr. Stiles' lectures (in archaeology) at the University of Nairobi:

- *'I did not know what to make of Esmond in the beginning. He always arrived alone dressed in a smart suit and silk tie with matching breast pocket silk hand- kerchief, incongruously carrying a battered leather satchel. He stood out in a crowd by his height, extravagant dress and the unruly shock of white hair on his head. He would take a seat and take out a large hard-backed notebook from the satchel for taking notes, which he did using an old-fashioned fountain pen. Esmond invariably asked knowledgeable questions and showed a deep interest in the subject under discussion, regardless of how trivial it might be'.*

---

[26] Steven Broad *Esmond Martin Obituary* TRAFFIC Bulletin, April 2018, Vol.30 No.1.

[27] See Shute (n.24).

[28] Aislinn Laing *Ivory trade investigator stabbed to death* The Times, 5 February 2018.

- *'I had to chuckle ... whenever he recounted his 'undercover' investigations to me. Esmond was not a person who could go unnoticed undercover. After observing him work in Cairo in 1998, where we coincidently collided, I saw that he obtained his information from brazen interrogations, with no thought as to the consequences that might befall him'.*
- *'He accomplished his sterling work with consummate determination, an iron will and superhuman stamina'.*[29]

**Rachel Katumwa 1993–2018:** This young ranger, an eco-guard in Virunga National Park, was only 25 years old when she was killed, not protecting the critically endangered mountain gorillas or the forest elephants who are slowly returning to the park, but two British tourists. They were driving towards the volcano when they were ambushed by four heavily armed men, who murdered her then forced the driver and tourists to march through the forest, until the driver, who had been injured, was released to give the alarm. It was a kidnap. Rachel was the first female ranger to be killed in the park.

Virunga, in DRC, is a deadly place to work. More than 170 rangers have been killed there in the past 20 years while protecting the park's wildlife. Situated in a particularly unstable part of the country, and extending over a very large area, it has to deal with many kinds of danger ranging from militias and rebel groups to local bandits and poachers. In an incident earlier in the year, five rangers and a driver were killed, and another ranger was injured.

The tourists were kidnapped near an army base, and the troops were slow to respond, provoking criticism. They were later released.[30]

## 38.2 NGOs

These come in all shapes and sizes. Wide-ranging specialist, large and tiny, all determined to do what they can, to play their part. We have met a few in our story, including the Environmental Investigation Agency which has provided most of our case studies. Part of the work of Elephants Without Borders is to

---

[29] Dan Stiles *Esmond Bradley Martin (1941–2018) A Tribute* SWARA, April–June, pp. 19–20.

[30] Patrick Sawer *Wildlife ranger killed trying to protect kidnapped Britons in DR Congo is named* The Telegraph, 13 May 2018. See https://www.telegraph.co.uk/news/2018/05/12/wildlife-ranger-killed-trying-protect-kidnapped-britons/. Accessed 25/10/2018; Jason Burke *Six Virunga park rangers killed in DRC wildlife sanctuary* The Guardian, 10 April 2018. See https://www.theguardian.com/weather/2018/apr/09/six-virunga-park-rangers-killed-in-drc-wildlife-sanctuary Acc 25/10/2018; and Jani Actman *Virunga National Park Sees Its Worst Violence in a Decade, Director Says* National Geographic, 14 June 2018. See https://news.nationalgeographic.com/2018/06/wildlife-watch-virunga-rangers-deaths-poaching-militia-gorillas/?user.testname=none. Accessed 25/10/2018.

count elephants. This NGO was established to protect elephants as they wandered across Angola, Botswana, Namibia, Zambia and Zimbabwe. Some, such as WWF, Save the Rhino and Save the Elephant are more familiar, their work well known. Another is TRAFFIC which may not have been known to you before reading this story. Born Free and IFAW, among their many functions, continue to provide invaluable data on the illegal wildlife trade. But there are so many others out there, also making valuable contributions, whose praises should be sung. So many reasons to be optimistic. So many seeds of hope.

Ian Player, Mark Shand and Princes William and Harry all set up NGOs.

**The Wilderness Leadership School and WRAP, the Wilderness Rhino Awareness Programme:** This was an early NGO and another of Ian Player's important achievements, the product of a fruitful partnership with his friend and mentor Magqubu Ntombela.

Magqubu's grandfather was a chief, and Magqubu, who was very familiar with the old conservation practices of the Zulus, knew that, in the 1860s, the Zulu King Shaka had carried out controlled hunting in the Imfolizi area of South Africa and realized that this historical and cultural knowledge could be used to put a programme in place to defeat the deadly poachers.[31] Another senior game ranger introduced Player to the American concept of wilderness, areas that could only be explored on foot, horseback or by canoe.

By 1958, half of the Hluhluwe-Imfolozi Game Reserve together with part of Lake St. Lucia had been designated as wilderness areas, and game rangers started taking groups of school children and young people along a network of trails developed in the two reserves. In 1963, these activities were incorporated into the Wilderness Leadership School. From small beginnings, the school has grown, and grown playing a vital role in educating children and young people all over the world about rhinos and conservation.

WRAP was another important development. It is all about rhino conservation, about *'creating a human culture that accepts its role of Trustee to the Earth'*. And it works with the local communities, those who live beside wilderness areas and poaching hotspots, so that a great deal of information and intelligence has been collected and fed through to antipoaching teams, an essential contribution if poachers are to be defeated.[32]

---

[31] Wilderness Leadership School 2015.
[32] Ibid, *Rhino Conservation*.

The school also works closely with employees of Pilanesberg National Park, collecting data. Rhinos are darted and captured. While sedated:

- Their ears are notched so that each animal can be identified.
- Microchips are inserted into their horns and rumps.
- The length of their horns and horn bases are measured.
- RHODIS data is collected, samples of tissue, horn and nails.

Everything is recorded.

Before we move on, there is just time to recount a very moving story. While trekking with another friend, Laurens van der Post, Player and Ntombela were deciding who should act as lookout for wild animals. The matter was swiftly decided. An enormous lion sprang out of the bush, followed by two white rhinos who settled down some 50 yards away. The rhinos would alert them to any predators; Ntombela had no doubts about that.[33]

**Elephant Family:** This modestly describes itself as *'a small dynamic NGO on a mission to save a forgotten species from extinction'* and has an enviable track record of helping Asian elephants. It is involved in projects in India, Thailand, Malaysia, Indonesia, Cambodia and Myanmar and specializes in human-elephant conflict. Working with a number of partners and providing funding, its aim is to promote *'grassroots conservation'*, so it empowers *'local experts rather than imposing Western ideas on operations'*, recognizing local communities as the key to success.[34]

**United for Wildlife and the Royal Foundation:** Represents a powerful collaboration. United for Wildlife means working with partners:

- Conservation International
- Fauna and Flora International
- IUCN
- The Nature Conservancy
- WCS
- WWF
- ZSL

---

[33] See (n.4).

[34] Elephant Family. See http://elephant-family.org/who-we-are and http://elephant-family.org/what-we-do/where-we-work. Accessed 07/03/2017.

And The Royal Foundation has made a long-term commitment:

- To scale up the response to conservation crises
- To engage 15–30-year-olds in the fight against the illegal wildlife trade

One way of achieving this of course is through education, and, in conjunction with ZSL, United for Wildlife regularly runs free online courses. The first of these was *Introducing Conservation*. Its aim was to inspire and connect with young people everywhere. And it succeeded, with students enrolling from many parts of the world, including Africa, India and South America.

**African Parks:** Another NGO that just keeps on delivering. Specializing in managing and restoring key national parks, many of them World Heritage sites, it is currently working with 15, although it plans to make this 20 by 2020. The work is challenging, but they are very good at it. Management priorities, as always, are concerned with developing infrastructure and strengthening law enforcement. Then key species can be reintroduced. The local communities are seen as essential to any solution; hence, there are educational programmes, opportunities for employment are created, and sustainable sources of income, in which sustainable tourism plays a key role, are established.[35] They have been responsible for turning round parks such as Garamba in DRC and Zakoumi in Chad, parks that were previously no-hopers.

## 38.3   Rhino Impact Investment Project

Because conservation costs so much money, there is now a shortfall in funding, and the gap is likely to widen, making it even more difficult for protected areas to secure threatened species. In other words, a new way of funding conservation was needed, and urgently. What normally happens is that donors give money to conservation organizations who use it to implement projects such as erecting fencing to protect a rhino sanctuary or to train rangers, reporting back on the success or otherwise of the project. In other words, short term contracts. Lack of funding makes even day-to-day management difficult.

---

[35] *World Heritage Site in Chad to be Protected Under New Agreement* African Parks. See https://www.african-parks.org/press-release/chad-ennedi-protected. Accessed 13/03/2018.

There had to be an alternative method of obtaining the requisite finances, a long-term solution. So the United for Wildlife partners have established the Rhino Impact Investment Project, whereby impact investors provide up-front working capital to fund long-term contracts, with the donor repaying *'the original investment plus or minus a percentage relative to conservation outcome'*. In other words, this is *'a commercial, impact investor-driven approach – conservation finance'*. It is *'Saving wildlife by paying for impact'*.[36]

The funding will go to selected sites, and a key performance indicator (KPI), the net rhino growth rate, will show how good the conservation outcome is. Secondary KPIs are used for short-term payment triggers. When the donor repays the investor, the amount will be the original investment plus or minus a percentage based on the actual conservation outcome.

A *'bespoke rhino gap assessment tool'* has been developed to *'assess and score'*:

- The *'sites' ecological, managerial and financial capacity to achieve additional impact outcomes'*.
- Any *'estimated funding required to enable sites to reach investment-readiness'*.[37]

The gap assessment results have enabled the African Rhino Specialist Group to identify 15 protected area sites containing key rhino populations in Kenya, South Africa and Zimbabwe that are willing to participate in this unique and imaginative experiment. Seven of them have been shortlisted and will now progress to the investment readiness phase of the project.

Rhino impact bonds will provide the funding. These are *'equity investments where return is pegged more to social returns than to maximum profits'*, and investors will be risking *'their profits and some of their capital to ensure that donor contributions are only used for successful projects'*. These should be available in 2 years' time. Meanwhile, a Finance Manager and a Performance Manager are being selected.

Funding for a pilot for a rhino impact bond was provided by the UK Government's Illegal Wildlife Trade (IWT) Challenge Fund (more later), and as part of the project, black rhinos in Tsavo, Kenya, are now having their ears notched and are being fitted with transmitters so that they can be easily identified and tracked.[38]

---

[36] Rhino Impact Investment Project (RII), p.4.

[37] Ibid, p.9.

[38] ZSL *Conservation > Tracking black rhinos in Tsavo* ZSL, 4 January 2018. See https://www.zsl.org/blogs/conservation/tracking-black-rhinos-in-tsavo Accessed 09/01/2018.

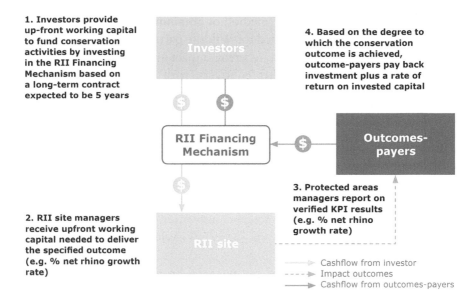

**Fig. 38.2**   Rhino Impact Investment funding model. Copyright Rhino Impact Investment Project (see RII (n.829), p.7)

This is a project with clout. It is funded by the Global Environment Facility (under the auspices of UNDP), the UK Government, UfW and ZSL, the latter also acting as the implementing agent. And its implementing partners include FFI, IUCN SSC African and Asian Rhino Specialist Groups, KWS, The Nature Conservancy, WWF-UK, Credit Suisse, DLA Piper and UBS.

A road map will be produced to reduce rhino poaching, and it can then be rolled out for other threatened species across the world.

## 38.4   Local Communities and Rangers

We have already seen the important role local communities play in protecting our animals, a role that has been formally recognized in, for example, the London Declaration. Further recognition came in autumn 2016.

The IUCN World Conservation Congress, best described as '*a unique global environmental parliament of governments and NGOs*', took place in Honolulu, Hawaii in September 2017. This global conservation summit laid down the global conservation agenda for the next 4 years and set the direction for the post-2015 sustainability agenda.

More than 10,000 people registered to attend, and more than 100 Resolutions and Recommendations were adopted. The outcome was 'the Hawaii commitments', a document called 'Navigating Island Earth', and among the Resolutions and Recommendations were two that were specifically relevant to our story and which we will briefly consider.

**Resolution WCC-2016-Res-025-EN:** Recognising, understanding and enhancing the role of indigenous peoples and local communities in tackling the illegal wildlife trade crisis…

This Resolution, among many matters, recognizes the essential role local communities play 'as sentinels of change and illegal activity within their environment' and the important fact that responses to the illegal wildlife trade (IWT) are not always beneficial for the communities themselves, which might make them less willing to 'contribute tangibly to combating IWT in the long term'. But the people who live alongside wildlife 'have a key role to play in conserving wildlife'. Thus, one way of combating IWT is by engaging local communities, for example, through outreach and education and incentivising them 'through financial and non-financial benefits'. We have already seen this at work in Kenya, Lewa, is an excellent example.

The IUCN has recognized the importance of rangers at several meetings. For example, the IUCN World Parks Congress in 2014 'officially pointed out that the actions of park rangers in the conservation of protected areas have been recognised as fundamental', so one of the strategies at the 2016 meeting was to strengthen their careers.

**Recommendation WCC-2016-Rec-103-EN:** Establishment, recognition and regulation of the career of park ranger:

Among other matters, the Recommendation considers that 'without sufficiently professional park rangers to implement development and management plans effectively, it will be hard for protected areas to fulfil their essential role'.

'Ranger' applies to 'all frontline protection staff regardless of title', including but not limited to 'wildlife warden, forest guard, forester, scout, watcher, game scout, and park guard'.

It recognizes the fact that they help to 'achieve several of the Aichi Biodiversity Targets' (which, as we know, is all about sustainability), that they are key workers 'for the development and management of all protected areas', that 'they assure the effectiveness of environmental protection' and that they 'maintain relations between public policies and the local communities in which they operate'.

To this end *'it is essential that all States give priority to the recruitment, training, equipment and well-being of park rangers'*, recognize what a *'high-risk occupation'* this is and protect them accordingly.

Prince Harry, in his speech *'In my view rangers are heroes'* to the trainee rangers at the Southern African Wildlife College, made it quite clear that he holds rangers in very high regard:

*After a summer working alongside rangers, I now fully understand the skill you bring to your work, the sacrifices you and your families make for you to do it, and the perseverance you demonstrate every day in the face of huge challenge. To those of you graduating today can I say congratulations – you are part of a profession that cannot be appreciated enough and that is dealing with very difficult circumstances. In my view: rangers are heroes. I hope you are incredibly proud of your achievements.*[39]

David Leto, a Maasai Mara Elephant Officer, is one of these exceptional people. This is his story.

It started with a tragedy. David's father was killed by an elephant. This left David, the eldest son, head of the family. But his mother still wanted him to go to university and study conservation, which he did, and this led to him becoming an elephant ambassador with WWF-Kenya dealing with human-elephant conflict.

David works with the local communities, setting up and running conservancies, extending and connecting *'areas of habitat for elephants and other wildlife'* and *'the community benefits from ecotourism and other sustainable development'*.

*It's my job to convince local people that revenge is not the right solution; to convince them that we can coexist. It's important for me to listen to the communities and relate to them. My personal experience and Maasai background help. They see I'm really on their side, dancing with them and sharing stories…*

*My dream is to see a secure and increasing wild elephant population, along with economic benefits for people. The elephants benefit from the community land, but the community also benefits from the elephants being here.*[40]

---

[39] HRH Prince Harry *'In my view: rangers are heroes'* A speech by Prince Harry at the South African Wildlife College, Kruger National Park, South Africa, December 2015. See https://www.royal.uk/speech-prince-harry-south-african-wildlife-col. Accessed 12/01/2017.

[40] David Leto *Good Neighbours* Action, Summer 2016 (WWF Magazine).

## 38.4.1   The Well-being of Rangers

Although we have glimpsed a few of the horrors many rangers are regularly subjected too, and the pressures they must experience as they contemplate what might happen to them and the animals they so courageously care for on a daily basis, *'more research has been conducted on post-traumatic stress disorder (PTSD) among elephants following a poaching incident than on the rangers protecting them'*, and this has to change.[41]

It is easy to forget that it is only in recent years, with the skyrocketing of poaching incidents and the muscling-in of organized crime, that rangers have been forced to become an army of protection for animals, forced to carry, and often use firearms and deliberately place themselves in situations of incredible danger. For they must run towards the armed gangs, and too often they have seen friends and colleagues die cruel deaths. A situation not unlike that in World War One.

Johan Jooste, head of antipoaching forces at South Africa National Parks (SANParks), pointed out that *'eighty percent of the work now, in many of the parks, is hard-core, 24/7 operations to combat poaching'*.[42] And surveys WWF carried out in 2016, *'the only extensive surveys to ever to examine rangers' working conditions'*, showed that:

* In Africa, 82% of rangers and in Asia 63% of rangers had faced a life-threatening situation in the line of duty.[43]

And many rangers are killed. Some 100–120 are known about, although there may be more as the data comes from a limited number of countries.[44] Added to that is the divorce rate, in some countries as high as 90%. Although Sean Willmore puts this down to the long separations from their families, surely a contributing factor must be the actual nature of the work itself.

However, if rangers are to be helped, more research is urgently needed, and it is finally being undertaken. Dr. Francis Masse of the University of Sheffield in the UK chose to examine the antipoaching forces in Mozambique for his doctorate, living with them for 5 months, so his knowledge was first-hand. And he is now carrying on postdoctoral research. One important finding was

---

[41] Jim Tan *Rangers face a 'toxic mix' of mental strain and lack of support* Mongabay, 24 May 2018. See https://news.mongabay.com/2018/05/rangers-face-a-toxic-mix-of-mental-strain-and-la. Accessed 29/05/2018.

[42] Ibid.

[43] See https://www.worldwildlife.org/publications/ranger-perceptions-africa.

[44] According to Sean Willmore who founded the Thin Green Line Foundation and is President of the International Ranger Federation.

the rangers' inability to deal with death threats, were these directed at their families as well, their families who would be without the ranger for sometimes as much as 5 months at a time. Another problem is caused by Mozambique's legislation. *'[Poachers] get arrested and in two weeks they are back in the communities right outside the park where the rangers have to pass in and out. I've seen rangers cry from the stress, from being afraid ... because of the strain and the stress and the anxiety that goes along with their work'.*[45] But in a country like Mozambique, having a secure though dangerous job is better than not having a job.

And in South Africa, SANParks runs Project Embrace, which offers psychological support to the Kruger Park rangers. This is not just down to Jooste supporting his workforce; legislation, in the form of the current South African Firearms Act, lays down a legal duty to provide *'psychological debriefing within 48 hours after experiencing any violent incident, discharging their firearm or witnessing a shooting'.*[46] The programme is run by the clinical psychologist Susanna Myburg-Fincham, whose aim is not just to prevent mental health problems before they occur but to fill in gaps in knowledge as well.

Let us leave the last word to the rangers Mongabay interviewed. Without exception, they wanted to be properly trained, properly equipped, fully supported and fairly paid. To be paid what they had earned. They all said that *'rangers were owed more than many of them currently receive'.*[47]

## 38.5   Conservancies

In Kenya, the award-winning conservancies of Lewa and Ol Pejeta and the Northern Rangelands Trust show how well different peoples can work together to protect both animals and livelihoods.

**LEWA Wildlife Conservancy:** One of the most successful private wildlife conservancies in Kenya, its importance has been recognized by the fact that it has been designated a World Heritage Site. And it started from small beginnings, with the original rhino sanctuary at Ngare Sergoi in 1983. Lewa, the land, was bought by the Lewa Wildlife Conservancy, which now owns thousands of other acres, as well as the family land, thus creating a number of different habitats.

---

[45] See Tan (n.41).

[46] Ibid.

[47] Ibid.

As well as managing the land for conservation and wildlife, Lewa has tried hard to guarantee economic opportunity and cultural stability to the communities who live within its boundaries. In fact, this has turned out to be one of its most important achievements. Education is a priority, and mindful of the fact that wildlife is key to their success, in August 2016, Lewa and Virry (the BAFTA award winning wildlife app) entered a partnership which children and parents around the world could use the app to watch Lewa's endangered wildlife.[48]

None of this, however, could be achieved without money. And although there is no trophy hunting in Kenya, it is forbidden by law, plenty of tourists flock to the country, with Lewa a very popular destination. People can stay there, go on safari there, even get engaged there just like Prince William and Kate Middleton. So many different opportunities present themselves. A successful model.

**Ol Pejeta:** Another successful conservancy in the same part of Kenya, this describes itself as *'a model for Wildlife Conservation'* and aims to demonstrate that *'conservation can be a sustainable and financially viable land use in the semi-arid areas of Kenya'*.[49]

It has an interesting history. Bought by Fauna and Flora International and the Arcus Foundation in 2003 because it is an essential part of the Laikipia ecosystem, it provides critical migration corridors and is home to black rhinos and other endangered species. The area needed protection. Two years later, they transferred their ownership to a Kenyan non-profit entity under a long-term management agreement, although both they and Lewa Conservancy, remain on the board.[50]

Because it was originally used for highly productive cattle ranching, wildlife was not tolerated, but in the 1970s, it underwent a radical transformation. There were three main reasons:

• More and more people started settling in the south of the country.
• In the north, elephant poaching increased.
• A hunting ban was imposed.

As a result, herds of elephants began to take up permanent residence, and game numbers began to increase. Fences and large numbers of elephants do

---

[48] Lewa Wildlife Conservancy *Kenyan Wildlife Conservancy Now Open for Children to Experience on Digital Platform* 26 August 2016. See wanjiku.kinuthia=lewa.org@cmail2.com.
[49] Ol Pejeta Annual Report 2017.
[50] Ibid.

not mix, so the fences came down and agricultural practice had to change. The farmers adopted the pastoralist ways of the Maasai. Cattle were kept in 'bomas' at night, to protect them from the wildlife, then herded to their grazing lands during the day, and although they were no longer so profitable, wildlife tourism was taking over, well able to make up for any shortfall in funding.

As well as being the largest black rhino sanctuary in east Africa, Ol Pejeta is now home to many other endangered species, and, of course, the very special northern white rhinos. And they are gradually expanding their landholdings to enable them to support the 130 black rhinos they are hoping to have by 2020.

Financially self-sustaining, it is managed to support both conservation and community outreach programmes, with the Mission Statement in its 2020 Map specifically stating that one of its objectives is 'to generate income through wildlife tourism and complementary enterprise for reinvestment in conservation and communities'.[51] In fact it is already 'one of the most productive conservancies in Laikipia', and 'the largest economic contributor', and because it operates on a not-for-profit basis, 'all surplus income is reinvested into conservation and our communities'. It also employs 900 people.[52]

Being on good terms with its neighbours, working together with them, brings benefits for everyone, including such diverse matters as improving quality beef genetics through artificial insemination services, improving water supplies, planting 'over 65,000 tree seedlings' and using the K-9 unit to help neighbours who have suffered thefts. It also enabled Ol Pejeta to get through 2017, a very testing year with extreme drought and 'dramatically reduced agricultural revenues', a year 'where the massive challenges of maintaining space for conservation were laid bare'.[53]

And every year both Lewa and Ol Pejeta continue to win awards.

## 38.6   Religion

We have already seen the uneasy relationship between some religions and ivory, so where can we find hope here?

---

[51] Ol Pejeta Conservancy *2020 MAP OPC-2020* Handbook – 261,114.pdf See www.olpejetaconservancy. org/uploads/assets/uploads/2016/01/OPC-2020-Handbook-261114.pdf Accessed 16/08/2016.

[52] See (n.49).

[53] See Richard Vigne CEO Ol Pejeta in ibid.

## 38.6.1  Pope Francis and the Environment

This Pope cares deeply about the environment, and under his papacy, the Vatican has already signed agreements regarding a number of global criminal activities including organized crime. In June 2015, he issued the first ever encyclical devoted to ecology.

**The Environmental Encyclical:** Welcomed by both religious leaders and campaigners, this encyclical is concerned with the loss of biodiversity. And it concentrates on ecology because the Pope believes there are significant ecological problems to be solved. He also believes there can be a fruitful dialogue between science and religion, which is perhaps why the encyclical is addressed, not just to the church's faithful followers, but to everyone living on the planet.

**A World Day of Prayer for the Care of Creation:** Another of the Pope's innovations, and inspired by the Orthodox Christians who have been praying for the environment every 1 September since the late 1980s, this date was adopted to be the World Day of Prayer for the Care of Creation, in the hope that it *'will offer individual believers and communities a fitting opportunity to reaffirm their personal vocation to be stewards of creation'.*

## 38.6.2  An Antipoaching Workshop

The Christian churches in South Africa wield a lot of influence, so when A Rocha, a Christian nature conservation organization, the Timbavati Foundation and the Responsible Tourism Foundation set up a Church Leaders Against Rhino Poaching workshop, it was almost certain to have a positive effect. The results, however, must have exceeded even their expectations.

The workshop, intended to raise awareness about the plight of the rhino, was held in Hoedspruit in the Protected Areas of the Lowveld, with representatives from 14 churches and denominations (with over 3000 members), all from communities on the periphery of the Timbavati Private Nature Reserve and the Manyeleti Game Reserve, attending.

A video showing rhino poaching was particularly powerful, moving some delegates to tears. One of the Pastors had been so concerned about the poaching, that she had preached to her congregation on the subject even before the workshop. Preached so successfully that two of them eventually confessed that they were involved.

She had good advice for the workshop delegates as well, suggesting that they too could talk about rhino poaching in church, as this would enable members of their congregations, family members or even friends to come forward and get help. Another Pastor pointed out that poaching would impact badly on their congregations, as so many of them worked in the tourist industry. Poaching led to fewer tourists.

And the outcome of the workshop? A booklet containing sermons about poaching and associated issues, which was to be distributed to other parishes across the country. The delegates also committed themselves to fight to prevent rhino poaching in their communities. And because of the churches' enormous influence in these communities, this was a very important contribution.

## 38.6.3   ARC (The Alliance of Religions and Conservation)

One body that has recognized the important role religions can play in conservation is ARC, an Alliance forged in 1995 by Prince Philip the Duke of Edinburgh. On the face of it, ARC might seem to be a religious organization, but this is misleading, as it is a secular body that *'helps the major religions of the world to develop their own environmental programmes, based on their own core teachings, beliefs and practices'*, its aim being to *'assist and enable the religions of the world to respond to the environmental challenges of the Twenty-First Century'*.[54]

It has enormous potential for good because 11 major world religions are involved, and, as we have seen, religious leaders have a great deal of power and influence over their followers. ARC's reach is global, for the religions include Baha'ism, Buddhism, Christianity, Daoism, Hinduism, Islam, Jainism, Judaism, Shintoism, Sikhism and Zoroastrianism, thus covering some 7% of the habitable surface of the earth, and its partner organizations include the China Taoist Association, WWF International, the World Bank and UNEP (the United Nations Environmental Program), further increasing its influence.

In 2012, WWF established its Sacred Earth programme *Sacred Earth: Faiths for Conservation*, which forges partnerships with both faith leaders and communities. These concentrate on a variety of environmental issues, including combating illegal wildlife trading.

That same year, Sacred Earth and ARC joined forces to launch an important project, the *Wildlife & Forest Programme*. Designed to help communities learn to protect endangered species and their habitats, not only does it look

---

[54]Alliance of Religions. See www.arcworld.org.

into traditional Chinese medicines and their connections with illegal wildlife trading but also at the trade in illegal luxury items. Elephant ivory and rhino horn. And it extends to the source countries in southeast Asia, India and sub-Saharan Africa as well. So here again is the potential to produce some positive results.[55]

## 38.6.4   Fatwas to Protect Wildlife

The first ever fatwa issued against the illegal wildlife trade stated that killing rhinos was against the will of God. In 1986, the Yemeni Foreign Minister met WWF, concerned about *'rampant rhino poaching to meet demand for dagger handles in Yemen'*. The result was an action plan. Then Yemen's Grand Mufti issued the fatwa. And to make sure everyone knew, it was written on a poster, and the poster was displayed all over Yemen.

Then, in September 2013, the Universitas Nasional, WWF-Indonesia and ARC,[56] together with Indonesia's Ministry of Forestry and the Indonesian Tiger Conservation Forum organized a field trip to Sumatra (Indonesia) for some Muslim leaders. While they were visiting a village that had been experiencing problems with elephants and tigers, the villagers questioned them on the status of animals in Islam. They were told that animals, like people, were creations of Allah, that it was haram (forbidden) to kill them and part of the worship of God was to keep them alive.

The result of this discussion came a few months later, when the Indonesian Council of Ulama issued what was probably only the second fatwa against illegal wildlife trading. It declared illegal hunting or illegal trading of endangered species to be haram. It explained and provided guidance on the Sharia law perspective on issues relating to animal conservation. It applied not only to individuals, but also to the government. Indeed, the Chair of the Council's environment and natural resources body, speaking about the fatwa, was under no illusions about the role of his government, *'noting that corruption can be an issue when wildlife, forests, and the interests of such industries as the palm oil business come into conflict'.*[57]

So what is a fatwa? It is a law, but it is a religious law. It is part of the body of Islamic Sharia law. And in both Yemen and Indonesia, which are Muslim

---

[55] See www.arcworld.org/projects.asp?projectID=578 Last accessed 06/12/2018.
[56] Including the distinguished Dr. Tony Whitten, tragically killed in a road traffic accident, who worked for many years in Sumatra. His obituary was in The Times on 3 January 2018.
[57] Ibid.

countries, Sharia law supplements national legislation. In Indonesia, there are some 200 million Muslims, so this law is binding on a great many people.

## 38.7   The Internet Again

Although the Internet continues to be responsible for so much illegal wildlife trafficking, there is also a positive side.

### 38.7.1   Online Petitions and the Social Media

We have already glimpsed a dark side to the Internet, but there is another side, a side that acts as some sort of counter-balance to the online illegal trading.

**Online Petitions:**  Earlier in our story, we saw that the ivory trade in Mong La, a town within Myanmar but on the border with China, was causing a high level of concern, so at the end of 2013, Vincent Nijman and Chris Shepherd returned there for 3 days, to carry out a very short survey. The results were not encouraging as the researchers counted 3300 pieces of carved ivory and 49 whole tusks for sale.

The full report was carried on the TRAFFIC webpage, but the news media also took up the story. Three days after the report was published, on 14 January 2014, an online petition was started by a stranger, Sue Lee, on the Care2 petition website. Its stated target was Asian governments throughout and near China, and it was *an effort to stop the untimely death of elephants being murdered for their ivory*. The petition and its progress were described in detail by the two researchers.[58]

**Sue Lee's Petition:**

*In China, it's illegal to kill elephants for their ivory and sell their tusks. But somehow, 30 tusks and thousands of pieces of ivory were recently discovered for sale at a market in Mong La. Investigations into wildlife monitoring and trafficking noted that this market could be one of the biggest unregulated ivory markets in Asia. How could this be if laws exist? Ivory is openly displayed at this market in the Eastern Shan State of*

---

[58] Vincent Nijman and Chris. R. Shepherd *Social media and the ivory ban: Myanmar and China cross-border trade* Pachyderm, No. 55, January–June 2014.

*China to cater to the many tourists that visit. There is obviously no form of government control to stop the sale of ivory throughout China and other Asian countries.*

*The problem has also reached epidemic levels in Myanmar, considered as the second largest elephant population area worldwide, just after India. It is believed that about 6000 elephants in the wild still roam the lands of Myanmar. It has laws forbidding trade in endangered species, but violations are rampant, especially in remote border regions. The laws need to be more strictly enforced!*

Readers were invited to sign the petition and share it worldwide *'in an effort to stop the untimely death of elephants just so their ivory can be sold worldwide. We need China and Myanmar to crackdown on the illegal sale of ivory in our efforts to protect the elephant'.*

The text made its readers sit up and take notice, a good start. There were a few mistakes. Some derived from the media itself, which was not always quite accurate, while others, such as including Mong La and the eastern Shan State in China rather than Myanmar, and confusion over the size of Myanmar's elephant population (Myanmar only has the second largest population of the rare Asian elephants) originated in the petition itself. And the occasional sweeping statement, such as *'no form of government control to stop the sale of ivory throughout China and other Asian countries'*. But mostly it was accurate.

The first day the petition went online, it was signed by 140 people. Slightly more the next day. Then the numbers gradually dwindled until, on 25 January, only two people signed. A total of 1019 signatures in all. Later that same day, the petition went viral, and 10 min later an additional 200 people had signed. And they kept on signing, so that by 4 February there were 50,000 signatures!

The investigators were also able to carry out a fairly detailed analysis of the signatories themselves. Here are some of their results:

- Most of the signatories were women (93.3% ticked the gender box and 73.6% of these were women).
- Of all the people who had signed, 2.9% had a doctorate.
- The signatories came from over 130 different countries.
- All the Asian elephant range States except Bhutan were represented, and 19 of the 37 African range States (the missing ones tended to be the francophone countries).
- 73 people from Hong Kong had signed, 51 from China, 47 from Taiwan and 8 from Macao. This represented 0.4% of the total. There were 4 signatures from Myanmar.

- 5.7% of those who had signed had added a comment; 0.5% of these wanted more legislation or better law enforcement.
- 37 of the comments linked the ivory trade with the trade in rhino horn.

**The Social Media:** Although Sue Lee's petition started off with a flourish, the effect soon began to die away, but it surged back into life once it was taken up by Twitter and Facebook, both of which had links to the petition site. So online petitions can deliver a powerful message.

### 38.7.2    Instagram, TRAFFIC and WWF

In November 2017, Abbas Allawi was found guilty of a number of offences regarding the illegal sale of endangered species and sent to prison for 14 months. The previous year, when members of the UK's Metropolitan Police's Wildlife Crime Unit had raided his house, they had discovered 3 rhino horns, 2 elephant tusks and 4 hippo teeth which he had been trying to sell for cash, at £60,000 kg, on Instagram. He could have made about £2 million.[59]

Although he claimed he didn't know this was illegal, evidence was produced showing Allawi was only too aware his actions were criminal. However, it prompted Instagram to team up with TRAFFIC and WWF, and by December 2017, Instagram had installed an alert. Users searching for hashtags associated with the illegal wildlife trade (or other abuses of animals) would trigger *'a content advisory warning pop-up informing them that Instagram prohibits the sale of endangered animals and animal abuse'.*[60]

### 38.7.3    The Global Coalition to End Wildlife Trafficking Online

In March 2018, a major new initiative was launched to control the online trade in endangered species, when 21 of the world's leading e-commerce and social media companies from four continents,[61] decided to work with TRAFFIC, WWF, IFAW and Tencent to make it *'inoperable'* for the traffickers

---

[59] The Times *Rhino horns for sale on Instagram* 11 November 2017.

[60] TRAFFIC *Instagram users to receive alerts about threatened wildlife* Wildlife Trade News, 4 December 2017    See    http://www.traffic.org/home/2017/12/4/instagram-users-to-receive-alerts-about-threat. Accessed 16/01/2018.

[61] Africa, Asia, Europe and North America.

to use online platforms and apps to carry out their illegal trade. The Global Coalition to End Wildlife Trafficking Online has the potential to be a game-changer because, according to Crawford Allan, TRAFFIC's Senior Director of Wildlife Crime, *'inconsistent policies across the web invariably create a "whack-a-mole" effect, where ads may be removed from one site just to pop up somewhere else'.*

The companies, which include eBay, Google and Microsoft, will receive policy guidance, educational information, training materials and assistance, sharing information on best practice and lessons that have been learned. They have pledged to work together and *'each company will develop and implement policies to help end wildlife trafficking online'.*[62]

The Global Coalition has been a beneficiary of the UK's IWT Challenge Fund (see later).

### 38.7.4    eBay Calls for a Ban

*We are proactively doing what we can to protect native, endangered and threatened species. Plus we are also following laws, government regulations and international treaties on animals and wildlife.*

*eBay supports a ban on re-export of raw ivory and supports all steps taken to protect elephants and rhinoceroses.*

*eBay takes the alarming decline in elephant populations throughout Africa very seriously and works closely with NGOs and law-enforcement agencies globally to fight international ivory trafficking.*

eBay already has a ban on any form of ivory sales, so when Kristen Foster, eBay's government relations director for Australia, New Zealand, Japan and ASEAN, appeared before the Australian Parliament's Joint Committee on Law Enforcement, it should have come as no surprise when she made the above statement, at the same time calling for a complete ban on sales of elephant ivory and rhino horn in Australia. The country was considering whether to follow the UK's example with its ivory ban.[63]

---

[62] See https://www.traffic.org/news/leading-tech-companies=unite=to-stop-wildlife-t. Accessed 04/11/2018; and WWF *Global Coalition to End Wildlife Trafficking Online.* See https://www.worldwildlife.org/pages/global-coalition-to-end-wildlife-trafficki Accessed 04/11/2018.

[63] Tom McIlroy *eBay joins calls for Australian ban on elephant ivory and rhinoceros horn sales* Afr.com, Financial Review, 4 July 2018. See https://afr.com/news/ebay-joins-calls-for-australian-ban-on-elephant-ivory-and-. Accessed 10/07/2018.

## 38.7.5  Resource of the Month

In autumn 2018, Chester Zoo's Learning Resource Hub released a video *'We can help reduce the demand for the illegal trade of wild animals by being more conscious of what we "like" and share on social media'* which has been specially designed for use with students.[64]

---

[64] Chester Zoo.

# 39

## Countries

More seeds of hope come from countries that have had a troubled past. We start with encouraging news from Tanzania, which, as we saw earlier, has experienced severe difficulties with poaching.

## 39.1 Tanzania's Mkomazi National Park: Back from the Brink

Only designated in 1951, this 'new' national park is an important home for species such as the black rhino and African wild dogs. Remote and inaccessible, it never attracted large amounts of funding, so the result was degradation. By 1988, its black rhinos and elephant herds had been killed by poachers. The land had been overgrazed, some had been deliberately burned and illegal hunting was rife. There were fears that it might have lost its status, that its importance had been overlooked and that the land would be released for subsistence farming. Its future looked bleak.

The park, which is actually a 3500 km² game reserve, is located in the southern tip of the Sahel in northern Tanzania, where it shares a border with Kenya's Tsavo National Park. In terms of ecosystems, *'it is a classic dry country reserve of grey-green nyika bush, ancient baobab trees and isolated rocky hills'* with the bush giving way to open savannah woodlands and shallow valleys of grassland.[1] And before it fell into decline, it was an important migration route for

---

[1] GAWPT *Mkomazi National Park* 2016. See http://www.georgeadamson.org/mkomazi Accessed 07/01/2016.

© Springer Nature Switzerland AG 2019
B. Martin, *Survival or Extinction?*, https://doi.org/10.1007/978-3-030-13293-4_39

the vast herds of elephant that moved north from Mkomazi, Tanzania, into Tsavo, Kenya.[2]

Home to other charismatic and endangered animals such as giraffes, cheetahs, lions, leopards, oryx, hartebeest and Grant's gazelle, as many as 78 species of mammals have been recorded, and over 400 species of birds. Something had to be done fast, and in 1989, its reclamation began. The Tanzanian government, recognizing just how important a wilderness it was, redesignated it a National Priority Project. This would ensure its complete rehabilitation, which included the reintroduction of endangered species.

However, a special kind of assistance would be required if this was to be achieved, so the government was fortunate when it managed to prevail on Tony FitzJohn and the George Adamson Wildlife Preservation Trusts (GAWPT) to become its main partner. Roads, boundaries and airstrips had to be cleared, an urgent priority, as was communication. So an entire radio network was installed. And water. Further works ensured supplies could be maintained. Staff, particularly rangers, had to be recruited, and antipoaching and aeroplane patrols set up.

As part of the government's policy on endangered species, a rhino sanctuary, the Mkomazi Rhino Sanctuary, was constructed and stocked. It is sizeable, some 45 km$^2$ in extent and is protected with an electric fence, for within its boundaries live not only black rhinos but also the equally endangered African painted dogs.[3]

Then, in 2012, three black rhinos from the UK, from the Aspinall Foundation's Port Lympne wildlife park, were translocated there, a male, Monduli, and two females, Zawadi and Grumeti. Prince William met them before they left for their new home. Prince Harry visited them during his 3 months as a volunteer, fed Zawadi her favourite carrots and sent his brother a photo.[4] Exciting news came in early 2015 that Grumeti had mated with Jamie another translocated rhino, who had been born in a zoo in the Czech Republic. Mobo, their calf, arrived in May 2015, the first eastern black rhino to be born back in Africa to parents raised in zoos. Prince William was informed of Mobo's birth.[5] A tiny seed of hope!

Other partners are involved. Chester Zoo provides funding for both guard/tracker dogs and the sanctuary fencing,[6] two key components for keeping the

---

[2] Tusk.

[3] See GAWPT (n.1).

[4] Valentine Low *Harry's despair in killing fields of Africa* The Times, 3 December 2015.

[5] Ben Webster *Relocation breeds success for royal rhino* The Times, 6 June 2016.

[6] Information given to the author by Chester Zoo.

**Fig. 39.1**   A grazing herd of African savannah elephants. Copyright Chester Zoo

rhinos safe, and both are very expensive. Tusk has been one of the main sponsors for well over a decade, providing funding for salaries, equipment and vehicles, as well as helping to build an environmental education centre and buy a school bus to carry local children to the reserve.[7]

So Mkomazi has been pulled back from the brink, once again a home for black rhinos and many other species. And the elephants are back as well. At the park's lowest ebb, they numbered only 11. But now the herds are back. *'Today, during the rainy season, close on 1,000 elephants range freely across the reserve, including many herds of breeding females with their young',*[8] and in autumn 2018, just before the London Wildlife Conference, Prince William visited the park to promote efforts to save the black rhinos.[9]

Two more countries that are making tremendous efforts to protect their animals are Nepal and Cameroon, two more beacons of hope.

---

[7] Tusk *Mkomazi National Park, Tanzania* See http://www.tusk.org/mkomazi-national-park Accessed 07/01/2016.

[8] See GAWPT (n.1).

[9] Photograph in The Times, 1 October 2018.

## 39.2   Nepal and Its Rhinos

Nepal has a remarkable and enviable record in rhino protection. In January 2017, it celebrated the fact that no rhinos had been poached there in the last 1000 days. Indeed, only three have been poached in the last 5 years. Not only that, their numbers have continued to grow, to the extent that some are even being relocated. Now, more than 605 of them live in Chitwan National Park, a haven not just for the greater one-horned rhino but for about 800 other species of animal as well, including elephants and tigers. It is a haven for wildlife. But back in 2002, things were very different. Thirty-seven rhinos were poached in that year alone. So how has this turnaround been achieved?

For a start, the park is heavily protected. According to the Deputy Warden 'more than 1,000 Nepalese soldiers guard the reserve round the clock, reinforced by 150 park employees and four sniffer dogs', and there are 'over 50 guard posts … embedded within the sanctuary, providing bases from which foot and bicycle patrols are launched'.[10] Drones are employed, and the rangers use SMART. The local villagers supply intelligence. All in all, a formidable team!

But there is more. There is political will at a high level, which is combined with 'the active involvement of the park authorities, Nepal Army, Nepal Police, conservation partners and local communities', in other words, 'a coordinated national response'.[11] Everyone working together.

And now there is the zero poaching framework. Nepal has been one of the driving forces behind this project, which comes with a zero poaching toolkit. The framework has six key pillars:

- Assessment
- Capacity
- Community
- Cooperation
- Prosecution
- Technology

The toolkit, which can be accessed by both governments and protected area authorities, contains 'the best available antipoaching tools and resources'.

---

[10] Denis D. Gray Nepal's bold moves have Rhino poachers running scared Nikkei Asian Review, 2 June 2016. See http://asia.nikkei.com/magazine/20160602-THE-BUSINESS-of-H Accessed 30/08/2016.
[11] WWF No rhinos poached in Nepal for past two years See http://wwf.panda.org/wwf_news/?266870/No-rhinos-poached-in-Ne… Accessed 30/08/2016.

Protected from the animals by buffer zones, the local communities who co-manage the park are an integral part of this success story. They benefit from tourism, both from employment opportunities and because they receive 50% of the annual revenues generated by it. In 2014, this amounted to almost USD1,000,000, useful money for building schools and roads, creating secure supplies of drinking water and providing veterinary support.[12]

The Nepalese regard their rhinos as a national icon, even putting them on their banknotes,[13] and despite the poverty and corruption in their country, factors that normally work in favour of the poachers, they keep their rhinos safe. A formidable achievement!

# 39.3   Cameroon's D'ja Conservation Complex

The Great Elephant Census discovered 8 dead elephants for every 10 live ones in parts of northern Cameroon, and *'with only 148 left, scientists believe they will soon be locally extinct'.*[14] We also know that the future of forest elephants of Central Africa is bleak, severely threatened by poaching and the illegal trade in their ivory. But there are grounds for optimism.

We saw earlier how an English student, supervised in part by the Zoological Society of London, had successfully carried out *'a rapid assessment of the status and distribution of (these elephants) in southeast Cameroon'*, a practical method of providing essential data. In fact ZSL has been working in Cameroon since 2007,[15] in the D'ja Conservation Complex which is rich in biodiversity. Forest elephants, critically endangered lowland gorillas and the common chimpanzee all live within its boundaries, and southeast Cameroon represents one of the remaining strongholds of the forest elephants, so its importance cannot be overemphasized.

At the heart of the Complex is the D'ja Biosphere Reserve, and in its periphery lie forestry concessions, agricultural land, mines and community forests. It is yet another UNESCO World Heritage Site in crisis, and the

---

[12] Dalberg for WWF *Not For Sale Halting the Illegal Trade of CITES Species from World Heritage Sites 2017 – Chitwan National Park*, Unpublished data, email to Dalberg Global Development Advisors, December 2015 and Dalberg and WWF, Protecting People Through Nature, 2016, http://awsassets. panda.org/downloads/wwf_dalberg_protecting_people_through_nature_1r_singles.pdf.

[13] See Gray (n.10).

[14] Alastair Leithead *Why elephants are seeking refuge in Botswana* BBC News, 31 August 2016. See https:// www.bbc.co.uk/news/world-africa-37230700 Accessed 04/09/2018.

[15] *Welcome to our new partner*. Zoological Society of London (ZSL). See http://pfbc-cbfp.org/news_en/ items/zsl-enen.html. Accessed 05/07/2016.

threats, particularly from poachers, are too many and too difficult for the Cameroonian government to deal with unless they have outside help.

But it is a 'seed of hope' because help has been forthcoming. Other partners and sponsors have rallied to the call, and a project has been designed to *ensure improved and sustained protection of the Dja landscape and help secure its status as a key stronghold for the great apes and African elephant*.[16] The aim is to help the government's Ministry of Forestry and Wildlife and its Conservation Service. Eco-guards are being equipped and trained, and while camera traps continue to provide vital information, particularly of the movements of elusive species such as golden cats, leopards and white-bellied duikers, SMART technology has been introduced, and antipoaching patrols make arrests whenever possible, seize ivory and ammunition, and send it to the capital, Yaounde. Prosecutions are encouraged.

And the local communities, the indigenous Baka who live in the reserve, are being made aware of how important this is. Because they are very poor, the temptation to kill elephants for their ivory is very high.[17] They need a tangible benefit.

## 39.4    Malawi and the '500 Elephants' Initiative'

*The Journey of Giants* aims to translocate 500 elephants over 350 km to a new home and secure the long-term future of Malawi's elephants. It is an ambitious project. Malawi's Department of National Parks and Wildlife (DNPW) is working with African Parks, which works with governments and local communities to rehabilitate protected areas, which they then continue to manage long term.

Three national parks, all managed by African Parks and DNPW, are involved: Liwonde National Park and Majete Wildlife Reserve in southern Malawi and Nkhotakota Wildlife Reserve in the north. Both Liwonde and Majete are havens for photo-tourists, bursting with wildlife of all kinds. Something for everyone. But too much for some local people. With over 800 elephants, there are now high levels of human-elephant conflict in Liwonde, and the ecosystem is beginning to suffer. While Nkhotakota, the country's

---

[16] *Africa > Dja Conservation Complex* ZSL. See https://www.zsl.org/conservation/regions/africa/dja-conservation-c. Accessed 26/07/2016; and https://www.zsl.org/blogs/conservation/how-to-survive-a field-expedition-in-the-dja-. Accessed 17/08/2017.

[17] Lauren Redmore *Protecting forest elephants in Cameroon* Blog on August 12th 2014. Protecting forest elephants in Cameroon/Zoological Society of Lon… See https://www.zsl.org/blogs/protecting-forest-elephants-in-cameroon. Accessed 05/07/2016.

oldest park and once home to more than 1500 elephants, now has less than 100. Almost poached out!

When, in 2015, African Parks started to manage Liwonde and Nkhotakota National Parks, two things had to be done urgently. Fences had to be erected round both parks, in Liwonde's case, to reduce human-elephant conflict, while Nkhotakota needed its new residents to be protected from poachers. And law enforcement had to be strengthened. Then the journey could begin. Phase 1 took place in summer 2016 when 261 elephants were successfully translocated to Nkhotakota; the remaining 239 were moved in 2017; and the park *is poised to be one of Malawi's most important wildlife sanctuaries*.[18]

But will the new home prove safe for these travellers? African Parks have a proven track record when it comes to the safekeeping of animals. Majete, which they started managing in 2003, has been an amazing success story, a case of 'back from the brink'. Most wildlife had been hunted out,[19] but, thanks to reintroductions, now it flourishes again, to the extent that the park has elephants 'to spare' to, in effect, seed another ailing park. The park is *Malawi's only Big Five reserve now home to a population of more than 400 elephants*, 'tourism has flourished and benefits are being derived by the local communities'.[20] At the beginning of 2017, the British Army sent experts in tracking and bush-craft to Malawi, more training for the rangers taking on the elephant poachers,[21] such a successful pilot scheme that more will be deployed in May 2018, so that all three parks will benefit.[22]

## 39.5   Chad and Zakouma National Park

Chad, another of the countries home to the endangered forest elephants, has suffered similar problems to DRC. Although it is Africa's fifth largest nation, and although it is oil-rich, it has been dogged by poverty, instability and violence. And the vast herds of elephants that once roamed there have been decimated by poachers sweeping in across Chad's borders. There have even been estimates of as many as 35,000 elephants killed annually since 2010.

---

[18] 500 Elephants *About Us What is '500 Elephants'* See https://500elephants.org/about/. Accessed 30/03/2017.

[19] African Parks *The Journey of Giants Prince Harry takes part in historic '500 elephants'* 28 October 2016. See https://www.african-parks.org/newsroom/press-releases/the-journey. Accessed 30/03/2017.

[20] Ibid.

[21] Ben Webster *Army helps to catch elephant poachers* The Times, 18 November 2016.

[22] *British soldiers deployed in fight against poaching* The Times, 17 February 2018.

But then African Parks was asked to help, to bring back elephants to Zakouma National Park which had once been home to thousands of elephants, until the Sudanese horsemen swept in to slaughter them for their ivory. The park, part of the Greater Zakouma Ecosystem, lies just south of the Sahara Desert and above the fertile rainforest regions, an ideal position for a safe haven for both Central and West African wildlife.[23] Only created in 1963 by Presidential decree, it enjoyed the highest legal protection given by the laws of Chad, but this did not help when the poachers attacked.

Working with the government of Chad since 2010, African Parks have managed to all but eliminate poaching, only 24 elephants in the last 8 years, and the wildlife has returned. Elephants have started to breed again; 127 calves under 5 years old were counted early in 2018, unlike 2011, when there was only 1. Indeed, surveys completed in April 2018 showed elephant numbers had risen to 559.[24] And satellite collars have been fitted to the herds so they can be tracked and monitored.

Local communities are involved as well, playing a key role protecting their wildlife by notifying the authorities of suspicious activities. They have also benefitted from Zakouma's employment opportunities. It is the largest employer in the region. Education is also a priority; the schools that have been built within the elephant migration zone are known as elephant schools. Community outreach programmes *ensure that about 5,000 Chadian children and villagers visit the park each year'*. Tourism, and its revenues, flourishes.[25]

Zakouma has been such a success story that in October 2017, the government greatly extended the area African Parks had been managing, bringing in even more vital habitats *for the benefit of local communities and wildlife'.*[26] And in February 2018, the Republic and African Parks signed another agreement, this time for the Parks to manage and restore Ennedi Natural and Cultural Reserve, another World Heritage Site described as 'a unique international treasure'[27] and *'an absolute global gem (which) requires protection to secure this historic and culturally significant landscape'.*[28]

---

[23] Zakouma/African Parks See https://www.african-parks.org/the-parks/zakouma. Accessed 06/03/2018.

[24] African Parks August 2018.

[25] Ibid.

[26] *Greater Zakouma* African Parks. See https://www.african-parks.org/the-parks/zakouma. Accessed 06/03/2018.

[27] Chad's Minister of Environment and Fisheries. See https://www.african-parks.org/press-release/chad-ennedi-protected. Accessed 13/03/2018.

[28] Head of Charities of the Dutch Postcode Lottery which is providing some of the funding, $3 m over three years. See https://www.african-parks.org/press-release/chad-ennedi-protected. Accessed 13/03/2018.

And then, after very careful preparation, in May 2018 the rhinos returned, four females, two males, six magnificent black rhinos from South Africa. It was a very special occasion because it meant that now there are 11 countries with wild rhinos. So they were welcomed by the Governor of Salamat, the South African Ambassador to Chad and the Chadian Minister of Environment, '*who was presented with a bronzed rhino statue on behalf of African Parks to commemorate the rhinos' return*'. Translocated by plane, which landed on a specially constructed airstrip, the park's staff unloaded them into the bomas they lived in until August, when they were released into a temporary location in the park.

It was a message of hope, made possible '*because of regional cooperation, a shared vision and joint will, to protect the rhino and change the trajectory that this species is on—from one of demise, to a brighter and more hopeful future*'.[29]

Then disaster struck (see later).

## 39.6   Gabon

Again situated in central Africa, this was another country where forest elephants thrived. Or at least they did until the poachers moved in. Between 2004 and 2014, about 80% of Gabon's elephants, some 250,000 animals, most of them living in the Minkebe National Park (close to the border with Cameroon), were cruelly slaughtered for their tusks. The profits almost certainly (compelling evidence) helped to fund Boko Haram, the terror group.

But now there are reasons to be encouraged. In 2009, President Bongo took power and appointed Professor Lee White, a British conservationist who had lived for many years in Gabon and has become a dual-Gabonese national, to run the 13 national parks, which he does from the headquarters of Lope National Park, in the centre of the country. His top secret security clearance means he can work with the country's intelligence agencies and their international partners, and one result of this has been to find the link between a Gabonese trafficking ring and Boko Haram, with the ensuing arrest of 9 people and knowledge of the kingpin's identity.

Fortunately, there are still about 45,000 elephants, but they are protected by unarmed eco-guards, unless the parks are so dangerous that soldiers patrol with them. So the President appealed to the UK's then Prime Minister David Cameron for military help, and it came in the form of elite troops from the British Army, mainly from the Second Battalion the Rifles, veterans from Iraq

---

[29] *The Return of the Rhino to Chad*, an email from Peter Fearnhead, CEO to subscribers to African Parks.

and Afghanistan. In fact, in the past 6 years, British soldiers have been deployed in the DRC, Ethiopia, Kenya, Rwanda and Tanzania, to poaching hotspots where they train the rangers.

The troops are there for training purposes, for '*tracking, planning and executing operations and making better use of intelligence*'. In Gabon, gun laws are strict, but eco-guards may soon be allowed to carry a shotgun, as this is the right of every citizen who can then go subsistence hunting for bushmeat. But they still lack much basic equipment, even compasses and head torches.

However, there is yet another battle to be fought, one that is equally important, the battle for the hearts and minds of the rural people. Because of human-elephant conflict, they hate elephants. Again, help is at hand. Three solar-powered electric fences that administer powerful electric shocks have enabled one village to have '*their best harvest in years*'. Tourism is of course, the other way to win them round, though the tourists have to feel safe. And there are magical experiences on offer, '*hippos surfing in the sea, and elephants on the beach while whales are breaching just off shore*'.

China's new legislation is also playing its part, forcing down the price for ivory, so that the poachers are only selling at £70 instead of £150 per kilogram. There is hope for Gabon's elephants.[30]

---

[30] Jerome Starkey *British troops join the war on ivory poachers linked to terror* The Times, 12 May 2018 and Jerome Starkey *The Elephant Man* The Times Magazine, The Times, 12 May 2018.

# 40

# Zoos, Ecosystems and Translocations

An eclectic collection, but all of them making essential contributions.

## 40.1  Zoos and Conservation

Zoos, good zoos, are another priceless asset. They tick all the right boxes. They carry out research. They educate. They take part in conservation/captive breeding programmes. And they raise awareness of the animals they care for. This is so important, because most people cannot get to see them in the wild, their natural habitat, and although many of us are fortunate enough to watch some wonderful programmes on television, it is not the same as seeing the real thing. No one can be prepared for the wonderful sight of young elephant calves joyfully playing together, rushing in and out of a pool of water, splashing each other (just like young humans). Or fail to be amazed by the prehistoric-looking rhinos. Their calves are delight as well, bouncing about and galloping around at high speed. Zoos, therefore, can show people that such animals really exist. Above all else though, they are arks. A repository of extraordinary 'seed banks'. A final sanctuary.

Zoos have been an important part of this story, participating in many of the programmes designed to save elephants and rhinos. Captive-breeding programmes of endangered species are most important. Chester Zoo's Malayan Sun bears were rescued from the illegal wildlife trade and are now part of the European Endangered Species Breeding Programme, whose purpose is to keep populations physically and genetically healthy. Northern white rhinos became extinct in the wild some years ago; they would not now be living on

© Springer Nature Switzerland AG 2019
B. Martin, *Survival or Extinction?*, https://doi.org/10.1007/978-3-030-13293-4_40

Planet Earth were it not for zoos. And when Najin and Fatu pass on, zoos will continue to play a key role trying to ensure the extinction is not permanent. They are a source of hope.

### 40.1.1  Cincinnati Zoo and the Sumatran Rhinos

Sumatran rhinos have always been difficult to breed in captivity, but, in 1995, with very few rhinos available, Cincinnati Zoo made a breakthrough, mainly because of two dedicated scientists, Ed Maruskar, the Zoo's Director, and Terri Roth who were determined to help save the species. The Zoo itself has a long history of captive breeding, and has an entire research facility specializing in it, CREW.

By this time, there were only three of the rhinos left in America. Ipuh, a breeding male at Cincinnati, was joined by two females, Rapunzel and Emi, and ultrasound showed that Rapunzel would never breed as sadly, she had a large tumour in her uterus. That left Emi.

It took time and chance to work out that female Sumatran rhinos are induced ovulators. Summer 1997 and Emi was still not pregnant because, as the ultrasound showed, she was not ovulating. So Roth and her team decided to take a chance, knowing full well that other attempts to mate this species had ended in fights, some vicious, and no pregnancies. After doing everything they could to relax the couple, they introduced them, though always watching them just in case. Luckily they mainly ignored each other. This went on for 42 days, and then they started mating though unfortunately it never led to a calf. But 2 days after their first encounter, the ultrasound showed Emi had ovulated for the first time.

Now knowing what to do, and that her cycle was about 21 days, Emi conceived on her next encounter although she lost that baby and the next four. In 2000, when she was pregnant for the sixth time, the team decided to try a progesterone supplement, and it worked. Andalas was born 16 months later, the first of his kind to have been born in captivity in 112 years. Suci was born in 2004 and Harapan in 2007, both without the progesterone.

Both Emi and Suci died of iron storage disease, which can only be diagnosed after death, Emi in 2009 and Suci in 2014. Ipuh was euthanized in 2013. He had cancer and had come to the end. And as the Sumatran rhino program was improving, it was decided that perhaps it might be better for the rhinos to be in managed sanctuaries over there, so, in 2007, Andalas was sent to Indonesia, with Harapan making the same journey 8 years later.[1]

---

[1] *The Great Rhino U-Turn* Mongabay, September 2018. See https://news.mongabay.com/2018/09/the-great-rhino-u-turn Accessed 25/10/2018.

## 40.1.2   The Chyulu Hills Programme

One of Chester Zoo's most important activities with regard to our animals is its work to ensure the future of the critically endangered eastern black rhinos. To this end, it supports the Big Life Foundation's Chyulu Hills programme, which is trying to ensure that at least some of these rhinos can continue to live in the wild. The Chyulu Hills contains *'the only truly free-ranging population of eastern Black rhinos in Kenya'.*[2] Part of the Amboseli-Tsavo ecosystem, the Hills, an area of about 65,000 acres of habitat, lie to the north of the Tsavo East National Park in Kenya, some of them in the Chyulu Hills National Park and the rest in the Maasai-owned communal land bordering it.

Nowadays, most black rhinos live within fenced enclosures, one of the best ways, possibly one of the only ways to keep them safe from poachers. It follows therefore that wild-living rhinos require a great deal of protection, including guarding them for 24 h each and every day. Even then some fencing is required, along the eastern boundary which is densely populated by humans. Extra water points are planned as well, so that in the dry season, the rhinos do not need to stray from the protected zone.

The fact that these rhinos live on communal land as well as in the National Park has led to an *'an extraordinary level of cooperation between Big Life Foundation and the Kenya Wildlife Service'*, a *'significant achievement at a time when mistrust often dominates the relationship between government bodies and NGOs'.*[3] In practical terms, it has meant far more rangers being deployed and has resulted in far more effective protection. Essential, because now there are only eight rhinos. Luckily, the animals are *'incredibly shy and live in almost impenetrable bush'*, which again has contributed to the significant drop in successful poaching attempts.

The Maasai must be involved for the project to succeed, so the Big Life Foundation has made sure of their active participation in many conservation projects. Education of course plays a key role, so Chester Zoo helps fund the *'Maasai Olympics'*, *'a conservation themed athletics event based on traditional warrior skills'* which is very popular and *'Rafiki wa Faru'* (Friends of the Rhinos), which *'improves conservation awareness in the surrounding communities'.*[4]

---

[2] See   http://www.chesterzoo.org/conservation-and-science/where-we-work/africa/black-rhir   Accessed 29/03/2017.

[3] Craig Miller and Jeremy Goss *The most exciting rhino space in Africa* The Horn, Autumn 2015, re-printed by Save the Rhino International 2016.

[4] See   http://www.chesterzoo.org/conservation-and-science/where-we-work/africa/black-rhir   Accessed 29/03/2017.

Eight rhinos are hardly a viable population, so there are plans (part of the national action plan) to bring in others through translocations. However, under Kenyan legislation, this can only be permitted if the area has been designated an Intensive Protection Zone, where rhinos are protected round the clock, 24/7, by armed rangers, and this can only happen when certain criteria have been met.

### 40.1.3 Captive Breeding the Critically Endangered Eastern Black Rhinos

We have already met Kitani, the beautiful eastern black rhino who lost her baby only 6 weeks after he had been born. Nothing could have saved him from this congenital heart defect. Time passed. Nearly a year. She could be mated again. She is an important part of an essential conservation breeding programme, for there are now only 650 of these rhinos left in the wild. But was Kitani ready?

Chester Zoo has a magnificent male Black rhino called Magadi, and early in 2016, viewers of Channel Four (a UK television channel) were privileged to see their courtship. This is what happened.

The first thing that must have struck the viewers was the overwhelming kindness and concern of the keepers. Kitani was there to breed, but only when she was ready and they worried about this. As one keeper said, *'she needs to be happy rather than to make a baby,'* and they hoped it would help lift her out of her depression, for some months on she still looked sad and spent a lot of time just lying in her bed. Could all this change?

Rhinos mating is a dangerous activity at the best of times as it involves an element of fighting, and such large animals have the capacity to inflict potentially lethal injuries. So the two of them had to get used to each other, and Kitani was provided with an escape route.

Magadi began regular visits, initially very short, lasting about 10 min and with no physical contact. Although Kitani quickly got bored and would wander back to her bed, Magadi was definitely interested, and finally she started to notice him, even started looking out for him. One day, he brought her a leafy branch, *'a peace offering'*. She seemed to be coming out of her depression.

Only now was it time to risk a meeting, though again everyone was concerned because it was hard to predict exactly what the outcome would be. They need not have worried. Inevitably, it started off with some sparring, but this quickly settled down, and finally Magadi was moved into the same enclosure. The pair got on. The result? On 19th June 2017, Kitani gave birth to Hazina, a healthy girl.

The European stud book, a key element in this captive breeding programme, is managed by Dr. Mark Pilgrim, Chester Zoo's Chief Executive Officer, and his colleague Becca Biddle. They can determine who should mate with whom to produce the best result in terms of genetics, and Gabe is one of the happy results. His mother, Ema Elsa, was mated with Kirfaru who came to Chester from Hanover Zoo. Dr. Pilgrim described him as being of *'genetically…very high importance to the breeding programme so on paper he's a perfect male to breed from'.* The two rhinos were attracted to each other at once, with Gabe arriving 14 months later. Dr. Pilgrim again, *'Gabe is going to play a very important part in the Eastern black rhino programme as he will add to the genetic diversity of Eastern black rhinos in zoos across Europe. A birth of any critically endangered animal is significant but this arrival was truly extra special'.*[5]

One reason for the zoo's success has been its cutting-edge research. Monitoring the female's hormone levels not only enables the keepers to work out the best time to introduce the animals to each other but also diagnose pregnancies and determine when the baby might be born, normally privately, at night. Recently however, they were 'wrong-footed' when Jumaane was born in the middle of the day watched by visitors entranced by the sight. All went well, except that the astounded keepers missed it![6]

This programme is particularly important because in February 2017, Cardiff University dropped a bombshell. Researchers in the School of Biosciences published their discovery of a massive decline in the genetic diversity of the southern-central black rhinos in Namibia, a totally unexpected result and one that makes their future look bleak unless urgent action is taken. The problem has resulted from a combination of hunting and habitat loss, which has *'reduced (their) evolutionary potential dramatically over the last 200 years'*, so that *'a new strategy of conservation must be adopted to save (the animal) from extinction'.* But there was hope as well. The new data would help identify populations requiring priority conservation, thus *'giving us a better chance of preventing the species from total extinction'.*[7]

[5] Dr. Mark Pilgrim *Z Magazine* Chester Zoo, Spring 2016, p.2.

[6] *Z Magazine* Chester Zoo, Autumn 2018, p.6.

[7] Professor Mike Bruford, Co-Director Sustainable Places Research Institute, in *Decline in genetic diversity of black rhino* Cardiff University News, 8 February 2017. See https://www.cardiff.ac.uk/news/view/567269-decline-in-genetic-diversity-of-black-rhino Accessed 25/10/2018.

## 40.1.4 'Never Forget'

Anyone connected with Chester Zoo was devastated when two of its best loved young animals, Bala and her little playmate Hari, delightful baby elephants, succumbed to EEHV, endotheliotropic herpesvirus, a deadly Ebola-like disease that attacks the critically endangered Asian elephants. Occurring both in zoos and in the wild, it has been recorded in Cambodia, India, Indonesia (Sumatra), Laos, Myanmar, Nepal and Thailand. Young elephants, typically around 2 years old, their weaning age, are its victims, and the virus is so virulent that, at the moment, even with the earliest diagnosis and the best veterinary care available, only about one in five animals survives. Tarli was one of the lucky ones. She lives in Woburn Safari Park in the UK.[8]

The disease is mysterious. While it lies dormant and undetectable in many elephants, in others, it swiftly moves in to kill in a matter of days. Researchers are currently working on improving the vaccine and determining which are the most effective drugs. *Never Forget* fundraises so that the Zoo can contribute to a number of these projects.[9]

## 40.1.5 Fighting the Illegal Trade

This is an important function of Chester Zoo's Act for Wildlife team, which describes itself thus:

> *WE ARE ACT FOR WILDLIFE*
>   *We are leading the way in saving endangered wildlife around the world. In the fight for conservation we're a force to be reckoned with. We work tirelessly to ensure species are protected.*

The Zoo is particularly active in the fight against the illegal wildlife trade and has worked with both TRAFFIC and the National Wildlife Crime Unit (NWCU) for a number of years. Recently, it has provided homes for trafficked songbirds. Toni and Milli their Malayan Sun bears, taken from the wild as cubs, were rescued from cruel captivity by conservationists from the NGO Free the Bears, Toni from a wildlife trader in Cambodia in 2004 and Milli from a military general. As a cub, she was a pet, but once she grew larger, she was chained to a tree and fed only occasionally. The chain had to be cut from her neck to free her. After being nursed back to health by Free the Bears and

---

[8] The Times, 16 February 2018.
[9] *Z Magazine* Chester Zoo, Spring 2016, pp.10–11.

undergoing initial rehabilitation by the team at the Rare Species Conservation Centre in Kent, UK, they are now living contentedly in Chester and have just produced their first cub, a female, Kyra 'Sun Goddess'.[10] Another happy ending.

Messages about the illegal wildlife trade are scattered throughout the Zoo, with the final focus found on *Islands*, which was opened in 2015 and is where the animals and plants from southeast Asia live. In 2016, the storyline was strengthened to emphasize not only the field conservation work the Zoo carries out but also to highlight the issue of illegal wildlife trading. Visitors follow in the footsteps of the field teams, passing various exhibits en route, such as notices warning of poaching, of what happens to poachers, carefully designed to attract interest.[11]

Eventually, the message gears up. Inside the Tapir House are three WARNING NOTICEs:

- The first defines *'wildlife'*.
- The second, *'wildlife trafficking'*.
- The third, *'wildlife trade'*.

*Illegal harvesting* is also defined, as is *poaching*—the illegal capturing or killing of wild animals.

And once one enters the *Sun bear tunnel*, all becomes clear. Some 30-ft. long, the 'tunnel' is more like a big gallery, light and airy, with plenty of space to convey the message, which, overall, is positive. The story starts with the 'Interactive crate', which is a model of the end of a transport container. As well as the many simple interactions there is also a short film about the illegal trade.

This is followed by a map.

In the final display cabinet, opposite the Sun bears home, Act for Wildlife tells us how we can help:

*There's something YOU can do today to help STOP illegal wildlife trade*
   *When you're ONLINE*
   *When you're ABROAD*
   *When you're OUT AND ABOUT.*

---

[10] Information provided for the author by Alex Knight, Chester Zoo and also Chester Zoo's *Z Magazine*, Autumn 2018, p.3.

[11] '*Islands Extension Brief*', Chester Zoo, 9 September 2016, p.3. and other information, all kindly provided by Rowena Hamilton.

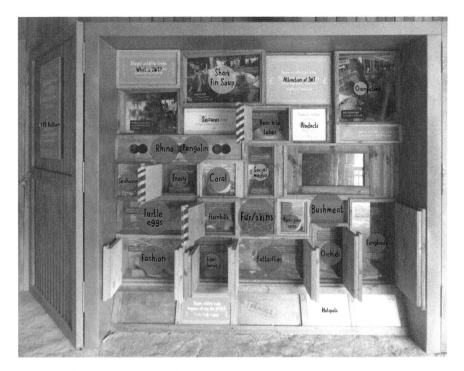

**Fig. 40.1** Chester Zoo's Interactive Crate. Copyright Chester Zoo

The online site is excellent, clearly explaining, with examples, what wildlife crime is, including illegal trading in elephant ivory and rhino horn, and the practical steps you can take if you witness a crime or have information about a potential crime or a crime that has occurred. As well as educating the public, this is an exercise in data gathering, possible evidence that may at some time lead to a successful prosecution. It gives specific advice on suspect trading on the internet. *'Try and provide full internet addresses and take a screen shot of what you have seen. Offenders often remove posts and they then can't be retrieved at a later date'.* An online form is provided for anyone wishing to report a (possible) criminal offence.[12]

The *'Dos and Don'ts of Reporting Illegal Wildlife Trade'* are clearly set out, in list form to make them easier to understand. They are comprehensive, examples include:

- *'DON'T put yourself at risk! Be sensible and discrete; don't make your report directly in front of the witnessed trade and only take photos if it's safe to do so'*; which is emphasized again in

---

[12] Act for Wildlife *Report it* Chester Zoo. See https://www.actforwildlife.org.uk/what-we-fight-for/conservation-challenges/our-campaigns/illegal-wildlife-trade/report-it/ Accessed 13/09/2018.

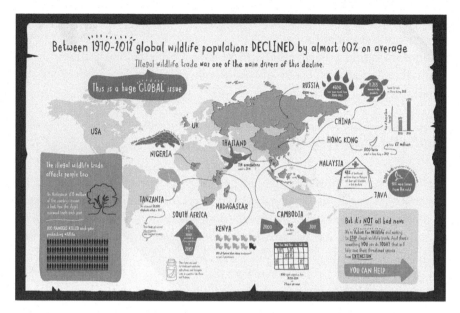

**Fig. 40.2** Map showing the decline of global wildlife populations between 1970 and 2012. Copyright Chester Zoo

- *'DO be careful not to arouse suspicion, if you think something illegal is taking place follow the staying safe tips and then make a report;*
- *DON'T point out illegal items to the sellers or staff selling the products'.*
- *DO look out for the different types of illegal wildlife trade, for example: Poaching – taking wild animals from their natural habitat is illegal so if you're travelling to or near any forested areas keep your eyes peeled for any suspicious activity. If you spot any traps, snares or nets these are sometimes a sign that poaching is taking place. Don't touch them! Just record where you've spotted them, and make sure you include a description of the surrounding area.*[13]

There is also a free app to download for British travellers to southeast Asia, a Wildlife Witness app. Using it to take a photo *when safe to do so*, this also pins the exact location the possibly illegal trade took place. The details are sent to TRAFFIC who can follow it up. The app is supported by zoos all over the world. Yet again the importance of personal safety is emphasized as members

---

[13] Act for Wildlife *Dos and Don'ts of Reporting Illegal Wildlife Trade* Chester Zoo. See https://www.actfor-wildlife.org.uk/what-we-fight-for/conservation-challenges/our-campaigns/illegal-wildlife-trade/dos-donts-reporting-illegal-wildlife-t Accessed 13/09/2018.

of the public may be unaware of just how dangerous it can be trying to prevent illegal wildlife trading.[14]

The positioning opposite the Sun bears' home is deliberate, their stories vividly illustrating the cruelty of this illegal trade. But again hope, for they were rescued and now they are safe.

## 40.2   Ol Pejeta's 'Boma' System

This boma system provides an unusual form of sustainable use which has enormous potential to benefit both farmers and wildlife. It is innovative because it uses cattle as a 'keystone' species to develop a different kind of eco-system, making a radical departure from the frequent conflicts between pastoralists and wildlife. So how did this come about?

In the 1950s and 1960s, cattle were the life-blood of the pastoralists who lived there, which left little room for wildlife. However, in the late 1980s and early 1990s, poaching became such a severe threat to the black rhinos that something had to be done. Sweetwater Sanctuary was established to protect them, and because, at that time, cattle and rhinos didn't mix, the Sanctuary was completely fenced off to keep out both cattle and poachers. This divided Ol Pejeta into two parts, the western side, which was for agriculture and the eastern side where there was wildlife, and hence tourists. Two income streams, but they were completely separate.

In 2005 everything changed again. This time, a radically new approach was taken. Cattle and wildlife were integrated. Cattle cleared away the old stale grass, regenerating new sweet grass, the sort of pasture wild herbivores love to graze on. Of course, it all depended on getting the correct cattle. Borans proved ideal. They bred well, produced good beef, and there was a 25% off-take.

The system works by dividing the 7500 cattle into small herds of 150 animals, each herd looked after by 2 men. During the day, the cattle graze on good grass under the watchful eyes of their herders, while at night, they are collected into their bomas which protect them from predators. Safely inside, the Borans graze on the old grass, which takes 2 years to clear, and at the same time, their dung fertilizes the land, thus enabling sweet new grass to grow.

---

[14] Zoo News *Zoos and aquariums call for an end to illegal wildlife trade* Chester Zoo, 7 September 2018. See https://www.chesterzoo.org/whats-happening/zoo-news/2018/19/call-for-an-end-to-illegal-wildlife-trade Accessed 13/09/2018.

Because the bomas are mobile, the cattle can be moved around, gradually regenerating the old grass and 'gardening' it for wildlife.[15]

Data obtained from the new system has shown that:

- The layer of manure should not be too deep, so the bomas need to be moved to prevent this.
- The accumulated *'hot spots'*, where the grass has been improved, result in *'glades'*.
- The bomas increase biodiversity.
- Reduce tick densities.
- Revegetate bare patches (in other words achieve restoration).
- The improved visibility they create makes it easier to see predators.[16]

Although some cattle are lost to both predators and tick-borne diseases, the percentage is tiny, some 3%, and the benefits far outweigh any losses. In fact everything benefits, including the wildlife:

- The system makes full use of the available resources.
- More people are employed, an estimated three to four times more than 15 years ago.
- All this brings prosperity which trickles down to the neighbouring communities who then receive better support.
- Because the system benefits both beef production and wildlife, the area benefits from two income streams, and should one fail for any reason, there is still the other as a backup.

Such an innovative form of sustainable use and what better way of involving the local communities the wildlife are so dependent on. Cattle and wildlife graze together. Pastoralists benefit not only because their cattle flourish but also from the extra tourists who flock in to see the increasing wildlife, which has also benefitted. Truly, a win-win situation.

The boma system can provide a more sustainable use of land in other pastoral areas too, where people are still struggling to see any benefits from conservation. It has meant that, in Ol Pejeta, rhinos now graze with cattle. Whereas in the past, land was taken from the farmers and used to keep the rhinos safe, a practice that must surely have generated a great deal of resentment and ill-will towards the animals, this is no longer necessary.

---

[15] Jerry Sellanga *Online Seminar*, Ol Pejeta, 2016.
[16] Ibid, Caroline Ng'weno.

The African Wildlife Foundation is attempting to set up a similar system in Tanzania, in the Maasai Steppe, as a possible strategy for reducing human-lion conflicts. The Maasai also collect their cattle into bomas for the night, to protect them from hungry lions. Unfortunately though, their bomas, which were made of wood from acacia trees, were not strong enough to do the job, but their two new mobile bomas are predator-proof.[17]

## 40.3 Translocations

Sometimes more dramatic action has to be taken. WWF has described translocation as *'one of the most successful techniques we've come up with in rhino conservation – and it's a technique we've perfected over almost 40 years'.*

Moving such large animals as elephants and rhinos from one location to another was never going to be either straightforward or easy though. It posed considerable risks to the survival of the animals involved, so initially, translocations were regarded as a last-resort activity, one with no guarantee of success. However, over the years, various techniques have been developed. There are always teams of qualified people involved, the receiving habitat is always well prepared, and now translocation works. Or almost always works.

There are various reasons for using this technique, one being the removal of animals to a safer place. This is particularly important if the animals are rare and, for example, in danger from poachers. It happens quite regularly to rhinos in both South Africa and Namibia, the animals being moved to small reserves that offer better protection.[18] It has also been important in Zimbabwe. In 1987, there were more than 1775 black rhinos in the Zambezi Valley, but by 1995, there had been a catastrophic fall in numbers, down to 315. A translocation programme moved some of the remaining rhinos to private game ranches and conservancies that had better resources to take care of them, and 49 went to the southeast lowveld of Zimbabwe, to what is now the Save Valley Conservancy.[19] Another reason is to improve genetic diversity, while a third is the repopulation of areas that can once again sustain a viable population. Solio Rhino Sanctuary

---

[17] African Wildlife Foundation *Maasai Steppe Predator – Proof Bomas Ending human-carnivore conflict in Tanzania* See https://www.awf.org/projects/maasai-steppe-predator-proof-bomas Accessed 16/08/2016.
[18] Nikki Le Roex Africa: *How Translocating Rhinos Promotes Genetic Health and Keeps Them Safe* The Conservation (Johannesburg). See http://allafrica.com/stories/201511232909.html Accessed 26/11/2015.
[19] Tusk *Save Valley Conservancy*. See http://www.tusk.org/save-valley-conservancy Accessed 13/09/2016.

in Kenya bred so many black rhinos that most of Kenya's other rhino sanctuaries started with a founder population from there.[20]

And we have already looked at one of the most ambitious attempts at translocating elephants to date, the *'500 Elephants'* Initiative in Malawi.

So how is it that in this current poaching epidemic that is threatening the very existence of the animals, there are 500 elephants available to be moved? We already know the answer. It is the reason for the one-off sales of elephant ivory, right back at the beginning of our story. Nature is not consistent. While elephants are extremely rare in some places, in others, there are just too many. Different populations, different threats, different numbers.

Malawi has worked very hard to rid itself of poachers, successfully clearing them not only from Nkhotakota Wildlife Reserve, where the elephants are moving to, but also from Liwonde National Park and Majete Wildlife Reserve, which enabled those elephant populations to flourish. And it is from these two locations that the elephants are being translocated.[21]

The Reserve is also being repopulated with other animals, zebra, buffalo and impala, to attract tourists and the income they bring, to one of the poorest countries in sub-Saharan Africa. Once again we see how wildlife can bring benefits to humans, the link between conservation and sustainable use.

India has also used translocation to help its rhinos, the greater one-horned rhinos. They used to live in Manas National Park, in Assam, but sadly, by 1999 there were none. However, in 2009, the population started to rebuild when five rhinos were reintroduced. Since then the numbers have grown wonderfully. Eighteen adults were translocated from other reserves in India and between 2013 and 2015, 13 calves were born. Even more extraordinary was the fact that two of the females gave birth twice, within about 2 years, a very rare occurrence. WWF, who were in part responsible for this unexpected success story, described it as *'providing a ray of hope for the future of this species'.*[22]

Nepal's rhino population continues to grow and has now increased so much that it too is translocating rhinos. Between 2016 and 2018, the plan is to move 30 rhinos from Chitwan National Park to Bardia National Park, with its 25 rhinos, and Shuklaphanta Wildlife Reserve, which only had 5. All three sites lie within the Terai Arc Landscape, and the object is to create a second

---

[20] Wildlife Extra News *Largest Ever Rhino Translocation in East Africa*. See http://www.wildlifeextra.com/go/news/laikipia-rhinos.html#cr Accessed 13/09/2016.

[21] Anna Dubuis *Hundreds of elephants drugged and moved to help save species* The Times, 22 February 2016.

[22] WWF *Together, we did it! India You helped provide sanctuary to rhinos* Action WWF Magazine, Spring 2016, p.5.

viable population, eventually restoring the greater one-horned rhinos to their historic numbers in Nepal.[23]

One rhino has already gone to its new home in Bardia, and with four more to follow, WWF-UK's Head of Asia Programme made this telling statement: *'The commitments and efforts of the government, NGOs and communities has resulted in Nepal becoming a leader for conservation globally and we hope to see more countries follow in its footsteps'.*[24]

While moving rhinos to safer places within the same country or within the same continent might seem an obvious way to mitigate risk, moving a large group of rhinos to another continent entirely seems somewhat bizarre, yet despite some initial controversy, this is now happening.

## 40.3.1   The Australian Rhino Project

On its face, this seems a bizarre plan, but it was the initiative of a South African, Ray Dearlove, who, although he had moved to Australia many years earlier, was still deeply concerned about the fate of rhinos. The aim of the Australian Rhino Project is to *'establish a breeding herd of rhinos in Australia as an insurance policy for the species'.*[25] Working with partners, this will be accomplished by translocating 80 rhinos over 4 years to *Wild Africa*, Monarto Conservation Park, an 120 acre site which has been carefully developed to resemble the Kruger National Park.[26] By June 2018, their housing facility had been completed. And New Zealand's Orana Wildlife Park had joined in, building a new rhino facility to house more of the animals, while at the same time, acting as a quarantine facility for rhinos moving on to Australia.[27]

Although the project has attracted considerable controversy, the idea is to return the rhinos once their future is more secure, and some of the private park owners are keen not only to see their animals moved to a safer place but

---

[23] Barker WWF *Rhino translocation off to a positive start in Nepal's Terai* 2 March 2016, *Rhino translocation off to a positive start in Nepal's Terai | Stories* … See http://www.worldwildlife.org/stories/rhino-translocation-off-to-a-po Accessed 09/08/2016.

[24] WWF *Successful Start for Nepal's rhino translocation expedition* 2 March 2016; Successful start to Nepal's rhino translocation expedition – WWF UK See http:// www.wwf.org.uk/about_wwf/press_centre/?unewsid=7818 Accessed 09/08/2016.

[25] Nosipho Mngoma *Plans in motion to relocate SA rhinos to Oz* The Mercury, 21 June 2018. See https://www.iol.co.za/mercury/plans-in-motion-to-relocate-sa-rhinos-to-oz-15598715 Accessed 26/06/2018.

[26] Jane Flanagan *Airlift to Australia offers hope for rhinos* The Times, 23 April 2018.

[27] Jonathan Guildford *Orana Wildlife Park to build new $500,000 rhino facility to help save animals from extinction in 2024* Stuff, 14 February 2018. See https://www.stuff.co.nz/environment/101449780/orana-wildlife-park-to-build-new-50 Accessed 08/03/2018. Australia's strict quarantine requirements mean that New Zealand will hold the rhinos for a year.

also to be relieved of the crippling expenditure currently spent on protecting them from poachers.[28]

## 40.3.2   Increasing Genetic Diversity

Genetic diversity is vital, so once the Mkomazi Rhino Sanctuary was established, the decision was taken to bring in new black rhinos to achieve this. We have already met the three captive-bred rhinos from Port Lympne Wild Animal Park in the UK who were duly translocated to their new home and their new companions. They settled well and are now breeding. Eric is the latest eastern black rhino to travel to Africa, this time to Tanzania, to strengthen the gene pool in the Singita Grumeti Reserve, an Intensive Protection Zone, for these precious animals need maximum protection.[29] A gift from San Diego Zoo, Eric is being carefully prepared for his new life, for he has to learn to live in the wild, how to be a wild rhino, before he can be released into his new home. There, he will find Laikipia, his proposed mate. She originally came from the UK, from Howletts Wild Animal Park, and was successfully 'wilded' into life in the Reserve.

## 40.3.3   A Dangerous Activity

Although the utmost care is taken when translocating any animals, sometimes things can go terribly wrong. This happened in July 2018 when 11 out of 14 black rhinos, about 1% of the entire world population, did not survive a move from their homes in Nairobi and Lake Nakuru National Parks to a new reserve in Tsavo East, a coastal reserve. And it was deeply shocking because the move had been planned so carefully. Tests eventually revealed they had been poisoned by the salt in the water.

Then, in early October, two of the six rhinos translocated from South Africa to Zakouma National Park were found dead, in different locations. The good news was that they had not been poached.[30] By early November, despite being closely monitored, two more had died. Post-mortem results showed that it was almost certainly due to starvation, although more tests were being run to check for any contributory factors. Despite all the care that had been taken to ensure their new habitat would be completely suitable for them to thrive, and reproduce, it seems they simply did not adapt to it.[31]

---

[28] Jane Flanagan *Airlift to Australia offers hope for rhinos* The Times, 23 April 2018.

[29] Jane Flanagan *Eric the rhino travels halfway round the world for love* The Times, 21 September 2018.

[30] African Parks *Statement from the CEO: Update on Rhinos in Zakouma*, in an email. A similar statement announced the deaths of the other two rhinos.

[31] Jane Flanagan *Rare rhinos starve after move to Chad* The Times, 7 November 2018.

# Part XIV

## Towards a Resolution

# 41

## Last Chance to Save

Eventually though, there comes a tipping point when deaths exceed births and populations start to decline. A first step on the road to extinction, and this is beginning to happen with some elephants and rhinos.

A major problem is the fact that it takes a long time for these animals to reproduce. Fortunately though, there are management plans in place, as well as breeding programmes, sometimes involving zoos like Chester Zoo, not only working with partners out in the field but also at home, actively participating in research, education and captive breeding programmes.

## 41.1 The Bornean Pygmy Elephant

A subspecies of the Asian elephant, these really are small animals, about a fifth smaller than their Asian cousins. Round and small. They have rounder faces, shorter trunks and generally look more rotund. Their tails are long, sometimes even reaching the ground, their ears are large, although this is merely an illusion, caused by their faces, and fortunately very few of them have tusks. Nor do they behave like mainland elephants.[1]

These are elephants of mystery, their origins obscure. Romantic stories abound, of the Sultan of Sulu in the seventeenth century, who was given a domesticated herd of the animals, and although they can be fierce, they are said to be more gentle and less aggressive than the mainland species.[2]

---

[1] World Land Trust *Bornean Elephant* See http://www.worldlandtrust.org/education/species/bornean-pygmy-el Accessed 0/08/2016.

[2] Ibid.

© Springer Nature Switzerland AG 2019

B. Martin, *Survival or Extinction?*, https://doi.org/10.1007/978-3-030-13293-4_41

When their DNA was examined by WWF, the results were surprising. The elephants were genetically distinct from other Asian elephants and must have been isolated from them about 300,000 years ago.[3] It is possible then that there were Javan elephants brought to Borneo by the Sultan of Sulu, and the Pygmy elephants are their descendants. Sadly, they are now endangered. With only about 1000 to 1600 of them left, they face a high risk of extinction in the wild, and they are of *great conservation interest* because Javan elephants on both Java and Sulu were hunted to extinction sometime in the nineteenth century.

Once again, the threat comes from people. And from the production of palm oil, that ubiquitous oil found in so many products, but which can and should be produced sustainably. Sabah has lost over 40% of its forests in the past 40 years, with the inevitable result that elephants go into palm oil plantations to feed and end up *causing considerable damage.*[4]

In January 2013, ten elephants in Kawang Forest Reserve were found dead, poisoned. But not by poachers, for they still had their tusks. A tiny baby was found *trying to wake her mother up.*[5] A grim example of what can result from human-elephant conflict, but the Danau Girang Field Centre working with their partner Elephant Family had been satellite tracking a number of collared elephants, and the data they obtained enabled them to set up the Kuamut Forest Reserve which has also linked together two other protected areas.[6]

Worse was to come, however. Four years later, the poaching started. In January 2017, the skeleton of Sabre minus his tusks was found, and 1.5 km away, another dead elephant, this time newly killed. His face had been hacked off to remove his tusks. Both had probably died within a month of each other. Sabre, named after his sabre-shaped tusks, had been special. Rescued from a palm oil plantation a few months earlier, fitted with a satellite collar then released into the wild, conservationists and members of Sabah's Wildlife Rescue Unit had thought him safe. Deeply upset and grieving for the two elephants, they were also very worried by the possible implications.[7]

---

[3] WWF *Borneo Pygmy Elephant* See http://www.worldwildlife.org/species/borneo-pygmy-elephant Accessed 30/08/2016.

[4] Elephant Family 2017 See http://elephant-family.org/what-we-do/where-we-work Accessed 27/02/2017.

[5] *Rare pygmy elephants 'poisoned' in Borneo* BBC, 29 January 2013.

[6] Elephant Family *Securing a Future for Borneo's Elephants* 201 See http://elephant-family.org/project-update-securing-a-future-for-borneos-elephants?ct= Accessed 11/08/2016.

[7] *World's smallest elephants killed for ivory in Borneo* The Guardian, 3 January 2017. See https://www.theguardian.com/environment/2017/jan/03/wborneo-p Accessed 28/02/2017.

These small elephants are losing their forests to palm oil plantations and logging. Human-elephant conflict is increasing. To make matters worse, now the poachers have arrived. We know that elephants are megagardeners, keystone species. Borneo and its forests need their elephants. We must not let them become extinct.

## 41.2 The Northern White Rhino

Last chance to save needs to be undertaken while there is still a fighting chance of saving the animals, and although we have seen the cutting-edge research that is going into the project, it may already be too late for the northern white rhinos. They might be gently slipping into the past.

At one time, northern white rhinos used to range widely throughout Central Africa, but over the years, their numbers and range reduced, and by 1984, only 13 were left, all of them living in the unstable DRC, in Garamba National Park, which borders with Sudan. The poachers who came for the rhinos came mainly from Sudan, and they were heavily armed.

Then the situation improved. Garamba was given trained staff, vehicles, a light aircraft, permanent guard posts and mobile patrols all in contact with each other, so that when Douglas Adams[8] and Mark Carwardine, the conservationist, visited the park in 1990, the numbers had increased. Up by 9 to 22. But not for long. Soon the population dropped to four, the last four northern white rhinos living in the wild.

In 2005, negotiations to take five of the rhinos from the park for a captive breeding programme failed. A year later, there was what turned out to be the final sighting of rhino spoor, and in 2007, this subspecies of white rhino was declared extinct in the wild.

That left seven, all of them in captivity. Najin, Fatu, Suni and Sudan, who were living in Dvur Králové Zoo in the Czech Republic, had been collected by a Czech who had devoted his life to the animals. In December 2009, they were transferred to Ol Pejeta Conservancy in Kenya. Back in their native habitat, perhaps they would breed. But it did not happen. Things only got worse. In Autumn 2014, two of the remaining males died, Suni, aged 34, of a suspected heart attack, and Angallifu, aged 44, who lived in San Diego Zoo, California. Now there were five.

---

[8] Author of *The Hitchhiker's Guide to the Galaxy*.

In November 2015, Nola died. She was 41 years old. She too had lived in San Diego Zoo, where she had been lovingly cherished, but she had been suffering from poor health for some time and finally the time arrived for her to be put down. A sad and very difficult decision. Only three left. Last chance to save!

Sudan, the last male northern white rhino died of old age on 19 March 2018. His death made headline news for he had thousands of friends and well-wishers throughout the world, and he will be greatly missed. He was called Sudan because that was where he came from back in 1975, when he was only 3, when there were still about 1000 of these rhinos ranging through Central Africa.

In 2017, Sudan turned to the dating app 'Tinder' for help. The idea was to fundraise a contribution to the target sum of USD10 million needed for the attempt to save this species. We have seen how well the research is progressing, and this money can be used to fertilize one of the female's eggs with his sperm and implant it in a southern white rhino surrogate mother. Sudan fathered two females, and the day before he died, his genetic material was collected, signalling a hope for the future.[9]

Sudan's profile on Tinder was a delight. *'I don't mean to be too forward, but the fate of my species literally depends on me'*. He described himself as *'one of a kind'*. The accompanying smiley face and rhino emoji said it all. And over USD85,000 was raised.

In July 2018, Nature Communications published the latest research findings, and they are encouraging, for the team has fertilized the eggs (oocytes) of a southern white rhino with northern white rhino sperm, thus generating hybrid rhinos. They have done this by developing a technique whereby eggs can be repeatedly recovered from live southern white rhino females, matured, fertilized and finally developed using assisted reproduction techniques (ART). They have established embryonic stem cell lines.[10]

Southern white rhinos did come back from the brink, so can we help the northern whites do the same? Until we have an actual live birth with the calf surviving and able, when adult, to reproduce, extinction continues to remain a bleak and obvious possibility.

---

[9] Press Release *The Last Male Northern White Rhino Dies* Ol Pejeta Conservancy 20 March 2018.

[10] Thomas B. Hildebrandt, Robert Hermes, Sylvia Colleoni, Sebastian Diecke, Susanne Holtze, Marilyn B. Renfree, Jan Stejskal, Katsuhiko Hayashi, Micha Drukker, Pasqualino Loi, Frank Goritz, Giovanna Lazzari & Cesare Galli *Embryos and embryonic cells from the white rhinoceros* Nature Communications 9, Article number: 2589 (2018), published 4 July 2018.

**Fig. 41.1**   A rare picture of a Sumatran rhino. Copyright Save the Rhino

## 41.3   Sumatran Rhinos

*'The ultimate goal (of captive breeding) is to keep remaining animals in human care until the threats have been addressed and the population is large enough to re-establish them in the wild'.*[11] Captive breeding programmes can play a major role in saving species, and although it could not save the northern white rhinos, there is still a chance of saving Sumatran rhinos, and a contingency plan is in place to do just that. But despite so many efforts, it never seems to be completely put in action. And the time clock ticks on, inexorably.

According to the International Rhino Foundation, in 2013 there were fewer than 200 Sumatran rhinos living in the wild. Most of these were located in Bukit Barisan Selatan, Gunung Leuser and Way Kambas National Parks in Sumatra, Indonesia, although a few were found in Malaysia, in the Tabin Wildlife Reserve, in Sabah.[12] By 2015 the situation had deteriorated consider-

---

[11] Margaret Kinnaird and Jon Paul Rodriguez *There is still a chance to save the Sumatran rhino (commentary)* Mongabay Series: Asian Rhinos, 3 May 2018. See https://news.mongabay.com/2018/05/there-is-still-a-chance-to-save-the-sumatran-rhin Accessed 10/05/2018. M. Kinnaird is WWF Wildlife Practice Leader and J.P. Rodriguez is IUCN Species Survival Commission Chair.

[12] International Rhino Foundation *Sumatran Rhino Conservation Program* See https://rhinos.org/where-we-work/sumatran-rhino-conservation-program/ Last Accessed 06/12/2018.

ably, and possible numbers zigzagged between 87 and 179, with subpopulations from 2 to 50 rhinos.[13] While a few of these are in Kalimantan, in the part of Borneo that belongs to Indonesia, they are now extinct in Malaysia. The Sumatran Rhino Sanctuary has five captive rhinos, three more live in a similar location in Sabah, Borneo, and one lives in Cincinnati Zoo in the USA.[14]

It is the same sorry story. Sumatran rhinos used to be widespread, but they are the only species of Asian rhinos that has two horns, and they have paid the price. The records show that thousands of their horns were sold and exported during the nineteenth and twentieth centuries, to be carved into beautiful objets d'art or used in medicine. Now we are left with only remnant populations of these attractive animals, the most ancient of all the rhino species.

Although poaching, of course, is still an ever-present problem, habitat destruction, particularly in the form of palm oil plantations, such an easy way to make money, is resulting in the fragmentation of the remaining rhino populations, and a steady encroachment of human settlements only adds to their difficulties.

A team of researchers from the University of Massachusetts Amherst and the Wildlife Conservation Society's Indonesia Program were determined something had to be done. The first thing was to gather information about the animals. Were they actually living where they were supposed to be living and how many were there? Not easy to find out with such rare animals. *Challenging* was the word the team used!

The team focussed on three areas. By noting any signs of rhino and determining the probabilities of them being in that particular location, they were able to develop a habitat model *'which predicted that rhinos now only occupy 13 percent of the surveyed area'* and *'with so many unknowns on how to manage Sumatran rhinos in the wild, or captivity, our study definitely shows where we must protect them at source'.*[15]

Five specific areas were identified by the team, who recommended that they should be designated *Intensive Protection Zones*, which means protection from poachers. Although intensive law enforcement should ensure that poaching is reduced to zero, unfortunately new roads had been planned in two of the five areas, which could be a disaster. Even with increased protective measures it

---

[13] UMass Amherst *Saving the Last Groups of Wild Sumatran Rhinoceros, Wildlife Conservation Society use enhanced population survey techniques* 16 September 2015 See https://www.umass.edu/newsoffice/article/saving-last-groups-wild- Accessed 05/05/2016.

[14] Jani Actman *Can Anything Save the Sumatran Rhinoceros From Extinction?* National Geographic 30 September 2015, interview with Rasmus Gren Havemollerr, University of Copenhagen.

[15] Pusparini 2015.

would make it far easier for potential poachers. Not surprisingly, the team recommended that these roads should not be built.

The other key recommendation was collect up *all remaining small populations and scattered individuals of healthy rhinos* and *consolidate them*. But to do this, the law would have to be changed, because under current legislation it is forbidden. Which was the reason why, in 2013, the Bandar Lampung Declaration, a conservation strategy designed to increase the rhino population, came into existence. It would remove this obstacle. But it has never been implemented and cannot be, without the consent of the President of Indonesia, which is needed to amend the law.

Translocation would not only improve the rhinos' chances of mating, which is not easy to achieve if males and females rarely encounter each other, it would also result in greater genetic diversity.[16]

By 2015, the Intensive Protection Zones had never materialized, and the numbers of wild rhinos had fallen below 100. Fortunately the nine captive rhinos were still alive. A team from the University of Copenhagen, Denmark, considered the latter to be essential if the rhinos were to be saved, *a last-chance solution to save the species*. They wanted all the remaining rhinos (wild and captive) to be *managed in a single program across national and international borders in order to maximise overall birth rate*, for a similar intensity of activity put into Sumatran rhino conservation as that put into Tiger conservation and *a similar high level intervention by President Joko Widodo of Indonesia could help pull the Sumatran rhinos back from the brink*.[17]

April 2015, Sabah Borneo. The Minister for the Environment stated that there appeared to be no rhinos left in the wild. This was a blow. That only left three rhinos, two females and a male, Iman, Puntung and Tam, in the Borneo Rhino Sanctuary in Tabin Wildlife Reserve. Could they be assisted to reproduce? As we have seen earlier, if female rhinos go too long without breeding, they are liable to end up with severe problems in their reproductive tract. This is what had happened here. It shows how essential it is to have a viable breeding herd. In the wild, these poor rhinos had not met mates.

The idea, therefore, was to try something similar to the northern white rhinos. To combine Iman's and Puntung's oocytes (immature eggs) with Tam's sperm to produce embryos which could then be implanted either back into one of the females or a rhino of a different species. And, as usual, there was a funding problem, where could the Borneo Rhino Alliance find enough money

---

[16] Ibid.

[17] Peter Dockrill *The Sumatran rhino is now considered extinct in the wild in Malaysia* Science Alert, 21 August 2015 See http://www.sciencealert.com/the-sumatran-rhino-is-now-considered-extinct-in-the-wil Accessed 31/03/2016.

to undertake the task?[18] Sadly, time was not on their side. In 2016, Puntung died, and in December 2017, Iman was fighting for her life.[19] Fortunately, this time she survived.

But there was some good news too. As the result of ground-breaking research, a great deal of patience and a determination not to give up, scientists from Cincinnati Zoo and Botanical Garden were pleased to announce a birth, Andalas, the first Sumatran rhino born in captivity in 112 years. Before she died, Emi, his mother, had two more calves, then it was decided to translocate Andalas to the Sumatran Rhino Sanctuary where, with a little help from his friends from the zoo, he was able to father a calf, Andatu, 'gift from God', the first captive-bred Sumatran rhino born in Sumatra. Delilah followed in 2016.[20]

And there was a discovery. At the end of March 2016, after 40 years, suddenly there appeared a Sumatran rhino in Borneo. She, because it was a female, was captured by conservationists and her location kept secret. Camera traps also showed there were possibly another 15 rhinos in the area.[21]

Finally, in 2017, consensus was reached, as time was fast running out. The difference between the official estimate of numbers still alive, and the minimum estimate was stark. Officially there were 100 rhinos, but when the minimum estimates of the four known wild populations were added up, the total was 30, with another 9 in captivity.[22] The only way to save these rhinos was to capture and consolidate the remaining wild animals in intensively managed captive breeding facilities. Indonesian government officials and rhino experts from around the world all agreed.[23]

By May 2018, a suitable female rhino had been identified *for immediate capture*. Living in Borneo, in West Kutai, not only was breeding impossible as she was on her own, but her forest home was under threat of clearance making it highly likely she would be poached. So she will be captured and translocated to a protected place. And the Indonesian government has given the

---

[18] John R. Platt *Extinction Countdown The Last 3 Bornean Rhinos Are in a Race against Extinction* Blogs, Scientific American, 27 April 2015. See http://blogs.scientificamerican.com/extinction-countdown/the-last-3- Accessed 05/05/2016.

[19] Dave Lawrence *Update on effort to save Iman the Sumatran rhino in Malaysia*. See http://hawaiipublicradio.org/post/update-effort-save-iman-sumatran-rhino-malaysia Accessed 09/01/2018.

[20] Jason Bittel *The Unprecedented Plan to Save the Sumatran Rhino* National Geographic, 20 September 2018.

[21] BEC Crew *The world's smallest rhino has just been sighted in Borneo for the first time in 40 years* See http://www.sciencealert.com/the-world-s-smallest-rhino-has-been-sighted-in-borneo-f Accessed 31/03/2016.

[22] *Worst-case scenario: There could be only 30 wild Sumatran rhinos left* Mongabay, November 2017, see https://news.mongabay.com/2017/11/worst-case-scenario-there-could-be-only-30-wil     Accessed 06/02/2018. This is the first article in a 4-part series *Is anyone Going to Save The Sumatran rhino?*

[23] See Kinnaird and Rodriguez (n. 11).

IUCN Species Survival Commission permission to work with a number of partners, a coalition of international conservation organizations, on a plan to boost the wider captive breeding programme.[24]

With the situation so desperate, an even more radical suggestion has been put forward, though not without some reluctance. That there should be an attempt to interbreed the Bornean rhinos with the Sumatran rhinos even though the two subspecies are entirely different and have not mixed in about 300,000 years. Which makes it a particularly hazardous enterprise. First proposed in 2012, although conservationists accepted it, it was never acted on, but advances in genetics at least means there is a possibility that sometime, somewhere, some development will occur. Possibly by creating two distinct groups, one keeping the original western Sumatran rhino, the other, a Sumatran-Bornean mix, which, if it worked, could be 'reintroduced' into Borneo.[25]

Then, on 20 September 2018, hopes were raised when the plan, the Sumatran Rhino Rescue, was announced. As many as possible of the remaining wild rhinos will be captured and placed in sanctuaries, the Sumatran Rhino Sanctuary at first, to be supplemented by two further sanctuaries in Sumatra and possibly one in Borneo. Wild rhinos will widen the narrow gene pool, and there is sufficient expertise for captive breeding to be successful, but it will not be easy, and it will certainly be very expensive in financial terms.[26] Only to be dashed about 3 weeks later, when once again Indonesia postponed the programme because Iman has stopped producing viable eggs.

Indonesia is also refusing to make Tam's sperm available. This is especially upsetting and frustrating because of all Cincinnati Zoo's hard work and the recent breakthrough in the white rhino IVF programme. Although there may still be as many as 100 rhinos left in the wild, there may be only 30,[27] with such small and scattered populations accurate numbers are hard to come by, and what is certain is that with such low numbers, there is an ever-increasing risk of genetic defects. According to the executive director of the Borneo Rhino Alliance, Malaysia has offered to send Tam to Indonesia, to see whether he will mate with one of the captive females at the Sumatran Rhino Sanctuary,

[24] Ibid.

[25] Jeremy Hance *Geneticists: It's time to mix the Sumatran rhino subspecies* Mongabay Series: Asian Rhinos, 29 May 2018. See https://news.mongabay.com/2018/05/geneticists-its-time-to-mix-the-sumatran-rhino-s Accessed 26/06/2018. See also Jessica R. Brandt et al. *Genetic structure and diversity among historic and modern populations of the Sumatran rhinoceros (Dicerorhinus sumatrensis)* Journal of Heredity (2018) DOI: https://doi.org/10.1093/jhered/esy019.

[26] See Bittel (n.20).

[27] See https://news.mongabay.com/2017/11/worst-case-scenario-there-could-be-only-30-sumatran-rhinos-left/ Accessed 25/09/2018.

but the Indonesian government had not responded. If these ancient animals are to be saved, Indonesia must start working together with Malaysia.[28]

And we must also protect their habitats. Sumatran rhinos are the closest rhinos we have to those living in prehistoric times, linking us to the iconic megafauna of the Ice Age and reminding us of the woolly rhinos our ancestors would have hunted. If these rhinos are to survive, they must have somewhere to live. And where Sumatran rhinos live, so too do other critically endangered species, not only Asian elephants but also orangutans, clouded leopards and Sumatran tigers.

## 41.4   Javan Rhinos

By 2016, there were an estimated 58–61 Javan rhinos left, all of them living in the wild on the Indonesian island of Java,[29] (since the last 1 in Vietnam was killed by a poacher/s sometime in 2009 or 2010) and confined to a tiny area, two National Parks, Ujong Kulon and Gunung Honje, on the westernmost point of the island. And the population remains stable. In April 2018, data estimated a minimum of 68 animals including 15 juveniles—2 calves were born earlier in the year while an adult male died of natural causes.[30]

The rhinos have been continually monitored for the last 30 years by the Rhino Observation and Activity Management (ROAM) programme, a partnership between WWF-Indonesia and Ujong Kulon National Park. Local people monitor the park for signs of illegal activity and all incursions, of which there are many these days, and camera traps are used to survey the population.[31]

As with the Sumatran rhinos, habitat loss has played a critical part in reducing their numbers, but in Ujong Kulon there has been another important factor as well, the presence of an invasive non-native species, the Arenga Palm. These trees block out sunlight, and this, in turn, stops the plants the rhinos feed on from germinating. It was essential to control them, and this is now beginning to happen.

---

[28] Basten Gokkon *Indonesian government puts off Sumatran rhino IVF program* Mongabay Series: Asian Rhinos, 10 October 2018. See https://news.mongabay.com/2018/10/indonesian-government-puts-off-sumatran-rhino-ivf-program/ Accessed 14/10/2018.

[29] Bibhab Kumar Talukdar *Conservation Insight Javan Rhino* BBC Wildlife, Spring 2016, Volume 34 Number p.61.

[30] Donny Iqbal (adapted by Basten Gokkon) *In protecting the Javan rhino, locals gain a 'more meaningful life'* Mongabay, 10 August 2018. See https://news.mongabay.com/2018/08/in-protecting-the-javan-rhino- Accessed 14/08/2018.

[31] Ibid.

However, the parks are adjacent to each other, in the shadow of an active volcano and *'recent research suggests a large tsunami could wipe out virtually the whole species'.* In December 2018, Arak Krakatoa erupted causing tsunamis in both Sumatra and Java,[32] so the plans to redistribute some of the rhinos, to translocate them to another suitable forest area in Java, or even Sumatra, might need to be brought forward. There are even more plans, perhaps still at the 'hopes' stage, plans to initiate a captive breeding programme. The local people love their rhinos. In the words of one rhino protector, *'if you think about it, Javan rhinos have helped us provide food and livelihoods for our families … Our lives become more meaningful'.*[33] They will make sure they seize any 'last chance to save'.

## 41.5   Garamba National Park

From animals to habitats. If animals are to be saved, they must have somewhere suitable and safe to live in. A very important habitat that has been under constant threat for many years is Garamba National Park.

When, in 1938, Garamba was designated a national park, it must have been a wonderful place for the wildlife that lived there. Located in the northeast of DRC, it covers an enormous area, for Americans readers, it is roughly the size of the State of Connecticut. It is a UNESCO World Heritage Centre, and it was managed by the Congolese Institute for Nature Conservation.[34]

The heart of the park consists of 490,000 hectares of land, with a further 752,700 hectares added by three adjacent hunting grounds, the entire protected area covering 1,242,700 hectares. Within it are two distinct biogeographical zones, the dense tropical forests of the Congo basin and the Guinea-Sudano savannas, divided by a transition zone. With an ecosystem extensive enough to support considerable populations of large mammals and maintain them on their seasonal migration routes, in its heyday, Garamba provided an outstanding sanctuary for its special inhabitants.

In fact, in its early days, it must have been a kind of earthly paradise. A paradise that at times housed elephant herds of more than 550 animals, something unimaginable now. As well as giraffes, it was also home to the northern white rhinos. And it buzzed with all kinds of other life. Now the rhinos have gone extinct in the wild, the elephants are under tremendous threats, and the

---

[32] See https://news.mongabay.com/2017/09/can-the-javan-rhino-be-saved-before-disaster-strikes/
[33] See Iqbal (n.30).
[34] A UNESCO World Heritage Centre. See http://whc.unesco.org/en/list/136 Accessed 13/04/2016.

giraffes are just hanging on. In fact, Garamba is the final stronghold for elephants and giraffes in the whole of the Congo. So what has gone so horribly wrong?

Unfortunately, it is the same old story. The colonial masters left, and there was a fight for power. The Belgian Congo was a vast, sprawling country with artificial boundaries, so there was bound to be trouble. There was, and there still is in places. The Congo was always a place that housed great wealth. It still does, and its wildlife and natural beauty are only part of it. Too many people and too many animals have lost, continue to lose their lives as a result of the multitude of conflicts that have ensued. And now Garamba is on the List of World Heritage in Danger.

Although the rhinos have gone, there are still elephants. And these are very special elephants. These are mixed populations of elephants, forest elephants, savannah elephants and elephants that appear to be hybrids of the two. And we know that forest elephants are at particular risk, the main threats coming from the militias, insurgent and rebel groups based in neighbouring countries. The very boundaries that protected Garamba when it was first designated are now providing a quick means of entry and exit to the marauding bands who are ransacking the area of its treasures.

2015 was a very bad year. In March, at about the same time the Kasane Conference was taking place in Botswana, raiders from north Sudan swept into the park, spread out into small groups and, over a period of 15 days, slaughtered 30 elephants. By June, African Parks warned of *a poaching onslaught*, 68 elephants had been killed in just 2 months and 10 in just 1 day. Their Director of Conservation explained that Garamba was being attacked by Sudanese militia because they were not finding enough elephants left to kill in CAR and northern Congo. He described the poachers as *very experienced elephant hunters*, for in 2012, they had killed some 300 elephants in 10 weeks in Bouba N'Djida National Park in northern Cameroon. What was even more disturbing was that some of them *wore government uniforms and carried army identity documents.*

People were dying too. In one incident, a ranger and two soldiers were killed by poachers, while in October, a 10-man patrol team was attacked by poachers from the Lord's Resistance Army who come into the Park from Uganda. The team, which had tracked down a poachers' camp, lost three park rangers and an army officer from the Congolese Armed Forces (FARDC). The attack must have been ferocious, because the helicopter that was taking part in the operation was so badly damaged by gunfire that it was put out of action. Fortunately though, not before it had flown in and picked up the six surviving members of the patrol, who had been forced to scatter in order to escape. The bodies of the dead rangers were later retrieved by a reinforced patrol team.[35] Erik Mararv, the park manager, was convinced the attacks were taking place because the whole rather than part of the park was being protected and that this had led to *a number of confrontations with various armed groups*.[36]

But it seems it was not just marauding gangs of militias who came a-poaching, other raiders were believed to be from the South Sudanese army, the Congolese army and the Ugandan army. Soldiers who found time to do some poaching while hunting down the Lord's Resistance Army.[37] A potentially deadly mix for everyone and everything.

2016 also started badly. In April, four rangers and Eric Mararv were flown in by helicopter to look for the body of a dead elephant that had been carefully hidden under branches and grass. As they approached it, the poachers opened fire, the rangers returned it, and in the ensuing gun battle one ranger was killed immediately. Four others were seriously injured. Dr. Mararv, who had been hit in the leg, managed, by reflecting light off a mirror, to attract the pilot's attention. So most of the injured were picked up and taken to base, where the helicopter refuelled. It then returned to the battle, dropped off reinforcements and evacuated the last of the casualties.[38]

This incident shows yet again the need to work together. The helicopter was provided by AFRICOM. Back at base, the park's headquarters, two members of the US Special Forces, working there as liaison officers, organized the medical treatment. The casualties were flown to a US military base in Sudan and then transferred to a United Nations military hospital in the CAR, where, sadly, two more rangers died of their wounds. But the fourth ranger and the park manager recovered.[39]

Since 2005, when African Parks took over Garamba, it has worked with the Congolese Institute for Nature Conservation to save the park, and although devastating attacks on both elephants and their protectors have continued, things are starting to improve. One reason for this is better training. There had been training, but it was only sporadic, priority going to operations on

---

[35] *Four rangers from the patrol team killed by poachers in Congo's Garamba National Park* See http://www. abc.net.au/news/2015-10-10/four-rangers-killed-by-poa Accessed 26/04/2016.

[36] *Poachers kill three park rangers and an army officer in DRC* BBC (Africa), 9 October 2015. See http:// www.bbc.co.uk/news/world-africa-34492072 Accessed 22/03/2016.

[37] *Poachers slaughter 30 elephants in Democratic Republic of Congo national park* News. See http://www.abc. net.au/news/2015-03-24/poachers-slaughter-30-ele Accessed 26/04/2016.

[38] Jerome Starkey *I'll be back to tackle poachers, shot ranger vows* The Times, 29 April 2016.

[39] African Parks .

the ground. However, in May 2006, DRC's two national parks, Garamba and Virunga, joined forces to offer high-quality anti-poaching training. And another partner, Rapid Response Facility, stepped in with funding. The fight-back had begun, but it continues to be both difficult and dangerous.

The first action was to upgrade the rangers' base in the park, Ishango Camp. Then a grant was awarded from the Zoological Society of Frankfurt, and 40 rangers from Garamba were given essential training. These men went on to form Garamba's elite ranger force responsible not only for protecting the world's remaining wild population of northern white rhinos but all the other wildlife from the poachers.[40]

Time to return to technology. We have seen just how vast an area the park covers, so there is no way the whole of it can be regularly patrolled. Many of the poachers use helicopters to locate elephant herds, before moving in to kill possibly all the animals. Speed is of the essence, so they bring chainsaws to hack out the tusks from the elephants' faces. The rangers acting need extra help which geospatial analysis can provide.

More working together. This time the partner was Digital Globe. Using a technique known as pattern analysis, their analysts examined former sites where poaching had occurred and used their findings to identify possible new places where attacks might happen in the future. In this instance, they statisti-cally compared past poaching activity with geographic locations. The aim was to develop a profile of places that attracted poachers.

They took a number of features into consideration including:

- Elevation and land use
- Centres of population and ranger stations
- Road and waterway access

They also analysed the 56 poaching incidents in Garamba between 2011 and 2013 when 70 elephants had been killed, all of them in the southern part of the park and most of them in 1 of 4 identified poaching zones. It meant the focus area could be reduced from 4920 to 1818 $km^2$, much more manageable. It was then further reduced to 107 $km^2$.[41]

Geographic information systems (GIS) also play a key role. They enable African Parks to analyse and monitor:

---

[40] Rapid Response Facility. See www.rapid-response.org/?p=336 Accessed 13/04/2016.

[41] DigitalGlobe *Elephant Poaching in Garamba National Park* See microsites.digitalglobe.com/interactive/garamba Accessed 13/04/2016.

- *'The movement of 50 elephants that have telemetric collars that allow rangers to track their location;*
- *The seasonal movement of pastoralists in the region;*
- *How patterns of wildfire are changing at a regional scale; Where trail systems are expanding'.*[42]

This provides geospatial intelligence for the rangers enabling precious resources to be concentrated and used more efficiently.

Garamba is gradually being pulled back from the brink.

---

[42] David Gadsden *Defending Elephants in Africa Garamba National Park Fuels Anti-Poaching Efforts with Technology, New Tools* ESRI Blog, 7 March 2018. See https://www.esri.com/about/newsroom/blog/garamba-combats-poaching/ Accessed 13/03/2018.

# 42

## Living Together: Resolutions to Human-Elephant Conflict and Other Problems

This too must be sorted out if elephants are to be better protected, for where there is human-elephant conflict, humans tend to develop very bad feelings towards elephants. And for good reason. People can become quite desperate if, for example, hungry elephants rampage over their fields, destroying the precious crops that not only provide food for their families but possibly also their annual income. Elephants can damage property, even kill people. In the Maasai Mara alone, about 70 people have been killed or injured by elephants since 2000.[1] Savannah elephants migrate every year and cover large distances. Growing populations often result in growing pressures on land, and people may well have unknowingly settled on elephants' paths since the last time the animals were there. More occasion for conflict. Roads are built across migratory routes, but roads will not stop elephants.

Why should this concern us though, when we are investigating illegal trading and poaching? Because, unfortunately, it can lead to potential disaster, for where people feel bad about elephants and they don't have much money, some might be tempted into poaching. In some of the poorest regions, the temptation might already be almost overwhelming, so a bad experience with the animals could provide the tipping point. Financial benefit outweighing other risks.

In fact, it is only recently, with the FLoD initiative (more of this later), that local communities have been able to give voice to their grievances, their anger fuelled by the perception that animals get a better deal than they do. Even in Kenya, whose Wildlife Conservation and Management Act 2013 provides

---

[1] David Leto *Good Neighbours* WWF, Action, Summer 2016.

compensation for wildlife-related injury or damage, there is insufficient funding to meet their claims.[2]

Their voices are eloquent:

*If a rhino is killed, someone comes immediately to find out what happened. If a person is killed or injured, nobody comes.*

*If an animal kills you there is no compensation. If you kill an animal you get ten years in prison.*[3]

The point is also made by NRT when *'relatively high levels of elephant mortality as a result of conflict killing'* topped the list of challenges for 2017.[4]

So how can human-elephant conflict be prevented? We will look at solutions in both Africa and India.

In the Tsavo Conservation Area, Kenya, as part of their Elephants and Bees Project, Save the Elephants and their partners are collaring elephants, more precisely, 20 *at-risk* elephants, in order that they can understand their movements and behaviour better, monitoring them, to try to avoid human-elephant conflict and to give them better protection from poachers. Nine of the animals had *spear and arrow wounds resulting from human-elephant conflict.*[5] An urgent necessity, as there have been a number of elephants killed around Amboseli, in revenge for human deaths and crop raids.

India is a country whose population is burgeoning, overflowing out and encroaching into areas that were previously the home of wildlife. Not enough space for everyone and everything. Marooning elephants *in tiny pockets in a sea of people.*[6] One answer is corridors, elephant corridors to link up these tiny pockets of land. Roads specifically for elephants, not for people. To benefit both people and elephants.

And they work! Take, for example, the Kerala corridor, the product of a partnership between Elephant Family, the Wildlife Trust of India and the World Land Trust. The corridor was *opened* for use in March 2009, and monitoring has shown it to be a lifeline, not just for the 6500 elephants who are using it, but for other species using it as well.

Roads, especially busy ones, are another hazard. But special bridges and tunnels are being built for elephants to use. In India, at least two eco-bridges

---

[2] Briefing etc. IUCN IIED.

[3] Briefing etc. These were comments collected during fieldwork.

[4] NRT *The 2017 State of Conservancies Report.*

[5] Lydia Tiller *Collaring At-Risk Elephants in Tsavo* Blog posted 13 February 2018. See http://elephantsand-bees.com/collaring-at-risk-elephants-in-tsavo Accessed 15/02/2018.

[6] Elephant Family 2016. See also Rahul Joglekar, Hugh Tomlinson *Elephants are electrocuted on power lines* The Times, 29 October 2018, which vividly describes the elephants' plight and provides some worrying numbers of elephant casualties.

were due to be constructed over the Chennai-Bangalore highway, a six-lane motorway, which passes near two reserve forests. The National Highways Authority would be providing the animals *with a safe and convenient passage*, and to make them elephant-friendly, the bridges would have *trees, plants and shrubs to recreate the feel of the forest on the route that is taken by the pachyderms*.[7]

Although villagers have traditionally protected their fields by beating drums, setting off firecrackers even firing guns into the air,[8] this does not really work, so alternative methods are being tried out. One answer is to use chilli, which is effective because elephants react badly to the hot chemical, capsaicin, which makes them sneeze and cough. Some African farmers plant 'chilli fences' around their crops to protect them, and in South Africa, farmers are taught by the Elephant Pepper Development Trust how to smear rope fences with a mixture of red chilli and waste engine oil, adding cowbells as an extra deterrent. Meanwhile, in north-east India, they are experimenting with ghost chillies, one of the hottest chillies in the world. In southern India, however, farmers only use smoke, very unusual smoke though. It is produced by burning a mixture of dry hay, tobacco and dry red chillies wrapped up in newspaper.[9]

Beehive fences also work well, and they are very effective at stopping raids by large groups of elephants because swarms of angry bees rush out whenever the fences are touched.[10] Elephants do not like bees. If they hear them buzzing, they run away, and they alert their family to the potential danger. For not only are African honey bees aggressive, their stings are also extremely painful, even for elephants, especially if they are stung around their eyes or inside their trunks. So Save the Elephants designed the fences, which carry beehives every 10 m along their length. They were field-tested in Kenya in 2007 and had an 80% success rate. Crop yields increased, because of better pollination and 'Elephant-Friendly Honey' generated extra income.[11] The fences have now been extended to Tanzania, Mozambique and Botswana and in Asia, Sri Lanka.

---

[7] Deepa Ramakrishnan *Eco bridges on highway soon for safety of elephants* The Hindu, 22 June 2016. See http://www.thehindu.com/news/cities/chennai/eco-bridges-on-highway-soon-for-safety-of-elephants Accessed 11/08/2016.

[8] Shreya Dasgupta *How to scare off the biggest pest in the world* BBC Earth, BBC-Earth-How to scare off the biggest pest in the world. See http://www.bbc.co.uk/earth/story/20141204-five-ways-to-scare-off- Accessed 18/08/2016.

[9] Ibid.

[10] Ibid.

[11] See elephantsandbees.com/beehive-fence/

An extension of this for farmers with larger farms could be to use bee-scented socks. Researchers have discovered that elephants are so alarmed by the possible presence of bees that even *a crude approximation of the honeybee alarm pheromone blend*, a *semiochemical blend*, has a remarkable effect. They sprayed white socks with their blend, Splat, and then hung them from branches round a watering hole in Kruger National Park. Twenty-nine elephants approached the socks and 25 turned away, while the other 4 animals examined them but swiftly dropped them.[12]

Another way of avoiding conflict is to provide early warning systems for humans. Although elephants are rare in India, some are attracted to tea and coffee plantations, where chance encounters with the workers often result in humans being killed. Now, people with mobile phones can use a simple early warning system to text an elephant's location to those living within 2 km of it. And a system of flashing red LED lights triggered by mobile phones when an elephant has been spotted has been set up to warn those without phones, the *flashing lights* warning *people returning home in the dark, when elephants are not easily visible*.[13]

Ecosystems in India have been working with Chester Zoo on the Assam Haathi Project in India for a number of years with the aim of reducing human-wildlife conflict, concentrating on mitigation through community-based interventions and alternative livelihoods. And it has been successful, having:

- *'Assisted over 80 villages, reducing crop-raiding & property damage by up to 95%;*
- *Protected 1400 households with community-managed solar electric fences';* and.
- *'Engaged 23 communities and 600 people in new sources of income through sustainable livelihoods'.*[14]

However, although these methods work well for smallish areas, they are no use for migrating elephants. So can long fences bring relief?

In Laikipia, Kenya, 163 km of electrified fence is being erected by Space for Giants to provide a *separation zone* between elephant migration routes and local communities' farmlands. Perhaps the most ambitious project of its kind

---

[12] Oliver Moody *Bee-scented socks scare off elephants* The Times, 24 July 2018; Mark G. Wright et al. *African bush elephants respond to a honeybee alarm pheromone blend* Current Biology, Volume 28, Issue 14, PR778-R780, July 23, 2018.

[13] See Dasgupta (n.8).

[14] Chester Zoo *Assam Haathi Project* See https://www.chesterzoo.org/conservation-and-science/where-we-work/south-asia/assam-haathi Accessed 13/09/2018.

attempted so far, this fence is urgently needed, for in the past 2 years, more than 20 people have died in human-elephant conflicts in this area. And what a fence it will be. It will carry *a 7000 volt pulsing electric shock, with dandling metre long wire strips, to stop the elephants well before they can use their trunks to uproot posts as they did in the past.*[15] It should work.

But fences can come at a price. In Nepal, although the Chitwan National Park authorities are managing to protect their rhinos from poachers, some that have left the Park and entered the buffer zone where there are farmers, have been killed by the electric fences put there to protect the farmers' crops.[16] A spin-off from help that can go tragically wrong.

Sometimes there are so many problems that it seems almost impossible to solve them all. Take the case of Kaziranga National Park. Yet another World Heritage site, Kaziranga is in Assam, in India, and it is home to the world's largest population of greater one-horned rhinos as well as many other rare and endangered species.

Such an important place, but its problems include:

• Poaching: So severe that the park managers and guards are authorized to shoot on sight.

**Fig. 42.1** A magnificent Greater One-horned rhino. Copyright Save the Rhino

[15] *News/Kenya's-jumbo-ele-fence-to-stop-human-wildlife-conflict* News 24.See http://www.news24.com/ Green

[16] Shiva Puri *Rhino deaths* Nepali Times, 21 February 2017. See http://www.nepalitimes.com/blogs/theb-rief/2017/02/21/rhino-deaths/ Accessed 23/02/2017.

- Flooding: An annual event which causes havoc, killing wildlife, among other things. Like drought, flooding can result in food shortages.
- Erosion: Mainly caused by flooding and which is shrinking the area of the park.
- Habitat loss.
- Invasive species: Such as water hyacinth and mimosa. The latter plant, which is poisonous to herbivores, is spreading through the grasslands. Even fire does not kill it so it has to be controlled manually.
- Poor infrastructure and manpower: To the extent that it affects management and anti-poaching activities.
- Human interference.[17]

People encroaching into Kaziranga pose particularly intractable problems which have now become politicized. In September 2016, 350 families were evicted from three villages, the action authorized by a High Court order made in October 2015. In one village, bulldozers and elephants were used to demolish the houses. A protest turned violent, and two people were killed when the police forcefully broke it up. The dispossessed moved onto government land or the land of people in neighbouring villages. Before the eviction, two ministers had told the villagers compensation was available under the Right to Fair Compensation and Transparency in Land Acquisition, Rehabilitation and Resettlement Act 2013.

An important element of the problem was a dispute as to where the villagers came from. The Indian Government was concerned that the people were infiltrators from Bangladesh, but many of the villagers were undoubtedly indigenous peoples who had moved into the protected forests because they had lost their homes and land to erosion. It is hard to see a way forward without further conflict erupting.[18]

People and animals must find ways of living with each other, for it is becoming increasingly difficult in a world of growing populations, shrinking habitats and a planet that seems to be warming up, for them to avoid each other. Animals and humans are being forced into each other's territories, often with tragic results.

---

[17] *Problems of Kaziranga National Park* See http://www.assams.info/assam/kaziranga-national-park-problems Accessed 21/02/2017.

[18] Samudra Gupta Kashyap *After Kaziranga eviction, the question: Where next?* The Indian Express, 2 November 2016. See http://indianexpress.com/article/india/india-news-india/kaziranga-national-park-eviction-encroaches-assam-3069626 Accessed 22/02/2017.

Although conflicts can be reduced, they can never be eliminated, so what can be done? A task force can be set up. This happened in 2016. The IUCN-SSC Task Force on Human-Wildlife Conflict was established by the IUCN's Species Survival Commission to focus on the problem and provide guidance and specialist support. Chaired by Dr. Alexandra Zimmermann, the Task Force consists of experts from a wide range of disciplines, including social scientists, economists, anthropologists, social psychologists, peace-builders, political ecologists, specialists in poverty alleviation and biologists of course.[19]

---

[19] Act for Wildlife, 2016. See https://www.actforwildlife.org.uk/conservation-news/world-leading-conservationist Accessed 29/03/2017.

# 43

# Working Together Creates People Power

We have already seen the contribution Northern Rangelands Trust is making to the people in Northern Kenya. It provides an excellent example of sustainable use, with both people and wildlife benefitting. And of working together, for some of its many partners and donors from across the world include:

- USAID
- The Nature Conservancy
- US Fish and Wildlife Service
- Fonds Francois pour l'Environnement Mondial
- Kingdom of the Netherlands
- Embassy of Denmark
- Danida International Development Cooperation
- The Lundin Foundation
- Fauna and Flora International
- Tusk

  Kenyan partners include:

- Sera Wildlife Conservancy/the County Government of Samburu
- Lewa Wildlife Conservancy
- Kenya Wildlife Service
- Ol Pejeta Conservancy

© Springer Nature Switzerland AG 2019
B. Martin, *Survival or Extinction?*, https://doi.org/10.1007/978-3-030-13293-4_43

Zoos include:

- San Diego Zoo
- St. Louis Zoo
- Zurich Zoo

One of their projects is the Sera Rhino Sanctuary. Sometimes the support is purely financial, and the contribution can be small or large. For example, St. Louis Zoo provided some funding for the purchase and installation of a solar-powered water pump, while money from the more wealthy Lundin Foundation was put towards *'sanctuary fencing; construction of housing complex (for the rangers); purchase of a Toyota Land Cruiser; VHF digital radio equipment; GSM/UHF transmitters and receivers; training costs for rhino monitoring and security personnel; translocation and sanctuary running costs'.*

Myanmar is keen to protect its wildlife, but it needed funding to help its rangers. So now the government, its Forest Department, is working with Elephant Family and the Wildlife Conservation Society.

## 43.1   The Wildlife-TRAPS Project

The *challenge* is:

- The rapid escalation in poaching and trade of species between Africa and Asia, which is causing a conservation crisis
- Growing evidence that wildlife crime is also supplementing other types of highly organized and sophisticated transnational organized crime, for example, trafficking in drugs, arms or people
- Undermining state sovereignty and the rule of law

It is a recognition of just how serious wildlife crime has become. Not just the enormity of the near prospect of wiping out iconic species such as elephants and rhinos, although that is bad enough, but how it interlinks with so many other dangerous activities.

*The response*, which is ambitious, is the Wildlife-TRAPS Project. Designed to last 3 years, it will:

- Increase understanding of the true character and scale of the international response required, identify intervention points, and test non-traditional approaches; as well as.

- Devise, develop and deliver a suite of ground-breaking partnerships and pioneering approaches to tackle wildlife crime between Africa and Asia.

In so doing, it will *'protect global biodiversity from the threat of illegal wildlife trade through strengthening the knowledge base, resolve and cooperation of governments, inter-governmental organisations, the private sector and NGOs, in tackling wildlife trafficking between Africa and Asia'.*

Financed by USAID, implemented by TRAFFIC and IUCN, it started in 2016, focusing on trade between Africa and Asia, aiming to foster trust and facilitate dialogue. And there will be *'ground-breaking partnerships and pioneering approaches'* to tackle wildlife crime.[1]

Wildlife-TRAPS. What do the letters stand for? Wildlife Trafficking, Response, Assessment and Priority Setting. It is the ultimate attempt at working together.

## 43.2   Transfrontier Conservation Areas (TFCAs)

Defined by the Southern African Development Community (SADC) Protocol on Wildlife Conservation and Law Enforcement of 1999 as *'the area or component of a large ecological region that straddles the boundaries of two or more countries encompassing one or more protected areas as well as multiple resource areas'.*

The most important TFCA in our story is the Great Limpopo Transfrontier Park which crosses the borders of South Africa, Mozambique and Zimbabwe. The Protocol commits the Member States of the SADC to promote the conservation of wildlife resources through the establishment of TFCAs, while the Southern African Wildlife College (founded by the Peace Parks Foundation, itself a partnership, between governments and the private sector) trains park rangers specifically for work in them. But once again an excellent idea can have some unintended consequences. In the case of the Great Limpopo Transfrontier Park, although bringing down the boundaries made it easier for the elephants to migrate, it also made it easier for the poachers to come in from Mozambique.

---

[1] TRAFFIC *Wildlife-TRAPS – Wildlife Trafficking Response, Assessment and Priority Setting.* See http://pfbc-cbfp.org/news_en/items/traffic-TRAPS-en.html. Accessed 03/03/2016.

## 43.3   A Bridge Too Far

Had it not been stopped in its tracks, this would have been yet another good idea that would have had bad unintended consequences. We return to Borneo, to the pygmy elephants and the orangutans, their homes again under threat, this time by a proposal to build a bridge across the Kinabatangan River. And with the bridge would have come a road, and traffic. Easier access to the forest. It would have made life easier for people in the local village, Sukau, but this village is adjacent to the Kinabatangan Wildlife Sanctuary, and leading conservationists were in no doubt it would been seriously bad for the wildlife.

Steve Backshall described the forest as *'one of the single most important pieces of rainforest on Earth'* and *'fragmenting this habitat... would be catastrophic'*.[2] Sir David Attenborough wrote to the chief minister of the region, explaining the reasons for his deep-seated concern, and it was this letter or, more accurately, its publication by *The Guardian* newspaper that led to the plan being dropped, despite the fact that preparations had already begun.

The change of plan was announced by Sabah Forestry Department's chief conservator at a dinner in the UK, in London, held by the South East Asia Rainforest Research Partnership. The reason for the decision? *'That headline broke the camel's back...it made us understand that the issue of a proposed bridge across a protected area for wildlife is now the number one environmental concern not just in Sabah, but globally too'*. And the headline? *'David Attenborough attacks plan for Borneo bridge that threatens orangutans'*.[3]

## 43.4   CITES and the UNESCO World Heritage Convention

Treaties, too, must work together. CITES and the CBD are a good example, but perhaps the importance of CITES working with the World Heritage Convention is rather less obvious.

The World Heritage Convention, which has 193 Parties, focuses on World Heritage Sites, some of which have featured in our story. These very special

---

[2] Jeremy Hance *David Attenborough's 'Guardian headline' halts Borneo bridge* The Guardian, 21 April 2017. See https://www.theguardian.com/environment/2017/apr/21/attenborou. Accessed 27/04/2017.

[3] Jeremy Hance *David Attenborough attacks plan for Borneo bridge that threatens orangutans* The Guardian, 2 March 2017. See https://www.theguardian.com/environment/2017/mar/02/david-atte. Accessed 27/04/2017.

areas are treasures that the world should cherish. And some of them are not just sanctuaries but the only home for our animals, for example, the last Javan rhinos live in Ujung Kulon National Park. In fact, these sites are home to 40% of all African elephants.[4]

The sites, which are internationally recognized as areas of outstanding universal value, have *'cultural and/or natural significance which is so exceptional as to transcend national boundaries and to be of common importance for present and future generations of all humanity'*.

Although a number of criteria must be fulfilled for a site to qualify, we are only concerned with (x), which states that the site must be:

- One of *'the most important and significant natural habitats for in-situ conservation of biological diversity'* often containing *'threatened species of outstanding universal value from the point of view of science or conservation'*.

It follows that the parties have agreed *'not to take direct, or indirect action that damages the sites'*, and the Convention uses *'reactive monitoring, which addresses specific threats on the sites and is one of the most comprehensive monitoring systems under any international convention'*.[5]

When a site is in trouble, it is placed on the List of World Heritage in Danger. The Tropical Rainforest Heritage of Sumatra, which contains the national parks that are home to Sumatran rhinos, was put on the list in 2011 because of *illegal harvesting and industrialization*. Garamba National Park, which as we saw earlier is beginning to pull back from the brink, is likewise listed.

In 2017, Dalberg, commissioned by WWF, produced the report *Not For Sale: Halting the Illegal Trade of CITES Species from World Heritage Sites*. The executive summary makes it quite clear that there must be more 'working together' between the two Conventions because the current international approach to stop poaching is not working.

- *'Stakeholders must redouble their efforts and address all parts of the wildlife trafficking value train'*.
- *'Increased collaboration and integration between CITES and the World Heritage Convention at the national and site level in particular, could lead to a more coordinated , comprehensive response, and save time and valuable resources'*.

---

[4] Dalberg for WWF, 2017.
[5] Ibid.

The report sets out a range of options as to how best to achieve this and makes clear the grim reality, *'stakeholders recognize the urgency of the challenge, and must now take the steps required to prevent irreversible damage to some of the most iconic species and places in the world'*. Both the CITES Secretary General and the IUCN Director General contributed forewords. And even if only some of the 193 stakeholders take action, there should be some improvement.

It is impossible to overestimate the importance of working together. If we are to save elephants and rhinos from extinction, it will depend on all the people who love these animals, which is most of us, working together to educate and/or defeat those who only see them as 'a nuisance', 'an enemy' or 'as money'. Governments must act and there is no doubt that people power can be a potent persuader.

**World Wildlife Day** provides us with another example of working together. It was first celebrated on **March 3, 2014.**

## 43.5   Governments Slowly Moving Forward

Governments have been very active on behalf of elephants and rhinos; many of them, such as China, Viet Nam, the USA and the UK, passing new legislation to curb the illegal trade in elephant ivory and rhino horn. Unfortunately for South Africa, it was forced to abandon its moratorium on domestic sales of rhino horn, but it did not do so without a fight.

### 43.5.1   The Destruction of Ivory Stockpiles

This is a movement that has just grown and grown, with governments seeming almost to fall over each other in their anxiety to give a practical demonstration of how much they care about elephants, by destroying their ivory stockpiles.

It began in Kenya, in 1989, so many years ago, to draw attention to the plight of their elephants. They burnt 12 tonnes of their ivory stockpile. Then did it again some 22 years later, in 2011, this time burning 5 tonnes. Zambia followed Kenya's example, and in 1992 destroyed 9.5 tonnes of ivory, although since then, it has struggled to deal with its own poaching epidemic. In 2012 Gabon destroyed 4.8 tonnes of ivory. In 2013, two countries, the Philippines and the USA, both large consumers of ivory, destroyed 5 tonnes and 6 tonnes, respectively. Then in January 2014, China crushed 6 tonnes in Guangdong Province; a fortnight later, Hong Kong announced that over the next 2 years, it planned to burn 28 tonnes of ivory, and in February, just before the London Declaration, France destroyed 3 tonnes of their ivory. In March 2016, Rome hosted Italy's first public ivory crush of almost a tonne of seized ivory, which included both tusks and carved objects. Singapore joined in in June, crushing, then burning a massive 7.9 tonnes of illegal ivory. More than 2700 elephants' tusks, the result of four separate seizures.[6] Then in November, just before the Hanoi Conference, and after taking DNA samples as required by CITES regulations, the National Specimen Destruction Council destroyed over 2 tonnes of ivory and 70 kg of rhino horn (and other illegal specimens), by crushing and burning. They were implementing a Prime Ministerial directive.[7] And it carries on, though the pace has now slowed.

However some countries, such as Cambodia, still retain their stocks.[8] In April 2016, just before they went to the Giant's Club Summit in Nanyuki, Kenya, Botswana's Environment Minister explained movingly why his country did not burn tusks:

*For us, burning an elephant's tusks is like putting the final nail in the coffin of a once magnificent animal. We believe we should preserve and protect whatever remains of these creatures as a reminder of how mankind's greed leads to the extinction of our planet's flora and fauna. We cannot burn the shame associated with this and hope it will disappear in smoke – therefore we will not attend the ivory burn event that will follow the summit.*[9]

And Namibia's ivory stockpile of 67.4 tonnes is worth about N\$125 million, 39.4 tonnes of it illegal ivory, from seizures. This alone is worth N\$71.3 million. We have already seen that use of their natural resources to benefit their citizens is part of their constitution, so the government would like another one-off sale to achieve just this.[10] But would it really be sustainable use or would it fuel another poaching spree?

---

[6] News24 *Singapore destroys tons of illegal ivory* 13 June 2016. See http://www.news24.com/Green/News/singapore-destroys-tons-of-illegal-ivory-2016061. Accessed 11/08/2016.

[7] *Destruction of elephant ivory and rhino horns confiscated from illegal trade and trafficking in Vietnam.* See http://iwthanoi.vn/destruction-elephant-ivory-rhino-horns-confiscate. Accessed 22/03/2018.

[8] *Cambodia seizes shipment of ivory hidden in hollow logs* Phys.org, 6 December 2017. See https://phys.org/news/2017-12-cambodia-seizes-shipment-ivory-hidden.html. Accessed 06/03/2018.

[9] *Why Botswana won't burn illegal ivory The Independent*, 25 April 2016. See http://www.independent.co.uk/voices/campaigns/GiantsClub/why-botswana-won-t-bu. Accessed 28/04/2016.

[10] Albertina Nakale *Namibia sits on N\$125 million worth of ivory* New Era Live, 24 September 2018. See https://neweralive.na/posts/namibia-sits-on-n125-million-worth-of-ivory#.W6kGNyDCcQQ. Accessed 26/09/2018.

## 43.5.2    Pause for Some Celebration

Once again, hard choices had to be made as to which countries to include.

**Tanzania:** In October 2015, Tanzania voted in a new President, who came with an excellent track record for rooting out corruption, so important when we know how corruption feeds the illegal wildlife trade. One of his pledges was to stamp out poaching, which of course has had adverse effects on the safari tourism the country relies on to bring in foreign revenues. Even then it came as a surprise when, 2 months later, 4 Chinese nationals were each sentenced to 20 years in prison, and each fined £3 million (about ten times the value of the horns), after being found guilty of smuggling 11 rhino horns from Malawi into Tanzania.[11]

**Thailand:** This country provides *an outstanding example* of how being put into a NIAP can make a significant difference.[12] Earlier in our story, we saw that Thailand was causing grave concern because it was easy for illegal ivory sales to take place. The NIAP forced it to take action. It passed new legislation, the Elephant Ivory Act, to regulate domestic ivory markets and regulations to criminalize the sale of African elephant ivory. The result was astonishing. A 96% drop in the past 2 years of the amount of ivory openly for sale in Bangkok's markets. In 2013, *'the country was considered to have the largest unregulated ivory market in the world'*.

**China and Hong Kong SAR:** At CoP17 China mounted an exhibition, a side event co-hosted by TRAFFIC and WWF, to highlight recent actions it was taking to implement CITES. Liu Dongsheng, Deputy Administrator of the State Forestry Administration and Head of the Chinese Delegation, spoke of the importance of CITES and made some important points: *'Only if all Parties take ownership in their protection endeavours and respect the initiative and practice of every country, can we improve the levels of wildlife protection… Only by improving people's livelihoods and promoting sustainable economic growth in source countries, can we create a sound environment for the survival of endangered species'.*

---

[11] *The Daily Telegraph*, 19 December 2015.
[12] Kanitha Krishnasamy, Senior Programme Manager for TRAFFIC in Southeast Asia in TRAFFIC *Massive downturn in Bangkok ivory market as Thailand implements national ivory action plan* 29 September 2016. See http://www.traffic.org/home/2016/9/29/massive-downturn-in-bangko. Accessed 18/10/2016.

This was acknowledged by John Scanlon, the CITES Secretary General, who spoke warmly of China's deep commitment to the treaty, the *serious measures* it had taken against the illegal wildlife trade as well as work on *legal and sustainable trade*, describing it as an *extraordinary achievement* made during his tenure in office in the past 6 years.[13]

Although China continues to make gigantic strides in its efforts to protect elephants, in particular, Hong Kong was identified by ETIS in 2017 as still a *'country/territory of primary concern'* that had not done enough to comply with its NIAP.[14]

In August and December 2015, TRAFFIC surveyed Hong Kong's ivory markets, following this up with a rapid survey in November 2016 when most of the same shops were re-visited. It was a time of great uncertainty for the local ivory traders, who knew that their activities were unpopular with many people and that even though their activities were controlled by a licencing system, the government was minded to do something to curb their sales.

The results were interesting. Briefly summarized, during the 2015 surveys, although ivory was readily available, there were not many sales, so most of the sellers were supplementing it with other goods, ranging from mammoth ivory, stone and timber carvings to jewellery and antiques, and by 2016, 7 outlets seemed to have closed, and the market appeared to be consolidating.

Shops selling ivory had to display a licence on their premises, but only about a third of them did so, or even claimed to possess one. Furthermore, although 42% of the traders stated that adequate CITES documentation was needed with their purchase, about a third of the traders, including those with licenced outlets, had no qualms about selling to international buyers and encouraging them to smuggle the small artefacts out in their personal effects.[15]

Something curious was also happening. ETIS data showed that a *considerable quantity* of pre-convention ivory (pre-1990), more than half of it raw ivory, was imported into Hong Kong between 2010 and 2014. Most of it came from Europe. However, although all the declared re-exported raw ivory was destined for mainland China, some of the imported raw ivory remained in Hong Kong. Why was this, when there were *'no more ivory carving factories*

[13] TRAFFIC *China exhibition highlights action taken to implement CITES* 26 September 2016. See http://www.traffic.org/home/2016/9/26/china-exhibition-highlights-a. Accessed 18/10/2016.

[14] TRAFFIC *Hong Kong poised for critical ivory market debate* 23 March 2017. See http://www.traffic.org/home/2017/3/23/hong-kong-poised-for-critica. Accessed 25/04/2017.

[15] Lau, W., Xu, L., Guan, J. and Xiao, Y. (2017). *Closing Strategy: Ending ivory trade in Hong Kong.* TRAFFIC, Hong Kong.

*in Hong Kong and most carvers only operate on a part-time basis undertaking repair and bespoke work'?*[16]

In January 2016, the Hong Kong Legislative Council announced it proposed to phase out local ivory trade, but it was not until March 2017 that its Legislative Council Panel on Environmental Affairs met to consider the Bill, which was finally passed on 31 January 2018. This will bring an end to commercial ivory trading and increase penalties for wildlife crimes to a maximum of 10 years. Ivory dealers will not receive compensation. WWF was so concerned that it might not happen, that in June 2017, it invited Erik Mararv, the manager of Garamba National Park, and Josias Mungabwa, a senior investigations officer at the Zambia Wildlife Authority, to attend a Legislative Council public hearing on phasing out the trade.[17]

However, the key words turned out to be *'phase out local ivory trade'*; and although it became illegal almost immediately to import and export ivory, actual trading in the commodity can continue until the end of 2021, which raises concerns, because the ivory trade between Hong Kong and China is closely linked and China's ban on the sale of ivory started in January 2018. TRAFFIC commented in March 2017 that *'parity in policy with mainland China is crucial if the full impact of curbing ivory trafficking is to be realised'.*[18] Fortunately, the recommendations in their recent report *Closing Strategy: Ending ivory trade in Hong Kong* should, if followed, help to mitigate the situation.

**Kenya:** All countries need new roads, new airports and large-scale developments if they are to grow their economies and improve life for their citizens, but this can have an adverse effect on local wildlife. In Kenya, for example, elephants migrate, and their migration routes, which are passed on from generation to generation, can lead them into deadly conflict with humans, so their National Environmental Management Authority (NEMA), working with The Nature Conservancy and Save the Elephants, is using animal movement data in a pilot scheme in infrastructure planning.

The data comes from Save the Elephants, which has been tracking collared elephants in Samburu, North Kenya, since 1998. Now this data, gathered from 144 individuals, can be combined with other data and translated into maps that show three essential components: elephant routes/movements,

---

[16] Ibid.

[17] Sum Lok-kei *Call to end deadly trade The Standard*, 7 June 2017. See http://www.thestandard.com.hk/section-news.php?id=183703. Accessed 25/10/2018.

[18] The same as 908 TRAFFIC.

human settlements and planned infrastructure. Many community conservancies are situated in Isiolo County where the scheme will be piloted, as are wildlife reserves, and the region is *'a vital corridor for thousands of elephants'* moving from Mt. Kenya to open grasslands further north.

If it works, it *'has the potential to prove that animal movement data can be a key piece of the puzzle for Kenya's smart, sustainable development'.*[19]

**UK:** In autumn 2017, the UK government finally bowed to pressure and consulted on a UK ban on ivory sales. The reaction was overwhelming, with over 88% of the more than 70,000 responses in favour of the ban.[20]

Although it will be an almost total ban, the UK will follow the line taken by both the USA and China, allowing *'certain narrowly-defined and carefully-targeted exemptions for items which do not contribute to the poaching of elephants'.* As a result of the consultation, the proposed exemptions have been tightened *'but still provide a balance to ensure people are not unfairly impacted'.* They include:

- Items containing less than 10% ivory made before 1947
- Musical instruments containing less than 20% ivory and made before 1975 (the year Asian elephants were listed on CITES)
- Items assessed by the UK's most prestigious museums to be extremely rare and important, and at least 100 years old—with *'a specific exemption for portrait miniatures painted on thin slivers of ivory'* again provided they are over 100 years old
- Commercial activities to and between specially accredited museums

The government's Press Release concluded with two important statements which recognized the context of the ban: *'As profits become ever greater, the illegal wildlife trade has become a transnational organised enterprise, estimated to be worth up to £17 billion a year'* and *'the further decline of elephants would also deprive some of the poorest countries in the world of their valuable natural capital, affecting economic growth and sustainable development'.*[21]

Then in July 2018, the Secretary of State announced that the government would be consulting on a possible extension to the ban. The dangers posed by

---

[19] *Elephants Will Have A Say In Mapping Kenya's Future* The Nature Conservancy. See https://www.nature.org/ourinitiatives/regions/africa/elephants-will-have-a-say-in-map. Accessed 17/08/2017.

[20] GOV.UK *Government confirms UK ban on ivory sales* Press release from Department for Environment, Food & Rural Affairs, 3 April 2018. See https://www.gov.uk/government/news/government-confirms-uk-ban-on-ivory-sales. Accessed 10/05/2018.

[21] Ibid.

mammoth ivory had been recognized, and the ban on elephant ivory was beginning to put other species under pressure, so hippos, walruses and narwhals could also be protected if sales of their ivory were to be banned.[22]

In December 2018, the Ivory Bill became the Ivory Act 2018, although its provisions will not come into force until late 2019.

**Viet Nam:** This country showed its commitment by hosting the Hanoi Conference, and although it will not be easy to change its culture, it continues to work hard to educate its people.

## 43.6   People Power

This must never be underestimated. The Mali Elephant Project shows what people can achieve. Mali has elephants, only a few hundred, but its people want to keep them safe. Mali also has problems.

Throughout Mali, elephants are considered to be *'a national heritage and to be protected'*. The project began in 2003, with the first 3 years spent on *'scientifically understanding the migration route (of the elephants) and identifying the priority areas for action'*.[23] Then 3 years of community outreach and engagement followed, eventually resulting in a model of community engagement which brought together the various clans and ethnicities to make common cause.

Mali's legislation empowers communities to manage their own natural resources, so the elders formed *'a representative management committee who decided the rules of resource management'*, which included protecting the elephants' habitats. Government foresters worked with the younger men to enforce the rules.

All was going well. Pasture land, protected by firebreaks, was being set aside for the animals. Then, in 2012, trouble erupted. There was a coup, which led to *'a retreat of government, lawlessness and occupation by rebel and jihadist groups'*. Worse was to come. It soon became apparent just how easily the armed groups were managing to recruit the young men to their cause. And suddenly there were elephant poachers as well. Something had to be done, urgently, if the communities were not to lose both their young men and their elephants.

It turned out to be easier than they expected. The elders met and issued edicts that killing elephants amounted to stealing from the local community.

---

[22] *Ministers consider ban on hippo and mammoth ivory*, The Times, 5 July 2018.

[23] Susan Canney *The role of communities in tackling illegal wildlife trade – the Mali Elephant Project* SULi News, Issue 10 August 2015.

This was a social sanction. The community brigades were extended. Their patrols looked for the bodies of poached elephants and collected information to lead them to the poachers. This required considerable manpower, and more than 500 young men were recruited, despite the fact that there was no money to pay them. They were paid in food. They regarded the protection of their elephants and other natural resources as *a more "noble" occupation* than joining the armed groups, and *they gained pride and self-esteem in being able to provide food for their families and benefit the community*. To them, elephants represent good luck.

So protecting their elephants and their environment fortunately overcame radicalization. But problems remain. Mali is still unstable. And although the brigades are now paid, they do not get much money. A *small recognition* or *incentive payment* rather than a salary. But they are not greedy. They have a job that carries status and *they understand how their livelihoods are diminished by resource loss and degradation*.[24]

Another excellent example of people power comes from Nepal. We have already seen how greater one-horned rhinos are being translocated from Chitwan National Park to Bardia National Park. This is how WWF told it:

> *The translocation involved the help of 33 elephants, who help herd the rhino, and a team of about 250 people. It was led by Nepal's Ministry of Forests and Soil Conservation with the support of WWF, National Trust for Nature Conservation, the Nepal Army and local communities. The translocation was funded by WWF, USAID and the US Fish and Wildlife Service.*

WWF also explained Nepal's success story with its rhinos: *'Nepal's growing rhino population reflects a top level commitment to species conservation, a well-coordinated protection response between the national parks, Nepal Army, Nepal Police and local communities, and the support of conservation partners such as WWF and the National Trust for Nature Conservation'.*[25]

## 43.7   The First Line of Defence (FLoD) Initiative

However, although people power and working together can provide an unstoppable combination, and the role of local communities is recognized as essential to both combatting wildlife crime and the long-term survival of

---

[24] Ibid.

[25] WWF *Rhino translocation off to a positive start in Nepal's Terai* 2016.

wildlife, more efforts are required if they are not to become disillusioned. Many of the people are pastoralists, small farmers and even foresters. Some live along the boundaries of the protected areas, some even inside them. For many reasons, land is at a premium, essential to both people and animals who may well be in competition.

We have already looked at some problems arising from human-wildlife conflict and the ways these are being resolved, but the First Line of Defence (FLoD) Initiative is concerned with the illegal wildlife trade. It is a collaboration of the IUCN and the International Institute for International Development (IIED), who have been working with a number of partners since 2013 *'on initiatives to highlight the important role that local people play in conserving wildlife, to combat the IWT (illegal wildlife trade) and to better understand the perspectives of local people who are engaged in anti-IWT projects'*. Local communities are *rarely consulted* about the trade or how they would tackle it, a fact that is emphasized in the briefing paper's heading *'Local communities: the overlooked first line of defence for wildlife'*.[26]

The FLoD Initiative, which develops and refines the earlier Beyond Enforcement Initiative whose draft Theory of Change (ToC) tackled the illegal trade through community-level action, was field tested initially in 2016 in three Kenyan conservancies.[27] The following year, with the assistance of WWF South Africa and WWF in Namibia, the FLoD Initiative's methodology was further refined by scoping work carried out in southern Africa, while a consultative workshop with 35 Kenyan conservancies from the Kenya Wildlife Conservancies Association collected additional experiences and insights as well as identifying key policy issues and policy intervention entry points.

Four pathways were identified for community-level action:

- Pathway A: Increase the costs of participating in IWT
- Pathway B: Increase incentives for wildlife stewardship
- Pathway C: Decrease costs of living with wildlife
- Pathway D: Increase non-wildlife-based livelihoods[28]

---

[26] Briefing *Local communities: the overlooked first line of defence for wildlife* IIED, IUCN, CEESP, Species Survival Commission, SULi, AFRICAN ELEPHANT SPECIALIST GROUP, May 2018. See http://pubs.iied.org/17455IIED Accessed 13/06/2018.

[27] The Olderkesi Conservancy (adjacent to the Maasai Mara National Reserve), Kilitome Conservancy (adjacent to Amboseli National Park) and Olkiramatian& Shompole Group Ranches.

[28] See Briefing (n.26).

Working so closely with the communities, really paying attention to what they are saying, has produced rich dividends. The results have been revealing and lessons have been learned, which will enable policies to be improved and put into action. Particular attention was paid to Pathway B, for *'the target communities agreed that it was critical to have a mechanism that allowed them to benefit from wildlife, as an incentive not just to prevent poaching but also to secure land as wildlife habitat'.* High-end tourism operators purchase land leases so that they can access wildlife-rich land excellent for photo tourism. The quid pro quo: money for access to land. Community wildlife scout programmes were also included, where the quid pro quo was jobs, training and equipment in exchange for patrolling, providing intelligence and collaboration with state-led enforcement efforts.[29]

The lessons learnt so far, highlight:

- *'The importance of local rights to own, use and manage wildlife in order to benefit from it; and*
- *The importance of putting the right enabling conditions in place to achieve this'.*[30]

Not only will this help tackle the illegal wildlife trade, it will also benefit sustainable wildlife management more broadly.

---

[29] Ibid.

[30] Cooney, R; Roe, D; Dublin, H; and Booker, F (2017), *Wildlife Wild Livelihoods.* UNEP Nairobi, in Briefing (n.1006).

# 44

# Survival or Extinction?

We have finally arrived at crunch point.

## 44.1 The Great Elephant Census

In 2016, *'more than 25,000 elephants were killed across Africa and 668 rhinos were killed in South Africa alone – a 5,000% increase on 2007',*[1] and the 2014 Great Elephant Census estimated there had been a 30% decline in savannah elephants since 2007. The cause? Mainly poaching.

The first attempt to estimate the number of elephants in Africa was carried out by Dr. Iain Douglas-Hamilton (Save the Elephants) some 45 years ago. It found about 1.3 million animals, and although this figure was challenged, it provided a necessary starting point. Ten years later, the number had plummeted to 600,000 elephants.[2] That happened during the previous poaching crisis.

In 2014, the Great Elephant Census began, and now 95% of the known savannah elephant range has been surveyed. The census, which has been described as providing *'one of the best—quality wildlife surveys to date in Africa',* has made good use of advances in technology and thus enabled accurate results to be obtained. In some cases, these have been deeply shocking. Before that, because each elephant range State was responsible for counting its own

---

[1] TRAFFIC Wildlife TRAPS – Wildlife Trafficking Response, Assessment and Priority Setting. See http://pfbc-cbfp.org/news_en/items/traffic-TRAPS-en.html. Accessed 03/03/2016.

[2] Paul Steyn *African Elephant Numbers Plummet 30 Percent, Landmark Survey Finds* National Geographic, 2016. See http://news.nationalgeographis.com/2016/08/wildlife-african-elephants-population-de. Accessed 01/09/2016.

B. Martin, *Survival or Extinction?*, https://doi.org/10.1007/978-3-030-13293-4_44

population, the picture was very fragmented and therefore inadequate, because *a better understanding of the sizes of and trends in elephant populations is critical for setting conservation priorities*.[3]

The aim of the census, the first ever continental-scale survey of African elephants, was *to provide a coordinated, timely, high-quality survey of savannah elephants across the African continent*. Its goals were:

- To accurately determine the number and distribution of African savannah elephants over the majority of their range
- To provide a baseline on a continental scale for future surveys

At the same time, it provided an opportunity to determine two other vital issues:

- Whether, at a continental scale, elephants were doing better in protected rather than unprotected areas.
- How populations of savannah elephants have changed across Africa in the past 20 years.

The results to *serve as a baseline for assessing change in savannah elephant populations throughout Africa*.

Although, for comparison requirements, a few surveys took place in the wet season, counting in the dry season was much more satisfactory. This was when the deciduous trees had lost many of their leaves, so the animals were easier to see. And the weather was better. As there was potential for some animals to be counted more than once as they crossed from one country into another during their seasonal migrations, the surveyors in the countries involved timed their counts to coincide with each other.

All savannah elephants were to be counted, including dead ones. This was possible because *dead elephants remain visible for several years after dying*.[4] Most teams further divided the dead ones into *fresh* and *old* carcasses, which were bones only. That provided enough information to work out the *carcass ratio*, which is frequently used as an index of population growth rates.[5]

---

[3] Michael J. Chase, Scott Schlossberg, Curtice R. Griffin, Philippe J.C. Bouche, Sintayehu W. Djene, Paul W. Elkan, Sam Ferreira, Falk Grossman, Edward Mtarima Kohi, Kelly Landen, Patrick Omondi, Alexis Peltier, S.A. Jeanetta Selier, Robert Sutcliffe *Continent-wide survey reveals massive decline in African savannah elephants* PeerJ, 31 August 2016. See https://peerj.com/articles/2345/. Accessed 01/09/2016.

[4] Ibid.

[5] Douglas-Hamilton I, Burrill A. 1991. *Using elephant carcass ratios to determine population trends* African wildlife: research and management. Paris: International Council of Scientific Unions. 98-105.

## 44.1.1    Calculating the Carcass Ratio

Let us call:

* The number of live elephants seen **X**.
* The number of dead elephants seen **Y**.
* By adding the two numbers, we get **Z**. In other words: **X + Y = Z**.
* The carcass ratio is obtained by dividing **Z** by **Y**.

Obviously, the largest and densest populations would be easier to count, but they were also more important, having better genetic diversity,[6] being *'most likely to persist for the long term'*[7] and thus *'the ones most critical for'* the animals' future survival. Although the target was to count 90% of these elephants, the surveyors did even better. An estimated 93% were counted, with the actual percentage likely to be even higher.[8] Between 2014 and 2015, 18 countries were surveyed, the surveys in South Sudan and CAR were nearly completed and Namibia did not take part. But large-scale surveys were carried out in 2015.

The first survey in the Great Elephant Census was across Kenya's Tsavo National Park, described as *'an iconic elephant habitat… (its) sheer size and teeming wildlife make it an excellent barometer for elephant conservation'* and *'historically home to one of the largest elephant populations in Kenya'*. There the surveyors saw Satao, the *mighty tusker* whose death we mourned earlier in our story, *'as he trundled across his habitat'*. Their memories were *bittersweet* as *'it marked the last time anyone would (see him) alive'*.[9]

So what do the results tell us? Here are the broad details.

## 44.1.2    Protected Areas, Unprotected Areas and the Carcass Ratio

Across the survey, about 84% (295,978) of the elephants were found in protected areas, leaving 16% (56,262) in unprotected areas, although the proportions varied from country to country. Mali and Angola found most elephants in unprotected areas.

---

[6] Frankham 1996 in Chase et al. (n.3).
[7] Armbruster & Lande 1993 in ibid.
[8] Ibid, p.11.
[9] Great Elephant Census 2016.

Unfortunately, the teams recorded high carcass ratios across the board. It made little difference whether or not the area was protected or unprotected. It did vary, though, between countries.

The highest carcass ratios were in Cameroon, 83%; Mozambique, 32%; and Tanzania, 26%.

The highest fresh-carcass ratios were in Angola 10%; Cameroon, 10%; the W-Arly Pendjari (WAP) Ecosystem (on the borders of Niger, Burkina Faso and Benin) 3%; and Mozambique, 3%.[10]

These high carcass ratios *'are not simply due to elephants' being persecuted outside protected areas. Rather, carcass ratios of 12-13% suggest that deaths likely exceeded births over the 4 years preceding GEC surveys in both protected and unprotected areas'.* Notably high carcass ratios were recorded in Kenya, the northern section of Tsavo East National Park, 52%; Mozambique, Niassa National Reserve, 42%; and Tanzania, Rungwa Game Reserve, 36%.

They potentially indicate high levels of poaching. *'For protected areas, the clear implication is that many reserves are failing to adequately shield elephants from poaching and human-elephant conflict'.*[11] In other words, both protected and unprotected areas need better protection if elephant mortality is not to continue to rise.

### 44.1.3   Trends

Elephant populations between 2010 and 2014 declined at an estimated 8%. If this trend continues, *'GEC survey areas will lose half of their savannah elephants every nine years'.* Elephants in Mali, Chad and Cameroon, which have small, isolated populations may possibly become extinct.[12]

The cause? *'These dramatic declines in elephant populations are almost certainly due to poaching for ivory'.* As we saw earlier in our story, elephants from Mozambique and Tanzania are the primary sources of savannah elephant ivory, and elephant populations in Niassa and the Selous have decreased by over 75% in the past 10 years. The census trend model shows, as of 2014, these populations to be declining at 14% and 17% a year, respectively. For

---

[10] See Chase et al (n.3), p.9.

[11] For example: Gandiwa E, Heitkonig IMA, Lokhorst AM, Prins HHT, Leeuwis C. 2013 *Illegal hunting and law enforcement during a period of economic decline in Zimbabwe: a case study of northern Gonarezhou National Park and adjacent areas.* Journal for Nature Conservation 21: 133-142; Muboko N, Mupohshi V, Tarakini T, Gandiwa E, Vengesayi S, Makuwe e. 2014 *Cyanide poisoning and African elephant mortality in Hwange National Park, Zimbabwe: a preliminary assessment.* Pachyderm 55:92-94; and Booth VR, Dunham KM. 2016 *Elephant poaching in Niassa Reserve, Mozambique: population impact revealed by combined survey trends for live elephants and carcarasses.* Oryx 50:1-10 in ibid.

[12] Ibid, p.12.

similar reasons, the elephants in DRC are also struggling. This all makes grim reading. Their conclusion, with caveats, is that *'the illegal ivory trade appears to be the major driver of recent population trends in savannah elephants'*.[13]

Can we derive any hope from the caveats?

First, different ecosystems within the same country may not all be doing badly. We have already seen this in Malawi, with the current translocation of 500 elephants to a new home because their two former homes were 'full'.

Second, not all historical data was available, so *'trends in some areas were unknown'*. Unfortunately though, the declines in some of the other countries were so severe that this would be *'very unlikely to change any conclusions at the continental scale'*.

The third caveat again relates to *missing* historical data, but again this is unlikely to make a difference to the results.

## 44.1.4   Regional Differences in Elephant Population Status

There were large regional differences. Populations in western and central Africa were doing badly. Chad, Cameroon, Mali and DRC had small, isolated populations that were declining due to both poaching and expanding human populations.[14] The WAP Ecosystem had over 2000 elephants, and its populations have grown, but it now had high fresh-carcass ratios which might be a warning of things to come.

The news from Eastern Africa was mixed. Poaching had decimated elephant populations in both Mozambique and Tanzania, even though numbers were still relatively high, while Kenya, Uganda and Malawi showed more positive trends.

Southern Africa presented the best picture, with MIKE showing less poaching had taken place than in any other part of Africa.[15] There were relatively high populations in Botswana, South Africa, Zambia and Zimbabwe, populations that were either suffering *'mild or non-significant declines'* or even increasing. With, as we have seen, Angola the exception. Results from there showed *'extremely high carcass ratios and large numbers of fresh carcasses'* which suggested *'high levels of ongoing poaching'*.[16]

---

[13] Ibid.

[14] Bouche P, Douglas-Hamilton I, Wittemeyer G, Nianogo AJ, Doucet J-L, Lejeune P, Vermeulen C. 2011 *Will elephants soon disappear from West African savannahs?* PLoS ONE 6:e20619.

[15] CITES 2014.

[16] See Chase et al., p.13.

### 44.1.5   Caveats

The first was that despite every attempt to ensure the highest standards, the methodology came with built-in limitations, meaning that actual numbers of elephants were probably higher than those estimated by the survey teams.

The second was that the Great Elephant Census only applied to savannah elephants. The plight of many populations of forest elephants is known to be dire so there is an urgent need for these to be counted. We have already seen how a successful survey was carried out in the D'ja Conservation Complex in Cameroon.

### 44.1.6   Future Surveys

These need to be carried out regularly because elephant populations can change rapidly. They should:

* Measure trends in population.
* Gauge how effective conservation measures are.
* *'Identify populations at risk of extinction'.*

They might also detect emerging threats such as droughts. And they should be on the same scale as this Great Elephant Census.

### 44.1.7   Immediate Results

There have some of these. The governments of Mozambique and Tanzania were so concerned and moved by the preliminary results found in their countries that they implemented new measures to stabilize their populations of elephants.[17]

## 44.2   Botswana 2018

Although poaching occurred in the borderlands, very little took place inside Botswana itself, so when Elephants Without Borders and the Department of Wildlife and National Parks started their 2018 census of northern Botswana

---

[17] Lopez O, 2015 *The global fight for Tanzania's elephants* Newsweek; and Wildlife Conservation Society 2015 *Govt of Mozambique announces major decline in national population* > WCS Newsroom.

in July that year, the initial results must have come as a shock. They found 22 fresh carcasses *already* more than recorded in the entire 2014 census, and 33 others, all killed within the previous 3 months, a grand total of 55 poached elephants counted in a month, a total which had catapulted up to 87 a short time later.

The savagery of some of the attacks was shocking as well, even to those accustomed to seeing such horrors. In one act of unspeakable cruelty, a poor elephant, while still alive, had his spinal cord hacked through making it easier for the poachers to hack off his face while stealing his tusks. Some attacks had taken place close to game drive tracks, even safari lodges close to the Okavango Delta wildlife sanctuary, yet another UNESCO World Heritage site and a tourist hotspot.[18] 5 white rhinos had also been poached.[19]

This epidemic of poaching has been particularly tragic because, until now, Botswana has always been a safe haven for elephants. 130,451 were counted in the Great Elephant Census. And elephants, who *clearly have a cognitive ability to understand where they are threatened and where they are safe*, have been making their way there from the surrounding countries, so many of them seeking sanctuary from the poachers in Angola, Namibia and Zambia that numbers were becoming unsustainable.[20] Data from their tracking collars give details of their journeys. And because so many of the dead were old bulls, tuskers, there could well be a connection with criminal syndicates from south-east Asia.[21]

As elephant numbers continued to grow, so did the pressures on land, with the inevitable increase in human-elephant conflict. We have already seen that there are a number of ways to reduce such conflict, but in June 2018, the hunting ban on elephants in areas outside game reserves and national parks, especially in marginal rangelands, was lifted. The Assistant Minister of Presidential Affairs called on the residents of Chobe to shoot elephants on sight if they were caught damaging crops, although he was promptly rebuked by the Minister of Natural Resources Conservation, for Chobe is a town that relies on tourists and elephants are already declining in that region. And the new President is consulting on whether to lift the ban on elephant trophy hunting.

---

[18] Louise de Waal *Conservation: Elephant poaching on the increase in Botswana* The South African, 6 August 2018. See https://www.thesouthafrican.com/conservation-elephant-poaching-on-the-increase. Accessed 14/08/2018.

[19] Alastair Leithead *Dozens of elephants killed near Botswana wildlife sanctuary* BBC News, 3 September 2018. See https://www.bbc.co.uk/news/world-africa-45396394. Accessed 04/09/2018.

[20] Dr. Mike Chase in Alastair Leithead *Why elephants are seeking refuge in Botswana* BBC News, 31 August 2016. See https://www.bbc.co.uk/news/world-africa-37230700. Accessed 04/09/2018.

[21] Jane Flanagan *87 elephants killed in poaching frenzy* The Times, 5 September 2018.

Botswana has huge unfenced borders with its neighbours, which used to be patrolled by armed antipoaching units exercising a 'shoot-to-kill' policy, but such large numbers of elephants/ivory must always be a temptation to wildlife criminal gangs and there were slight indications as early as 2016 that some might be infiltrating the country. Dr. Mike Chase, who was *shocked* and *completely astounded*[22] by the scale of the poaching which was *'by far the largest I've seen or read about anywhere in Africa to date'*, took up the matter with the authorities, who were apparently unaware of what was happening, sending details of the dead animals, including GPS locations and photographic evidence,[23] but President Masisi allegedly responded *'nothing but hysteria'* to criticism of his government's antipoaching policy.[24]

Although we can only speculate on what is driving this disaster, we can observe that the antipoaching units were disarmed in May 2018 and that sustainable use is not working for the many people who currently feel alienated from or indeed are being adversely affected by the elephants. They do not derive sufficient benefit from the thousands of photo tourists who flock to their country, whereas they did, or think they did, benefit from trophy hunting. And elephants who have sought sanctuary continue to be horribly killed.

## 44.3   Akashinga

From despair to hope, and to another all-female group of rangers, only the second in the world, this time based in Zimbabwe. *Akashinga*, The Brave Ones.

Hope comes in more than one form with Akashinga. The rangers themselves are drawn from single unemployed mothers, who have suffered abuse and deprivation, vulnerable women. Women with no hope, until Damien Mander of the International Anti-Poaching Foundation met some of the Black Mambas while they were fundraising in New York and realized he could do something similar in Zimbabwe. The team is now a year old.

It is important to get an idea of size. The Akashinga operate within the 'buffer zones', vast areas of unfenced wildness surrounding the national parks and constituting almost 20% of Zimbabwe's land. These are game management

---

[22] Alastair Leithead *Dozens of elephants killed near Botswana wildlife sanctuary* BBC News, 3 September 2018. See https://www.bbc.co.uk/news/world-africa-45396394. Accessed 04/09/2018.

[23] Louise de Waal *Conservation: Elephant poaching on the increase in Botswana* The South African, 6 August 2018. See https://www.thesouthafrican.com/conservation-elephant-poaching-on-the-increase. Accessed 14.08.2018.

[24] *Botswana denies hand in elephant poaching* The Times, 10 September 2018.

areas, originally created for limited hunting, trophy hunting, some of the profits going to the local communities, to the estimated four million people living in the borderlands. Income that is rapidly drying up.

It became an urgent necessity to involve local communities, for them to become stakeholders again if these ecosystems were to survive. Hence Mander's initiative could not have come at a better time. The training is modelled on that of special forces, and it is very hard, but the women have proved themselves to be more than equal to it. Very few drop out. And according to Mander, the project is *'already putting more money per month into the local community than trophy hunting did per year'*.

Professor Muposhi, a conservation biologist, waxes lyrical, *'developing conservation skills in communities creates more than just jobs. It makes local people directly benefit from the preservation of wildlife'*, which, in turn, saves *'not only landmark species such as elephants but entire ecosystems'*.[25]

## 44.4    A Potential Major Breakthrough Using Elephant DNA

We saw earlier how Dr. Wasser and his team had developed a genetic map for African elephants. In September 2018, they published a paper in which they showed how this information could now be applied to help pinpoint key organized crime syndicates involved in illegal trading of ivory.[26]

They made a number of assumptions:

- That an elephant's (**A**) tusks frequently end up in different shipments (so a minimum of two seizures are required), having become separated between the killing and the export location
- That both tusks were probably 'owned' by the same cartel
- That the traffickers tended to use the same ports for export
- That the shipments were probably close together in time
- That the geographical origin of many the tusks in the two seizures containing **A**'s tusks would be the same (possibly/probably indicating a poaching hotspot)

---

[25] Jeff Barbee *Africa's new elite force: women gunning for poachers and fighting for a better life* The Guardian, 17 December 2017. See https://www.theguardian.com/environment/2017/dec/17/poaching-wildlife-africa-conservation-women-barbee-zimbabwe-elephant-rhino. Accessed 06/09/2018.

[26] S.K.Wasser, A. Torkelson,M. Winters, Y.Horeaux, S. Tucker, M. Y. Otiende, F. A. T. Sitam, J. Buckleton, B. S. Weir *Combating transnational organized crime by linking multiple large ivory seizures to the same dealer* Sci. Adv. **4**, eaat0625 (2018).

- That not only would a connection between the poaching operations on the ground and the cartel be indicated
- The size of the cartel

Applying this to their data from 2011 to 2014, three major cartels exporting out of Africa were identified and tied in to poaching hotspots, one of which was in Gabon.

This should make law enforcement more effective. It should be possible to identify not just cartels but some of their members, as they appear to operate using a similar pyramid-shaped hierarchy to that used by the rhino horn cartels (of course there might well be overlaps). And it should make it easier to prosecute, not just the poachers, which, as we have seen, is rarely satisfactory, but those higher up the chain, and for more serious offences, transnational organized criminal offences, which carry more severe penalties.

It would also enable enforcement to be more proactive, concentrating on the poaching source, as recent seizures would predict near-future poaching events.

However, a note of warning was sounded. For this to work well, they needed to expand their database, which meant more tusks from major seizures should be sent to them for analysis, something countries should be doing anyway under CITES Resolution Conf. 10.10.

## 44.5   London, October 2018

The action now returns to London.

### 44.5.1   The BIAZA Initiative

A new initiative, the first of its kind, was announced in advance of the next Illegal Wildlife Trade Conference that was held in London on 11–12 October 2018. On 7 September, BIAZA, the British and Irish Association of Zoos and Aquariums, a conservation wildlife charity, issued a *Position Statement on the Illegal Wildlife Trade* which it recognizes as having *'an increasingly devastating impact on biodiversity worldwide'* and as *'the major threat of extinction for a growing number of species including high profile animals such as elephants…'* that *'Europe, including the UK, is both a destination market and a hub for trafficking species in transit to other regions'* and they want to do something about it. Therefore its members undertake to:

- *'Work to add to information about the illegal trade within the UK and Ireland'.*
- Work on increasing captive breeding knowledge.
- Raise awareness of the IWT with the visiting public.
- *'Ensure that BIAZA standards are met on the transfer of animals, including due diligence on their provenance and their destination'.*

BIAZA also *'requests that UK and Irish Governments:*

- *Strengthen enforcement agency resources for the regulation of the wildlife trade including border controls for the import of species, particularly with steps to increase the identification of illegally traded species...and to increase the staffing of the UK Border Force CITES team particularly post Brexit)'.*

And for the UK Government, that it:

- *'Continues to fulfil its promises made earlier this year with the announcement of the Illegal Wildlife Trade Challenge Fund'.*[27]

We've already seen how Chester Zoo is educating and encouraging its visitors on how to take action, and since some 30 million members of the public pass through the gates of BIAZA's 117 members every year, even if only some of them decide to 'do something', it could make a big difference.

## 44.5.2    The Wildlife Financial Taskforce

This initiative was launched by Prince William in advance of and in response to UNODC's finding that only one in ten jurisdictions spends money on investigating the transnational organized crime gangs involved in the illegal wildlife trade. The Taskforce has been set up by United for Wildlife, and more than 30 global banks and financial organizations are signatories, as well as TRAFFIC and RUSI, with the UK government contributing £3.5 million to the funding. It will tackle money laundering.[28]

---

[27] BIAZA *Position Statement on the Illegal Wildlife Trade* 7 September 2018. See https://biaza.org.uk/news/details/biaza-position-statement-on-the-illegal-wildlife-trade. Accessed 13/09/2018; and BIAZA *First of its kind zoos and aquariums pact calls for an end to illegal wildlife trade* see https://biaza.org.uk/news/detail/first-of-its-kind-zoos-and-aquariums-pact-calls-to-an-end-to-illegal-wildlife-trade. Accessed 13/09/2018.

[28] Valentine Low *Banks join fight against illegal trade in endangered species* The Times, 10 October 2018; and GOV.UK *UK aid to crack down on criminal gangs driving the illegal wildlife trade.* See https://www.

### 44.5.3   'Sounding the Horn' and a Bombshell from China

Two important events in October once again made rhinos headline news. A new report exposed loopholes in UK legislation which potentially allowed modern rhino horn to be laundered into the system as antiques, and China announced it was 'adjusting' the 25-year-old ban on the use of rhino horn and tiger bone in traditional Chinese medicine.

We have already seen how the machinations of the Rathkeale Rovers were having a devastating impact on rhino poaching, despite the fact that the horns they were stealing came from long-dead rhinos. And how the 2016 survey of the ivory markets in the UK influenced the UK government's decision to put forward a Bill to ban most sales of elephant ivory. So, in 2017, researchers carried out a year-long survey into the sale of alleged rhino horn antiques in London auction houses, and Save the Rhino International published the results in 2018.[29]

Put at their most brief, these were disturbing, though not unexpected, exposing potential loopholes and revealing *trends that we consider render the trade open to exploitation by those trafficking modern rhino horn or cheap, older, ground-down or re-carved horn*.[30] Perhaps of equal importance is the fact that we finally have hard evidence about some of the sales of rhino horn in the UK, enough, hopefully, to persuade the government to consider bringing in new legislation similar to their Ivory Bill. Particularly important, because the researchers have continued their monitoring and, by July 2018, had found that *more lots had been offered than in the whole of 2017, with sales totalling a minimum of £1.2m*.[31] This huge figure does not of course include sales by antique shops, antique dealers and private sales.

Then came China's bombshell. In recent years, China had seemed fully aware of the need to protect the wildlife on this planet, playing a leading role when it banned the sale of elephant ivory, so it came as a complete shock when a few months later, the State Council announced that some medicinal use of rhino horn and tiger bones could be resumed. It did, however, impose conditions, making it clear that the products must come from

---

gov.uk/government/news/uk-aid-to-crack-down-on-criminal-gangs-driving-the-illegal-wildlife-trade. Accessed 11/10/2018.

[29] Brace, S. and Dean, C. (2018). *Sounding the Horn: A survey of rhino horn antiques sold in 2017 at auction in the UK*. Save the Rhino International, London.

[30] Ibid, p.4.

[31] Ibid, p.2.

farms, be in powdered form and could only be used by *qualified doctors* in *qualified hospitals*.[32]

The move was welcomed by doctors using traditional Chinese methods, who, fearing these were in danger of dying out, had been campaigning to save them.[33] WWF, however, made it quite clear that this would be a disaster that would lead to *devastating consequences* and be an *enormous setback* to the protection of wild rhinos.[34] The *South China Morning Post* stated that safeguards would be needed, with WWF's Margaret Kinnaird pointing out that the move would cause '*confusion amongst consumers and law enforcers and expand the market in other tiger and rhino items which should also be banned*'.[35]

## 44.5.4    London Conference on the Illegal Wildlife Trade (October 2018): Declaration

Unlike the earlier London Declaration (2014), this one is brief and to the point, the government representatives recognizing '*the significant detrimental economic, environmental, security and social aspects*' of the trade and politically committing to '*call upon the international community to act together to support and build urgent collective action to tackle the illegal wildlife trade as a serious crime carried out by organised criminals, and to close markets for illegally traded wildlife*'. They also reaffirmed the commitments they made in the 2014 Declaration and the Kasane 2015 and Hanoi 2016 statements, with any new participants now affirming their determination to combat the illegal trade.

In four key sections, the Declaration deals with the impact of the trade, tackling it as a serious and organized crime, working in partnership and demand reduction.

It ends with an acknowledgement of all the efforts that have been made so far by governments, the UN, CoPs, NGOs and the private sector and confirms '*our intention to build upon the successes already achieved*', committing themselves '*to enhance our individual and collective efforts to meet our existing commitments*'.[36]

---

[32] BBC News *Rhino horn: Alarm as China eases 25-year ban on rhino and tiger parts* 30 October 2018. See https://www.bbc.co.uk/news/world-asia-46027702. Accessed 30/10/2018.

[33] Didi Tang *Chinese doctors hail return of rhino and tiger remedies* The Times, 3 November 2018.

[34] See BBC News (n.32).

[35] See https://www.scmp.com/comment/insight-opinion/article/2171 1506/safeguards. Accessed 3/11/2018.

[36] See https://www.gov.uk/government/publications/declaration-london-conference-on-the-illegal-wildlife-trade-2018/london-conference-on-the-illegal-wild. Accessed 25/10/2018.

## 44.6    The Illegal Wildlife Trade (IWT) Challenge Fund

Established in 2013 by the UK Government, the IWT Challenge Fund provides money for practical projects around the world, projects that in some way help to combat the illegal wildlife trade. Concerned with:

- *'Developing sustainable livelihoods and economic development, to benefit people directly affected by IWT*
- *Strengthening law enforcement*
- *Ensuring effective legal frameworks*
- *Reducing demand for IWT products'*

Their subject matter is familiar, as are the NGOs who have applied for funding and the countries they will work with and the wildlife that will benefit. Money has already been allocated to 61 projects. Their range is wide, for example:

- *'Educational Children's Videos to Reduce Endangered Species Demand in Vietnam'*, to be made by Humane Society International and CITES Management Authority of Viet Nam.
- *'Technology and Innovation Against Poaching and Wildlife Trafficking'*, whereby the Stimson Centre will design a *gold standard* wildlife protection technological system as a pilot to protect rhinos in Tsavo West National Park.
- *'African Wildlife Forensics Network – capacity and coordination for law enforcement'*, whereby UNODC aims to reduce the disparity between arrests and convictions in prosecuting IWT-related crimes in eight African countries, including Botswana, Gabon, Zambia, CAR, DRC, Mali, Angola and Zimbabwe, by establishing a wildlife forensic network.
- *'Strengthening community anti-poaching and ecotourism in the Western Terai Complex'*, Nepal, ZSL will establish *'12 Community Based Anti-Poaching Units and a Rapid Response Network'* to improve intelligence and help *'local communities benefit from recovering wildlife populations through ecotourism'*.
- *'Developing investigation and prosecution capacity to save Angola's elephants'*, which is concerned with Angola's NIAP and National Elephant Action Plan. This Project aims to *'Strengthen criminal justice system for wildlife crime; investigate, analyse and reduce domestic and trans-boundary ivory trade;*

*and develop sustainable alternative livelihoods for ivory traders'*, thus reducing both elephant poaching and ivory trafficking in Angola.

IFAW has been awarded a grant to establish an operational Wildlife Crime Investigations Unit (WCIU) and community enforcement networks in Malawi (elephants to benefit). The IUCN—Eastern and Southern Africa—to strengthen local community engagement in combatting illegal wildlife trade. The EIA has been given money to make a training film to aid enforcement officers combatting the illegal ivory trade; Northern Rangelands Trust (NRT) to connect enhanced livelihoods to elephant and rhino protection; RUSI to detect and prosecute money-laundering in Kenya and Tanzania; and TRAFFIC International on combatting global wildlife cybercrime. The Mali Elephant Project has also benefitted.

Among the many other beneficiaries are ARC, Flora and Fauna International, Wildlife Conservation Society and INTERPOL. Most of the countries involved are African range States and countries in Southeast Asia.

## 44.7   The Theory of Scarcity Value

We have already seen how countries seem to have been tumbling over themselves in their haste to destroy their ivory stockpiles, sometimes their rhino horns as well. While this is good news, we need to sound a note of caution. It could lead to another problem. Let us examine a theory, the theory of scarcity value.[37]

Earlier in our story, we saw that some Chinese officials had stockpiled rhino horns because they were worth more than gold. We also saw that the result of this was the blizzard of poaching that occurred around the 1990s, when so many rhinos were so cruelly slaughtered. And now there appear to be discrepancies in Hong Kong SAR's ivory stocks. Add to this the burning of stockpiles. Fewer rhinos, fewer elephants, less rhino horn, less ivory, this is where danger might lie, the theory of scarcity might begin to operate. Fewer animals, more valuable stockpiles. So stockpiles could begin to grow again but at the expense of our animals.

While this might be just a theory borrowed from a detective story, it does expose potential dangers, so we need to think of workable solutions now. Just in case.

---

[37] Borrowed from Michael Innes *Appleby and Honeybath* 1983.

## 44.8   Rhino Numbers

Although there is no data corresponding to Great Elephant Census, we do have a certain amount of information. There is mixed news. We already know that there are now only two northern white rhinos and that the numbers of both Javan and Sumatran rhinos are causing deep concern. And although some rhino populations are increasing, and some of the measures to combat poaching are finally beginning to work, it's still not enough. The poachers, the organized crime gangs, the hangers-on, have not gone away. Reluctant to abandon such a lucrative trade, they will continue to kill rhinos until either demand ceases or all the animals are dead.

World Rhino Day 2016 should have been a day for celebration, but it was not. News came that 6 white rhinos had been killed. Six white rhinos were shot and dehorned in three separate incidents, all in Hluhluwe-Imfolozi Park in KwaZulu-Natal, the very park that had nurtured their comeback. It demonstrated an *alarming escalation* of the number of poaching incidents in the province in the previous few weeks and added to fears that poachers were now choosing this Park in preference to Kruger National Park whose (anti-poaching) measures had been tightened and appeared to be working well.

The Provincial Government acted immediately, setting up a high-powered task team which was given 6 months to review the situation, and then report back with recommendations. Its members, all experts, were drawn from Ezemvelo KZN Wildlife, the South African Police Service, an international expert in policing, a legal expert from the Ian Player Foundation and someone from the Premier's office. And its brief was to assess just how fit for purpose the criminal justice process was when dealing with poaching incidents, whether there were sufficient resources in the system to turn the situation around.[38]

Zimbabwe too was losing its rhinos, the numbers climbing steeply from 20 in 2014 to 51 in 2015, and with the two important black rhino populations in the Save and Bubye Valley Conservancies, monitored and managed by the Lowveld Rhino Trust also coming under attack. These, together with a third population in nearby Malilangwe, are key populations, offering Zimbabwe its best chance of keeping the species. But they need to increase in size, and, unfortunately, they are all, particularly the Save Valley Conservancy, close to the porous borders with South Africa and Mozambique where so many poachers come from.

---

[38] Tony Carnie *Six KZN rhino killed on Rhino Day*. See www.iol.co.za/news/south-africa/kwazulu-natal/six-kzn-rhino-killed-on-rhino-day-2072033. Accessed 27/09/2016.

Zimbabwe's parks and Wildlife Management Authority anti-poaching unit began the fightback. In August 2016, it arrested 10 suspected poachers, four of them in Save Valley Conservancy. Two of the suspects were on the police's 'wanted' list, and another was the nephew of a prominent Member of Parliament. Unfortunately in Zimbabwe, though, *'bribery is common with many high-profile cases failing to go to trial'.*[39] So will the change in regime make a difference?

Zimbabwe decided to take drastic action and dehorn all its black rhinos.

## 44.8.1   The Sera Rhino Sanctuary

In 2003, Kenya began to implement its third 5-year Rhino Conservation and Management Strategy, aimed at sustaining the 5% annual increase in the population of black rhinos that had taken place between 1986 and 2003.[40] Then came the next massive onslaught of poaching, and although Kenya's rhinos fared badly, by the beginning of 2013, there were still 631 animals left, *'just 19 … short of the 2010 target set out in the 2008 – 2012 National Strategy'.* The news in 2014 was even better, rhino poaching reduced by 40%.

However, the irony is this: too many rhinos living together can give rise to problems, including fighting. This was beginning to happen, so another National Rhino Strategy, 2012–2016, was developed to take account of the new challenges and enable rhino numbers to continue to increase. There were two strategic objectives:

- Diversification of the management and custodianship of rhinos by setting up sustainable financial mechanisms for community rhino conservation projects and establishing a viable rhino population (minimum 20 rhinos) in at least one community sanctuary
- Translocation of rhinos from productive populations to avoid overstocking and to diversify the gene pool

Black rhinos would be reintroduced onto land that had previously supported a population before they were poached to extinction. It would be a sanctuary, the first community-operated black rhino sanctuary in East Africa (possibly the whole of Africa), and the new rhinos would be an economic asset, providing

---

[39] Save the Rhino International *Poacher arrests in Zimbabwe's Save Valley* August 2016. See https://www.savetherhino.org/latest_news/news/1534_poacher_arre. Accessed 29/09/2016.

[40] Northern Rangelands Trust *Rhino Conservation Programme Phase 1* Report, November 2015.

both employment and trading opportunities for the local communities. A targeted tourism strategy would be developed. Sustainable use.

The plan was to split the revenue from tourists two ways:

- 40%: For conservancy operations
- 60%: For the local community to support '*projects in water, health, education and infrastructure*'[41]

By 2015, although 86% of the Conservancy revenues were coming from tourism, the financial sustainability plan recommended that trade in livestock and local products be developed further, just in case there was a slump in tourism. The prosperity and fate of rhinos and people were inextricably linked.

As KWS already had a good working relationship with NRT, they were the obvious partners, and this was the start of the Sera Rhino Sanctuary.

Three years after construction began, of the original 13 black rhinos released, 10 remained, 5 males and 5 females. Now, in 2018, there are 11.[42] Initially they were looked after by 24 full-time rangers, but that number has now grown to 33 and is supplemented with a mobile anti-poaching unit.[43] Every rhino has its own GPS device fitted to its horn, and they are well protected. Some 200 people patrol the Sanctuary, which is enclosed by a fence, which itself is guarded even at night.[44]

In March 2016, they celebrated the birth of the '*first black rhino calf to be born on community land in northern Kenya for over 25 years*'. The press release fairly fizzed with excitement, the birth demonstrating '*the strength of the growing community conservation movement*' and representing the community's '*hopes that the Sanctuary can nurture a viable breeding population of Black rhinos, that could eventually help repopulate other community conservation areas*'. A little female. This is a project that has merged '*a successful community model with Kenya's National Rhino Strategy, in a move that aims to benefit local communities as much as an iconic and critically endangered species*'.[45] And to make sure that everyone felt involved, '*Sera management and Board placed strong*

[41] Ibid.

[42] NRT State of Conservancies Report 2017.

[43] Tusk *Sera Wildlife Conservancy* 2018.

[44] Danae Mercer *Inside Africa's first locally owned rhino conservation programme: 'sometimes they do charge'* Independent, 1 December 2017. See https://www.independent.co.uk/travel/africa/rhino-tracking-kenya-s. Accessed 07/08/2018.

[45] NRT Press Release *Sera Community Rhino Sanctuary Celebrate First Black Rhino Birth*. See www.nrt. kenya.org/news-list/2016/3/21/press-release-sera-community-rhino-sanctuary-celebrate-first-b. Accessed 13/05/2017.

*emphasis on community engagements in 2016, with numerous meetings held with herders*[46] *and other community members'.*

The following year a male calf, Loijipu, was born. Sadly his mother rejected him, but he was welcomed by the Reteti Elephant Sanctuary where he is being reared by hand.

And now Sera has another venture, a walking safari, another income generator, another new experience for tourists to try out.

As well as prosperity, the Sanctuary has also brought peace. Three pastoral communities, the Rendille, Samburu and Borana people, live in this geographical area, and traditionally they had not lived together in harmony. Historic injustices still provoked retaliation. Other issues were water and grazing. If rhinos were going to live in the Sera Community Conservancy, something had to be done. The something was *'a massive peace drive'* with NRT's peace team assisting. It *'encouraged dialogue as a means of conflict resolution'*, and when relations became tense, peace meetings between the communities helped diffuse the situations. The Sera grazing committee, which includes representatives from all three communities, has also helped. Set up in 2007, at the very beginning of the Sanctuary project, conflict resolution is one of its functions. It mediates disputes and helps recover stolen livestock.[47]

## 44.9 Changing Minds/Demand Reduction

One of the things that has become clearer as we near the end of our story is that there are very different ways of viewing ivory and rhino horn. We know that ivory comes from elephants and that elephants have families. They are highly intelligent animals who exhibit familiar emotions. They grieve for their loved ones. They can even discriminate between humans who might harm them and humans who will certainly help them. They have learnt to trust some humans, and they will even make their way to them if they have been injured, seeking help which is lovingly given. We see wonderful creatures. What we don't see is ivory/money!

So how do we solve the problem of the people who only see elephants as money? They don't see an animal; they merely see tusks. A commodity to be exploited. They are blind to the killing, to the terror of the beasts, the agonies they endure. The pain and suffering of drawn-out death or the cruel steel of the chainsaws as they slice through the skulls to get at the teeth, the 'precious'

---

[46] NRT State of Conservancies Report 2016.
[47] NRT 2015.

tusks. They witness, but do not see the destruction of whole families, the frightened bewilderment of orphaned calves. No, what they see is money. Huge quantities of money. Money to make them wealthy beyond the dreams of avarice. These people do not want to change.

And there is a third group, an intermediate group. For some of the humans who kill elephants do so because they have themselves been injured by them in some way. Perhaps they have lost their crops, or relatives and friends have been killed. Populations are growing. More and more people are moving into lands where once only elephants roamed. The climate is changing and with it the elephants' habitats. Drought, the enemy of animals and people, is never far away. The opportunities for human-elephant conflict are growing. For some of these victims, poaching presents a real alternative to poverty. They have not benefitted from the animals, merely lost out to them. So they have had no love for them. Fortunately, we have already seen how some of these conflicts are being resolved and how people and elephants can live together in harmony.

With rhinos and rhino horn, although the first two groups apply, there are few human-rhino conflicts. So the third group is different. With them, joining the ranks of the rhino poachers is often prompted by poverty. The world must lift them out of poverty and teach them to love rhinos.

## 44.9.1  Minds Can Be Changed

We have already seen this happen. The local communities in Managalane lost their homes when Sabie Game Park first arrived. They became so disillusioned that many were prepared to join poaching gangs. But it all changed after the CBNRM, a community-based natural resource management, got under way, and they were able to participate in the sustainable use of their land. They derived an income from the wildlife and that helped the populations of wildlife to recover. It also reduced human-wildlife conflicts, and it *'encouraged the local ownership of wildlife resources'*. Lubilo expresses the effect with particular clarity: *'Just as if you kill or attempt to steal someone's cattle the community applies instant justice, simply because they have attached value and ownership, so people will stop rhino poaching if they are able to appreciate that they own the rhino'.*[48] People can be incentivised to become very protective of their animals.

---

[48] Lubilo 2015.

## 44.9.2   The Reteti Elephant Sanctuary

The same thing happened in the Northern Rangelands Trust. On 12 September 2016, the following announcement was made: *'Samburu Community to Open Elephant Sanctuary'*,[49] although Namunyak Wildlife Conservancy had officially opened the Reteti Elephant Sanctuary on August 20. Like Sera, Reteti is a first, the first community-run elephant sanctuary in Africa, and it is almost entirely run by local people.[50]

The sanctuary takes in orphaned and abandoned elephant calves and releases them, when they are ready, back into the wild herds adjoining the sanctuary in the remote Mathews Range. There, are found some 8700 elephants, the second largest population in Kenya. Fortunately very few calves, normally between five and ten, are rescued a year, and the intention is only to take the calves into care when everything else has failed. It has not been possible to reunite them with their herds at an earlier stage. So far though, the results have been excellent, and very few babies have needed to be reared by hand. At the close of 2017, there were just 12 elephants, and Loijipu, the baby rhino in the rescue facility, looked after by their dedicated keepers.[51]

It is a great success story. Both administration and training are good. In addition to working closely with the rangers and KWS, the mobile elephant rescue team also raises community awareness and helps with the mitigation of human-wildlife conflict, still a problem in the community conservancies.

And it happened because the local communities understood that wildlife was providing them with an opportunity to improve their livelihoods. They wanted their own sanctuaries, and everyone rallied round to make it happen. To quote NRT, *'Reteti is more than just a sanctuary. It is transforming the way local people relate to elephants, driving a sense of pride and ownership amongst a previously apathetic demographic'*.[52]

Once again, working together has made this possible. This time nine partners, including Namunyak Wildlife Conservancy, Samburu County Government, KWS, NRT, San Diego Zoo, Conservation International, Tusk Trust, The Nature Conservancy and Save the Elephants and several individuals are all working together to save elephants.

---

[49] Northern Rangelands Trust Press Release – *Samburu Community to Open Elephant Sanctuary* 12 September 2016.

[50] See (n.42)

[51] Ibid.

[52] Ibid.

### 44.9.3  British Antiques Dealers and Auction Houses Stand Up to Be Counted

October 2018 was a busy month. At about the same time, *Sounding the Horn* was published, and just before the London Conference, the British Antiques Dealers' Association (BADA) asked the UK government to introduce tough new legislation on sales of artefacts made of or containing rhino horn as the previous, strict regulations had been relaxed in 2016.

Then in Hong Kong, in November 2018, as the result of pressure from 37 NGOs and a petition of more than 10,000 names, Bonhams decided to follow Christies' example. It cancelled an auction selling rhino horn, 21 carvings, and there are to be no more of these sales. The following statement by its CEO was quite clear: *'All rhino carvings that have been offered at Bonhams have been antiques, with a known provenance and carried CITES licences … We do however, recognize that there are widely held concerns about this issue and have decided that the sale of the rhinoceros carvings … will not now take place. In future, Bonhams will not offer artefacts made entirely or partly from rhinoceros horn in its salesrooms'.*

Sotheby's was quick to follow. After withdrawing rhino horn artefacts from a Hong Kong auction, the chairman of Sotheby's Asia declared that, in the future, Sotheby's would not be offering such items for sale that it deplored *'any illegal slaughter and trading of endangered wildlife'* and that it *'strongly supports conservation efforts from the global community'*.[53]

### 44.9.4  China Has a Rethink?

Shortly after its shock announcement lifting the ban on trade in tiger bone and rhino horn, China's State Council announced that this had been postponed.[54] It would be encouraging to think that this was partly down to listening to the many people who had expressed their grave concern. Margaret Kinnaird welcomed the decision, *'this move helps maintain the leadership role China has taken in tackling the illegal wildlife trade and reducing market demand'*.[55]

---

[53] *Sotheby's and Bonhams stop selling rhino horn at auction in Hong Kong after wildlife group pressure* Antiques Trade Gazette, 26 November 2018. See https://www.antiquestradegazette.com/news/2018/sotheby-s-and-bonhams-st. Accessed 26/11/2018.

[54] *China's change of heart on rhino and tiger ban* The Times, 13 November 2018.

[55] Javier C. Hernandez *China, After Outcry, Reinstates Ban on Rhino and Tiger Parts in Medicine* The New York Times, 12 November 2018. See https://www.nytimes.com/2018/11/12/world/asia/china-rhino-tiger-ban.html. Accessed 14/11/2018.

## 44.10    People Do Care

Although the dreadful spectre of extinction has loomed like a dark cloud through-out our story, we have seen that there are green shoots, grounds for optimism, grounds for hope. Just how deeply people do care for their/our animals is illustrated by the following incident which happened in August 2016.

It was a bad year for the monsoon in India. The torrential rains led to severe flooding, and one poor elephant became separated from her herd and was swept down a swollen river all the way to Bangladesh. From Assam, in north-eastern India, a journey of some 60 miles. In Bangladesh, she became trapped by rising flood waters. However, the people were determined she should not die and desperate efforts were made to save her.

At one time, things seemed bleak. A group of veterinary surgeons riding a tamed elephant managed to reach her, but she was aggressive. She must have been terrified. So, unable to treat her and with their own elephant by then refusing to approach her, they were forced to retreat. She wandered aimlessly through the floods trying and failing to reach higher ground, gradually growing weaker and ever more distressed.

But the people would not give up. The vets eventually managed to tranquilize her, only to have her collapse into a nearby pond. There she would have drowned if the local villagers had not managed to drag her to safety. No easy task. According to one of the vets, when the villagers saw she had become uncon-scious, *'dozens jumped into the pond and helped us tie the animal with ropes and chains. Finally, hundreds of villagers helped us to pull her to dry ground'*.[56]

She was taken to a safari park to recover, to regain her strength and to wait for the flood waters to recede, when she would finally be allowed to return to her home, in India.

## 44.10.1    Refuse to Be Discouraged

We have seen that rhinos can become even more vulnerable when there are only a few viable populations, so, despite recent setbacks over translocation, more are being considered. Olmoti, a young female black rhino living in Yorkshire in the UK, will shortly travel to the Czech Republic to bond with four companions before the five of them fly off to Africa, to Rwanda, in June 2019.[57]

---

[56] Hugh Tomlinson *Elephant swept across border by monsoon is hauled to safety* The Times, 13 August 2016.
[57] *Getting outta here* The Times, 22 November 2018.

## 44.10.2   And Continue to Be Amazed

Recent research indicates that some elephants are rather good at solving mathematical puzzles. Authai lives in Japan, in Ueno Zoo. She is an Asian elephant and is 14 years old. Together with two other, slightly older elephants, she was taught to use a computer-controlled touch panel, and then the experiment began.

Cards showing pictures of bananas, apples and watermelons of different sizes, to make sure the animals were not simply choosing the cards with the biggest covered area, were displayed on the panel, two at a time, the numbers of fruit ranging from 0 to 10. Using the tips of their trunks to indicate, the elephants had to choose which one of the two cards contained most items, and to do this correctly, they had to count the items. When they were right, they were given a piece of fruit as a reward, but if they were wrong, the apparatus beeped.

Authai, who was better than the other two, got it right on an impressive 181 out of 271 occasions, a 66.8% rate of success. Artit only managed 49.73%, in other words, below the level of chance, while Surya decided it was not for her, and after two sessions, she refused to go to the experiment.

It seems that only Asian elephants can count, so the lead researcher, Dr. Irie, has suggested that because Asian and African elephants became separate species over 7.6 million years ago, this could well have led them developing different cognitive abilities. The team will carry on researching and is proposing to include other animals as well.[58]

# 44.11   Survival or Extinction?

Here are a few stark reminders of some of the problems still to be overcome. For example:

- The acting manager of the Madikwe Game Reserve noticed a new trend: poachers had started killing both babies and adults. Even more disturbing, sometimes body parts such as an eye or an ear, rather than the horn, were

---

[58] Mark Bridge *Elephant proves she is equal to jumbo maths puzzles* The Times, 24 October 2018; Asian elephants could be the maths kings of the jungle Heideberg, New York, 22 October 2018, published by Springer.   See   https://www.springer.com/gp/about-springer/media/research-news/all-english-research-news/asian-elephants-could-be-the-maths-kings-of-the-jungle/16208962 Accessed 11/12/2018. See Irie N. et al. (2018). *Unique numerical competence of Asian elephants on the relative numerosity judgement task* Journal of Ethology, DOI: https://doi.org/10.1007/S10164-018-0563-y for a detailed description of the research.

**Fig. 44.1** Authai doing her maths. Copyright Springer Japan and Dr. Naoko Irie

being taken.[59] In February 2017, the Thula Thula Rhino Orphanage in South Africa was viciously attacked: one calf was killed, another had to be euthanized and the staff were *beaten* and *brutalized*.[60] That same week, another attack took place on a privately owned game reserve, the victims were two hand-reared calves. Impi died but kept his horn. Vrystaat survived but lost his horn, receiving terrible injuries in the process. The veterinarians from Save the Survivors hoped to save him, although it would be many months before treatment could stop.[61]

- In March 2017, the unthinkable happened. Vince, a young white rhino who lived with two other rhinos in a zoo near Paris, France, was shot three times in the head.[62] The poachers, for poachers they were, fled with 1 of his horns leaving the other partly hacked off and people wondering what might

---

[59] Jamaine Krige *Poaching of rhino calves raises concerns* SABC News, 17 February 2017. See http://www.sabc.co.za/news/a/0902fd804019bc7ba4c9ef75889f6f7e/Poaching-of-rhin Accessed 21/02/2017.

[60] IFAW STATEMENT *Attack on Thula Thula Rhino Orphanage* 22 February 2017. See http://www.ifaw.org/united-kingdom/news/statement-attack-thula-th Accessed 11/05/2017.

[61] Elizabeth Claire Alberto *Little Rhino Refuses To Give Up After Poachers Stole His Horn* The Dodo, 24 February 2017. See https://www.thedodo.com/rhino-poaching-survivor-2282750064.html Accessed 11/05/2017.

[62] Katie Forster *Poachers break into Paris zoo, shoot rhino dead and steal its horn* The Independent Online, 7 March 2017. See http://www.independent.co.uk/news/world/europe/paris-zoo-thoiry-r Accessed 07/03/2017.

have happened. Were all three rhinos intended targets, had there been an equipment failure, why was this incident incomplete? Security appeared to have been good, cameras and five members of zoo staff living on site, but it hadn't been enough to protect Vince. Zoos have been warned. They will have to up their game.

- Poachers have managed to kill one of the rhinos in Chitwan National Park, Nepal, hacking off its horn.[63]
- Illegal trading in wildlife seems still to continue with impunity in Mong La's infamous market.
- Reopening the legal trade in rhino horn in South Africa will have consequences, but we can only speculate on what these might be. It will be an experiment, and we must be ready with possible solutions should it all go wrong and rhino poaching again start to skyrocket worse than ever. We need a Plan A and Plan B, perhaps even Plan C, ready in case of emergency. But at least there have been some successes catching poachers in Kruger National Park, with SANParks arresting 23 at the end of July, after one of their rangers had been shot dead while arresting suspects, and the Rhino 9 team arrested 27 in September. Serious charges will follow, including murder, attempted murder and firearms offences, and 16 unlicensed firearms were recovered.[64]
- And, as Dr. Daniel Stiles pointed out earlier in our story, we don't even have a Plan B should a sales ban on ivory be ineffective, or even make matters worse. We should be prepared for this as well.
- Then, in October 2018, China's shocking announcement that it had lifted its ban on the sale of powdered rhino horn and tiger bones came like a bolt of lightning from a clear blue sky.

At this point we can do no better than quote from Prince Harry's speech to the Southern African Wildlife College:

*'In the face of this, it would be easy to become despondent – but all of you here today know that we must not. We can win this battle. This is a test for all humanity and we cannot afford to fail. Nature needs us to fight her battles and in this case, protect her animals, some of which have been on this planet for tens of thousands of years'.*[65]

---

[63] The Times *Poachers kill rare rhino* 10 April 2017.

[64] See https://www.iol.co.za/news/south-africa/limpopo/sanparks-says-23-suspected-poachers-arrested-since-end-of-july-16428236 and https://www.iol.c0.za/news/south-africa/gauteng/27-alleged-poachers-arrested-firearms-and-rhino-horn-recovered-17092298. Accessed 20/09/2018.

[65] HRH Prince Harry *In my view: rangers are heroes*. A speech by Prince Harry at the Southern African Wildlife College, Kruger National Park, South Africa. See https://www.royal.uk/speech-prince-harry-south-african-wildlife-col. Accessed 12/01/2017.

**Fig. 44.2**   Dan and Sloan Stiles with Sudan. Copyright Daniel Stiles

And so the time has come. We have seen what is happening. Now we must decide. We, you and I dear readers, can make a difference. In fact, it is only we, humans, who will eventually determine the fate of our beloved animals. Their fate really is in our hands, and we must not let them down. The iconic picture below says it all. It shows Dr. Dan Stiles with his young son Sloan and old Sudan. Dan Stiles has dedicated so much of his life fighting passionately for the future of the animals he so obviously loves, for a world rich in wildlife that Sloan, our children and our grandchildren can inherit. And there is Sudan, the last male northern white rhino, who has now left us, but who has bequeathed us hope, that most precious of commodities, hope that we will carry on fighting to save our wonderful natural world.

So, dear readers, what is it to be? Survival or extinction? Our choice, but can we do it? It's a no-brainer, of course we can. All of us working together. We must!

# Further Reading

I have included this section, although the footnotes show where I got the material included in the book. By collecting some of the main sources together, I hope this makes it easier for those of you who might want to find out more.

## Case Studies

Environmental Investigation Agency *Back in Business: Elephant Poaching and the Ivory Black Markets of Asia* Jo Hastie, Julian Newman and Mary Rice 2002.

Environmental Investigation Agency 2010 *Open Season: The Burgeoning Illegal Ivory Trade in Tanzania and Zambia.*

*The Shuidong Connection: Exposing the global hub of the illegal ivory trade* EIA 2017.

## Surveys in Africa

Courouble, M., Hurst, F. and Milliken, T. (2003). *More Ivory than Elephants: domestic ivory markets in three West African countries.* TRAFFIC International, Cambridge, UK.

Martin, E. and Milliken, T. (2005) *No Oasis: the Egyptian Ivory Trade in 2005* TRAFFIC International, Cambridge, UK.

© Springer Nature Switzerland AG 2019
B. Martin, *Survival or Extinction?*, https://doi.org/10.1007/978-3-030-13293-4

Milliken, T., Pole, A., and Huongo, A. (2006) *No peace for Elephants: Unregulated Domestic Ivory Markets in Angola and Mozambique.* TRAFFIC International, Cambridge, UK.

Martin, E. and Vigne, L. *Illegal Ivory Sales In Egypt* TRAFFIC Bulletin, 30 January 2012 (Survey carried out in March and April 2011).

Esmond Martin and Lucy Vigne *LAGOS, NIGERIA: One of the Largest Retail Centres for Illegal Ivory Surveyed to Date* TRAFFIC Bulletin Vol.25 No.1 (2013).

Nkoke, S.C. Lagrot J.F. Ringuet, S. and Milliken, T. (2017) *Ivory Markets in Central Africa – Market Surveys in Cameroon, Central African Republic, Congo, Democratic Republic of Congo and Gabon: 2007, 2009, 2014/2015.* TRAFFIC. Yaounde, Cameroon and Cambridge, UK.

Moneron, S., Okes, N. and Rademeyer, J. (2017) *Pendants, Powder and Pathways.* TRAFFIC, East/Southern Africa Regional Office, Hatfield, Pretoria, South Africa.

## Surveys in Southeast Asia

Daniel Stiles (2008). *An assessment of the illegal ivory trade in Viet Nam* TRAFFIC Southeast Asia, Petaling Jaya, Selangor, Malaysia.

Chris R. Shepherd and Vincent Nijman (2008). *Elephant and Ivory Trade in Myanmar.* TRAFFIC Southeast Asia, Petaling Jaya, Selangor, Malaysia.

Daniel Stiles (2009). *The elephant and ivory trade in Thailand* TRAFFIC Southeast Asia, Petaling Jaya, Selangor, Malaysia.

Esmond Martin and Lucy Vigne (2011) *The Ivory Dynasty: A report on the soaring demand for elephant and mammoth ivory in southern China* Published by Elephant Family, the Aspinall Foundation and Columbus Zoo and Aquarium 2011.

Grace G. Gabriel, Ning Hua, Juan Wang *Making a Killing: A 2011 Survey of Ivory Markets in China* IFAW (The International Fund for Animal Welfare).

Doak, N. (2014). *Polishing off the Ivory Trade: Surveys of Thailand's Ivory Market* TRAFFIC International, Cambridge, UK.

## Surveys Carried Out in Japan

EIA *Japan's Illegal Ivory Trade and Fraudulent Registration of Ivory Tusks* EIA-GLOBAL.ORG 2015.

Tomomi Kitade and Ayako Toko (2016) *Setting Suns: The Historical Decline of Ivory and Rhino Horn Markets in Japan* TRAFFIC, Tokyo, Japan.

Tomomi Kitade and Ryoko Nishino *Slow Progress: A Reassessment of Japan's Ivory Market in 2018* TRAFFIC Briefing, September 2018.

## Surveys in the USA

Esmond Martin and Daniel Stiles *Ivory Markets in the USA* Published by Care for the Wild International and Save the Elephants 2008.

Daniel Stiles *Ivory Trafficking in California.* USA 2014.

*Bidding Against Survival: The Elephant Poaching Crisis and the Role of Auctions in the U.S. Ivory Market* IFAW 2014.

## Surveys in the UK

Lau, W., Crook, V., Musing, L., Guan, J., and Xu, L. (2016) *A rapid survey of UK ivory markets.* TRAFFIC, Cambridge, UK.

Brace, S. and Dean, C. (2018). *Sounding the Horn: A survey of rhino horn antiques sold in 2017 at auction in the UK.* Save the Rhino International, London.

## Crisis Surveys

Milliken, T. and Shaw. J. (2012) *The South Africa – Viet Nam Rhino Trade Nexus: A deadly combination of institutional lapses, corrupt wildlife industry professionals and Asian crime syndicates.* TRAFFIC, Johannesburg, South Africa.

UNEP, CITES, IUCN, TRAFFIC (2013). *Elephants in the Dust – The African Elephant Crisis. A Rapid Response Assessment.* United Nations Environment Programme, GRID-Arendal.

## The Internet and Social Media

IFAW *Killing with Keystrokes: An Investigation of the Illegal Wildlife Trade on the World Wide Web* 2008.

IFAW Wanted – *Dead or Alive: Exposing Online Wildlife Trade* 2014.

Vincent Nijman and Chris R. Shepherd *Social media and the ivory ban: Myanmar and China cross-border trade.* Pachyderm No.55 January-June 2014.

Xiao Yu and Wang Jia *Moving Targets: Tracking online sales of illegal wildlife products in China* TRAFFIC Briefing, February 2015.

Krishnasamy, K. and Stoner, S. (2016) *Trading Faces: A Rapid Assessment on the Use of Facebook to Trade Wildlife in Peninsula Malaysia 2016* TRAFFIC, Petaling Jaya, Selangor, Malaysia.

## Documents Produced for CITES

CITES 62nd meeting of the Standing Committee 23–27 July 2012 *Interpretation and implementation of the Convention, Species trade and conservation, Rhinoceroses* Report of the Working Group, SC62 Doc.47.1.

*Species trade and conservation, Rhinoceroses, Assessment of Rhino Horn as a Traditional Medicine.* A report prepared for the CITES Secretariat by Kristin Nowell on behalf of TRAFFIC, April 2012, SC62 Doc.47.2.

## Organized Crime and Terrorism

UNODC *Transnational Organized Crime in Eastern Africa: A threat assessment* 2013.

*Symposium on Combating Wildlife Crime: Securing Enforcement, Ensuring Justice, and Upholding the Rule of Law. The Proceedings.* Editors Kala K. Mulqueeny and Francesse Joy J. Cordon, 2014.

Tom Maguire and Cathy Haenlein *An Illusion of Complicity: Terrorism and the Illegal Ivory Trade in East Africa* Royal United Services Institute for Defence and Security Studies (RUSI), Occasional Paper, September 2015.

Julian Rademeyer *Beyond Borders: Crime, conservation and criminal networks on the illicit rhino horn trade (Part2 of a 2-part investigation into rhino horn trafficking in Southern Africa)* Global Initiative Against Transnational Organized Crime, July 2016.

Julian Rademeyer *Diplomats and Deceit: North Korea's Criminal Activities in Africa* The Global Initiative Against Transnational Organized Crime, September 2017.

Rossi, A. (2018) *Uganda Wildlife Trafficking Assessment* TRAFFIC International, Cambridge, United Kingdom.

# Ecology

Nasseri N.A., McBrayer L.D. and Schulte B.A. (2011) *The impact of tree modification by African elephant (Loxodonta Africana) on herpetofaunal species richness in northern Tanzania.* African Journal of Ecology, 49:133–140.

Campos-Arceiz Ahimsa, Blake Steve *Megagardeners of the forest – the role of elephants in seed dispersal* Acta Oecologica 37 (2011).

Cromsigt J.P.G.M., te Beest M. (2014) *Restoration of a megaherbivore: landscape-level impacts of white rhinoceros in Kruger National Park, South Africa.* Journal of Ecology, 102.

# Game Management

Pack, S.P., Golden, R.E., Walker, A. 2013 *Comparison of national wildlife game management strategies: What works where and why.* Consultant Report from the University of Maryland Sustainable Development and Conservation Biology Program to the Heirig Centre for Science, Economics and the Environment.

# Sustainable Use and Local Communities

Economists at Large 2013 *The $200 million question: How much does trophy hunting really contribute to African communities?* A report for the African Lion Coalition prepared by Economists at Large, Melbourne, Australia. Lead author – Roderick Campbell.

Phil Franks and Rob Small (2016) *Understanding the social impacts of protected areas: a community perspective* IIED Research Report, IIED, London.

Rodgers Lubilo *Sustainable use, local communities and tackling illegal wildlife trade in Managalane, Mozambique* SULi News, Issue 10, October

*Briefing Local Communities: the overlooked first line of defence for wildlife* IIED, IUCN, CEESP, Species Survival Commission, SULi, African Elephant Specialist Group, May 2018.

# Counting Elephants

Michael J. Chase, Scott Schlossberg, Curtice R. Griffin, Philippe J.C. Bouche, Sintayehu W. Djene, Paul W. Elkan, Sam Ferreira, Falk Grossman, Edward Mtarima Kohi, Kelly Landen, Patrick Omondi, Alexis Peltier, S.A. Jeanetta

Selier, Robert Sutcliffe *Continent-wide survey reveals massive decline in African savannah elephants* PeerJ, 31 August 2016.

Stephanie Brittain *A rapid assessment of the status and distribution of Loxodonta cyclotis in South East Cameroon* September 2013 (supervised by Dr. Marcus Rowcliffe & Paul DeOrnellas). To read, see https://www.iccs.org.uk/wp-content/thesis/consci/2013/Brittain.pdf

## Forensics and Data Collection

Hsing-Mei Hsieh, Li-Hung Huang, Li-Chin Tsai, Yi-Chen Kuo, Hsien-Huei Meng, Adrian Linacre, James Chun-I Lee *Species identification of rhinoceros horns using the cytochrome 6 gene* Forensic Science International 136 (2003).

S.K. Wesser, W.J. Clark, O. Drori, E.S. Kisamo, C. Maitland, B. Mutayoba and M. Stephens *Combatting the Illegal Trade in African Elephant Ivory with DNA Forensics* Conservation Biology, Volume 22, No.4, 2008.

Dr. Cindy Harper *RhODIS profiling and a DNA database as a tool to protect the Rhino* Peppin et al 2009.

Iyengar, A. *Forensic DNA analysis for animal protection and biodiversity conservation: A review* Journal for Nature Conservation (2014).

Critchlow R., Plumtre A.J., Dirciru M., Rwetsiba A., Stokes E,J., Tumwesigye C., Wanyama F., and Beale C.M. *Spatiotemporal trends of illegal activities from ranger-collected data in a Ugandan national park* Conservation Biology, Volume 20, No.5.

*Saving Rhinos Using DNA* Smithsonian's National Zoo & Conservation Biology Institute, 9 February 2018.

## Miscellaneous

Varun Vira, Thomas Ewing and Jackson Miller *Out of Africa: Mapping the Global Trade in Illicit Elephant Ivory* Commissioned by Born Free and carried out by C4ADS, published 2014.

TRAFFIC *Wildlife TRAPS – Wildlife Trafficking Response, Assessment and Priority Setting.*

Lau, W., Xu, L., Guan, J. and Xiao, Y. (2017) *Closing Strategy: Ending ivory trade in Hong Kong* TRAFFIC, Hong Kong.

Irie, N. *et al* (2018) *Unique numerical competence of Asian elephants on the relative numerosity judgement task* Journal of Ethology. DOI: 10.1007/s10164-018-0563-y.

# Articles

It is always worth looking out for interesting articles in newspapers and magazines. You can see how varied the sources can be by looking at the footnotes in the book. Those in National Geographic and Mongabay.com (the global conservation news service) often provide rather more depth. For example:

Bryan Christie *Blood Ivory* National Geographic, October 2012.

Jani Actman *Can Anything Save the Sumatran Rhinoceros from Extinction* National Geographic, 30 September 2015.

Bryan Christie *South Africa Just Lifted Its Ban on the Rhino Horn Trade* National Geographic, 23 May 2016.

Paul Steyne *African Elephant Numbers Plummet 30 Percent, Landmark Survey Finds* National Geographic, 2016

Christina Russo *Exclusive: This is Africa's New Elephant Poaching Hot Spot* National Geographic, 5 July 2016.

Jani Actman *Virunga National Park Sees Its Worst Violence in a Decade, Director Says* National Geographic, 14 June 2018.

Jason Bittel *The Unprecedented Plan to Save the Sumatran Rhino* National Geographic, 20 September 2018.

Or:

Mongabay Series: Asian Rhinos

A 4-part series which started in November 2017: *Is Anyone Going To Save The Sumatran Rhino?*

Jim Tan *Rangers face a 'toxic mix' of mental strain and lack of support* Mongabay, 24 May 2018.

*The Great Rhino U-Turn* Mongabay, September 2018.

# Index

© Springer Nature Switzerland AG 2019
B. Martin, *Survival or Extinction?*, https://doi.org/10.1007/978-3-030-13293-4